Springer Monographs in Mathematics

Springer Japan KK

AKIHITO UCHIYAMA

Hardy Spaces on the Euclidean Space

 Springer

Akihito Uchiyama (1948–1997)

Library of Congress Cataloging-in-Publication Data applied for

/

ISSN 1439-7382
ISBN 978-4-431-67999-8   ISBN 978-4-431-67905-9 (eBook)
DOI 10.1007/978-4-431-67905-9

Mathematics Subject Classification (2000): 46E30, 46E15, 46J15

Printed on acid-free paper
©Springer Japan 2001
Originally published by **Springer-Verlag Tokyo** in 2001
Softcover reprint of the hardcover 1st edition 2001

SPIN: 10843141

Foreword

This book is based on a type-written draft which Akihito Uchiyama (1948-1997) completed in September, 1990. The mathematics community was awaiting its appearance, but he passed away in the summer of 1997, before he could publish it. Because we believe that his book is still worthy, we have decided to proceed with its publication. The original manuscript was transformed into a TeX file with the help of Professors Shige-Toshi Kuroda and Akira Mizutani of Gakushuin University, along with their former students Ms. Makiko Kokubu and Ms. Motoko Oikawa. We thank them very much for their hard work.

We also send our hearty thanks to Professor Peter Wilcox Jones, Uchiyama's Ph. D adviser, who kindly wrote a very impressive essay about Uchiyama's personality and mathematical works.

Nobuhiko Fujii
Akihiko Miyachi
Kôzô Yabuta

March 2001

Recollections of My Good Friend, Akihito Uchiyama

Peter W. Jones

Akihito Uchiyama came to the University of Chicago in the early 1980's at a time of expansive progress in Fourier Analysis. The field had achieved explosive growth since 1970, when Charles Fefferman had proved H^1, BMO duality, and the influential paper by Charles Fefferman and E.M. Stein had appeared in Acta Math. Chicago was one of the centers of Fourier analysis, and the home of Alberto Calderón and Antoni Zygmund. Calderón had recently proved a major result: the Cauchy integral is L^2 bounded on Lipschitz curves with small constants. It was clear that this result had opened a new area of research in the theory of Hardy Spaces. Antoni Zygmund was old and infirm but made a point of coming to the department every day. He never missed a lecture in the Calderón-Zygmund Seminar. Robert Fefferman and William Beckner were the junior professors and there were always several Dickson Instructors in analysis. R. Narasimhan, the famous complex analyst, was also closely associated with the group. The atomosphere was lively, and the most talented young Ph. D.'s were always vying to get a job there. This was the scene upon which Akihito Uchiyama appeared.

I had been aware of his work for some time, and it was well known that he was one of the brightest stars of the new generation. Before coming to Chicago, he wanted to spend time with John Garnett's group at U.C.L.A., so John arranged this. To appreciate fully how Akihito Uchiyama was appointed at Chicago, one must first understand the spirit of that department. The mathematicians prided themselves on their daring and willingness to take chances. The practices of the mathematical community outside the department were certainly taken into account, but the department saw itself as a place where new ideas were to be fomented.

I approached Professors Calderón, Beckner and Fefferman and explained to them the importance of Uchiyama's work. A very brief discussion was held and it was decided to try to hire Uchiyama. Now we needed to convince the Chairman, Felix Browder. We explained to him that we had a brilliant researcher and wanted to make him an offer. We also told him that Uchiyama did not yet have a Ph.D. This made Felix Browder quite happy, but he had one more question. Could Uchiyama speak English. I replied that his English was very weak, and he might have problems teaching. "Excellent!", replied Browder. This was the kind of daring appointment the department loved to make - the only issue was the excellence of the candidate.

Before arriving for his appointment, Uchiyama came on a visit from UCLA, and his famous personal toughness was about to be revealed to me. I called him before his departure from Los Angeles and told him we were in the grips of a brutal Chicago winter cold snap - he needed to bring a heavy jacket! The instant reply was that this was unnecessary because the visit was only for a few days. I thought at first he had misunderstood me, but it rapidly became clear that this was not the case. Further arguments on my behalf were rebuffed and I eventually gave up. When he arrived, the temperature was −20° C with the infamous Chicago wind howling about us. I had arrived with an extra jacket, hat, and gloves for him. We then had a good spirited discussion that was repeated many times later. While I wanted very much to take care of my guest, Akihito did not want to make any impositions on his host. Freezing in the bitter cold was to him well worth the price of putting someone else to task for taking care of him. The fact that the task was so insignificant as my grabbing a jacket from the coat rack did not matter! (But I did finally succeed in getting him to accept the jacket.)

When Akihito Uchiyama arrived for his appointment at the university, he became an instant presence in the Mathematics Department. One of the first things that happened was that he became simply known as Akihito. Here I should explain a little about the customs of university life in America. To call someone Professor X, would imply either that you were unfamiliar with that person, or that he was quite old. Even Alberto Calderón was called Alberto within the department, and it was only at conferences that he was referred to as Professor Calderón. Akihito acceded to this custom with his usual grace and seemed to prefer this kind of collegial informality.

Akihito's daily routine remained fixed for the entire time of his stay. He was remarkably disciplined and preferred to lead a Spartan existence. His apartment was in the same complex where I lived, about twenty meters around the corner. Every morning he awoke well before me for his breakfast. Upon entering the kitchen I would look out the back window into the court-yard, and there was Akihito. Dressed only in a thin blue suit with dress shoes on, he would be doing calisthenics for about twenty minutes. You could set your clock to his arrival time, and he became somewhat of a local celebrity for the punctuality he showed. No weather condition except pouring rain (when he exercised in his apartment) could deter him. One day I looked out into a ferocious blizzard and saw Akihito's outline, standing in 25 centimeters of snow. I went out and asked him if this wasn't a bit extreme, but he replied with a grin that it was a beautiful day. His blue suit was all that was needed, and the snow was cold enough so that it wouldn't melt into his shoes.

After calisthenics, Akihito would walk to the department, which was about 500 meters away. Here too, the routine was unvarying. The entire morning was spent doing research, with no breaks. He was always happy, however, to receive a visitor and would do his best to answer any questions. But hard work was the order of the morning, and Akihito was soon back on the job. He had a light teaching schedule, and in the beginning of the year I would check in to see how the class was progressing. At this time his English was still uncertain, and I was quite worried that the students would rebel. What I saw astonished me, and the memory will stick with me for the rest of my life. The perfectly prepared lectures were written in impeccable handwriting on the blackboard, with barely a word spoken. He would stop to stress one or two of he words to clarify the main concept, but that was virtually the extent of his speaking. While I knew he was an accomplished lecturer, I was stunned by his clarity, and totally unprepared for another trick he had. Akihito still had a hard time understanding questions, especially from undergraduate students whose English was often imprecise. He prepared before each lecture a list of possible questions so that he would understand them when they came. Immediately upon the conclusion of the question, Akihito would turn to the board and write out his prepared answer. Far from having any problems, Akihito rapidly became one of the most popular teachers. His kindness and patience with the students made him very well loved. Some of the credit here must go to the undergraduates at Chicago, who are extremely sophisticated and are used to dealing with foreign instructors. But clearly the greatest credit goes to Akihito, who found his own novel way to deal with a tough situation, and win the day. As time went on, Akihito's English improved, and by the end of a year his lectures were delivered in standard (that is, spoken!) style.

Lunchtime meant that it was time to head off to Jimmy's Bar, about one kilometer away. Someone would be dispatched to tell Akihito - the crowd of analysts would not leave until he arrived. There was a lot of give and take at lunch, with many jokes and stories told. Akihito shed his usual reserve here and was known for his good sense of humor. While quiet on many issues, he had sharp opinions on the role of universities and researchers, and he defended his views with great conviction. While much of his character was very Japanese, these political views were much more in line with American ones. I think this aspect of Chicago was one that Akihito appreciated most. He was rapidly accepted by his colleagues and his views, while certainly open to debate among us, were widely respected. Another feature of Akihito's personality was also seen at Jimmy's. No matter how much he disagreed with someone's opinions, and even if the person was someone who was not present, Akihito would never attack the person. He would simply state his own views and then defend them. He held a special status at lunch because of this. While there were many pranks, practical jokes, and one-upmanship on display, Akihito was to be spared this.

Now it was back to the office for the rest of the day, where Akihito returned to his research. Aside from a visitor to his office, the only break came if there was an analysis seminar. Here Akihito was mostly a silent participant, but when he asked questions they were penetrating. He was always searching for new problems or methods that might be useful in his own research.

After the seminar, it was back to his office, where research continued until dinner. He preferred to eat a simple prepared dinner and did not want to have it take too much time. He would return to his office and work until 7 or 8, when he returned home. At that time, he would prepare his lessons in the manner I described and then turn to his English practice. Akihito confided that he had meager language talents, but that this was not an obstacle. His mastery of English would come by overpowering the subject with bulldog tenacity. He had several shoe boxes filled with filing cards for practice. Each card had a word on one side with the pronunciation and definition on the other. Twenty words were added every day, and there were regular reviews of the older cards. This took an enormous amount of time and energy, but in his characteristic style Akihito felt there was no need to inform others about his hard work. As far as I know, only I and my wife Lisa were aware of the closet filled with shoe boxes of cards. By the end of one year at Chicago, his English had advanced to an impressive level, but he never scaled back his training regime.

Akihito Uchiyama's contributions to mathematics were deep and wide. One could write several papers explaining his contributions and how they have been absorbed into the field of harmonic analysis, developed, and used in further breakthroughs. History will certainly record better than I his successes and wide influence. But no account of his work would be complete without comment on his crowning achievement on decompositions of BMO functions. The results here are published in Acta Mathematica. Charles Fefferman's duality theorem on H^1, BMO has an abstract corollary the conclusion that any function $\varphi \in \mathrm{BMO}(\mathbf{R}^n)$ can be decomposed as

$$\varphi = u_0 + \sum_{j=1}^{n} R_j u_j,$$

where $u_0, \ldots, u_n \in L^\infty$, and R_j denote the jth Riesz transform. (The converse direction, that any such φ lies in BMO is quite easy.) The difficulty with this result is that it relies on duality plus the Hahn-Banach Theorem to produce the bounded functions u_j. In other words, one has no idea where the u_j "come from" or behave locally. An additional complication is that the decomposition is far from unique. I found the first resolution of this for the one-dimensional case. There one has only one singular integral operator, the Hilbert Transform. The constructive solution I found used complex function theory and solutions of the dee bar equation. While admittedly useful in function-theoretic situations, my method was entirely constrained to one dimension. Worse, one did not get optimal local "smoothness" of the two bounded functions u_0, u_1. Technically speaking, this was because the method I used was L^∞-based, whereas the smoothness properties can only come from a study of local L^2 behavior. Akihito Uchiyama then studied the problem and came up with his brilliant solution to the general problem in $\mathbf{R}^n$. Furthermore, his method is L^2 based and so gives all possible smoothness results simultaneously.

The first major lecture Uchiyama gave on his result was at a large, international conference held at the University of Chicago. I had counted over 200 mathematicians in the audience. They would be witness to what later became known as "The Legend of Uchiyama". The fantastic idea of Uchiyama was to build a vector of functions $(u_0(x), u_1(x), \ldots, u_n(x))$ that had constant absolute value R. (The vector lies on the sphere of radius R in $\mathbf{R}^{n+1}$.) The strategy was to build this vector inductively, going through the dyadic scales and approximating via the "wavelet" decomposition of φ. A clever analysis of Fourier transforms allows one to pass from scale n to $n+1$. This trick actually allowed Uchiyama to solve a well known conjecture on exactly which families of singular integral operators could be used to replace the (first order) Riesz transforms.

The lecture was crystal clear, a hallmark of Akihito's style, but the methods used were so new that no one could fully understand the construction. At the end of the talk, one of the world's most famous analysts rose and hailed the result as one of the main breakthroughs of the decade. But could Professor Uchiyama just explain the main philosophy of the proof. The audience was hushed as it awaited Akihito's response. After a brief pause he replied, "The circle is round." This reply came to be known as The Legend of Uchiyama. He was of course referring to the appearance of the unit sphere in the construction. The deeper meaning was that as one passed from stage n to $n+1$ in the construction, the new vector was projected onto the sphere of radius R. The resulting error is quadratic and of size R^{-1}. The end result of the construction, assuming one starts with φ of BMO norm 1, is a decomposition that does not yield exactly φ, but has a BMO error of order R^{-1}. Taking R large enough allows one to iterate to a full solution, summing a geometric series corresponding to the error after repeating n times. Understanding the last four sentences requires several years of training for a graduate student. Only then can one appreciate the full depth of Uchiyama's statement that the circle is round.

Akihito Uchiyama completed in his lifetime research whose full effects will take decades to play out. His decomposition of BMO functions is considered to be the Mount Everest of Hardy space theory. But all these accomplishments do not come close to giving full measure of the man I knew. His gentle kindness, unflinching honesty, and independent spirit were felt by all who met him. I believe his greatest legacy is the vision he had for the future of mathematics. His good friends understood and will not forget that, to Akihito Uchiyama, great research was but one aspect to be strived after in the large and expanding mathematical world in which we live.

New Haven, March 2001

Preface

The foundations of the real Hardy space $H^p(\mathbf{R}^n)$ were laid by C. Fefferman and E. M. Stein "H^p spaces of several variables" (Acta Math., 129 (1972), 137–193) and by R. Coifman and G. Weiss "Extensions of Hardy spaces and their use in analysis" (Bull. Amer. Math. Soc., 83 (1977), 569–645). In this book, we will explain some of the important results on $H^p(\mathbf{R}^n)$.

Sendai, September 1990   Akihito Uchiyama

Contents

* These sections can be skipped.

0. Introduction

First, we prepare notations.

$\mathbf{R}$ = {the set of real numbers},

$\mathbf{C}$ = {the set of complex numbers},

$\mathbf{N}$ = $\{1, 2, 3, \cdots\}$, $\mathbf{N}_0 = \{0, 1, 2, 3, \cdots\}$,

$\mathbf{Z}$ = $\{0, \pm 1, \pm 2, \pm 3, \cdots\}$,

$[t]$ = {the greatest integer not exceeding a real number t },

$x = (x_1, \cdots, x_n)$, $\xi = (\xi_1, \cdots, \xi_n) \in \mathbf{R}^n$,

$dx = dx_1 \cdots dx_n$ = {the Lebesgue measure on $\mathbf{R}^n$},

$|x| = \{x_1^2 + \cdots + x_n^2\}^{1/2}$,

$S^{n-1} = \{\xi \in \mathbf{R}^n : |\xi| = 1\}$,

$\mathbf{R}_+^{n+1} = \{(x, t) : x \in \mathbf{R}^n, \ t \in (0, +\infty)\}$,

$B(x, t) = \{y \in \mathbf{R}^n : |x - y| < t\}$.

If $B = B(x_0, t_0)$ and $\delta > 0$, then

$\qquad \delta B = B(x_0, \delta t_0)$, $x_B = x_0$, $\ell(B) = t_0$,

$\qquad Q(B) = \{(x, t) \in \mathbf{R}_+^{n+1} : x \in B, \ t \in (0, t_0]\}$.

$I(x, t) = \left\{ y = (y_1, \cdots, y_n) \in \mathbf{R}^n : \max_{1 \le j \le n} |x_j - y_j| \le t/2 \right\}$.

If $I = I(x_0, t_0)$, then

$\qquad \delta I = I(x_0, \delta t_0)$, $x_I = x_0$, $\ell(I) = t_0$.

"Dyadic cubes" are cubes of the form

$$\left(k_1 2^h, (k_1 + 1) 2^h\right] \times \cdots \times \left(k_n 2^h, (k_n + 1) 2^h\right], \quad \text{where } h, k_1, \cdots, k_n \in \mathbf{Z}.$$

The "dyadic double" of a dyadic cube I is the dyadic cube $\tilde{I}$ such that

$$\tilde{I} \supset I \text{ and } \ell(\tilde{I}) = 2\ell(I), \text{ where } \ell(I) = \{\text{the side length of } I\}.$$

$\Gamma(x, \delta) = \{(y, t) \in \mathbf{R}_+^{n+1} : |x - y| < \delta t\}, \qquad$ where $\delta > 0,$

$\Gamma(x, \delta, h) = \{(y, t) \in \Gamma(x, \delta) : t < h\}, \qquad$ where $h > 0,$

$\alpha = (\alpha_1, \cdots, \alpha_n) \in \mathbf{N}_0^n, \ \alpha! = \alpha_1! \cdots \alpha_n!, \ |\alpha| = \alpha_1 + \cdots + \alpha_n,$

$D_{x_j} = \dfrac{\partial}{\partial x_j}, \ D_x^\alpha = D_{x_1}^{\alpha_1} \cdots D_{x_n}^{\alpha_n}, \ D_t = \dfrac{\partial}{\partial t},$

$\nabla_x = (D_{x_1}, \cdots, D_{x_n}), \ \nabla_{t,x} = (D_t, \ D_{x_1}, \cdots, D_{x_n}),$

$\nabla_x^m f = \left(D_{x_{j_1}} \cdots D_{x_{j_m}} f\right)_{j_1, \cdots, j_m \in \{1, \cdots, n\}} \qquad \left(\mathbf{R}^{n^m}\text{-valued function}\right),$

$\deg p = \{\text{the degree of a polynomial } P(x)\},$

$\mathrm{supp}\, f = \{\text{the support of } f\},$

$\chi_E = \{\text{the characteristic function of a set } E\},$

$\chi = \chi_{B(0,1)} = \{\text{the characteristic function of the unit ball of } \mathbf{R}^n\},$

$\overline{E} = \{\text{the closure of } E\},$

$|E| = \{\text{the Lebesgue measure of } E\},$

$\mathrm{av}(f, E) = |E|^{-1} \displaystyle\int_E f(x)dx,$

$(\phi)_t(x) = t^{-n}\phi(x/t),$

$P(x, t) = \{\text{the Poisson kernel of } \mathbf{R}_+^{n+1}\}$

$\qquad = \Gamma\left((n+1)/2\right) \pi^{-(n+1)/2} t \left(|x|^2 + t^2\right)^{-(n+1)/2},$

$\|\{\lambda_j\}_{j\in\mathbf{N}}\|_{\ell^p} = \left\{\displaystyle\sum_{j\in\mathbf{N}} |\lambda_j|^p\right\}^{1/p},$

$\ell^p = \{\text{the set of sequences of real numbers such that } \|\{\lambda_j\}\|_{\ell^p} < \infty\},$

$\|f\|_{L^p(E)} = \left\{\displaystyle\int_E |f(x)|^p dx\right\}^{1/p} \quad \text{if } 0 < p < \infty,$

$\|f\|_{L^\infty(E)} = \mathrm{ess}\cdot\sup_{x\in E} |f(x)|,$

$\|f\|_{L^p} = \|f\|_{L^p(\mathbf{R}^n)},$

$L^p = L^p(\mathbf{R}^n) = L^p(\mathbf{R}^n, \mathbf{R})$

$= \{\text{the set of real-valued measurable functions on } \mathbf{R}^n$

$\qquad\qquad\qquad\qquad\qquad \text{such that } \|f\|_{L^p} < \infty\},$

$L_{\mathrm{loc}}^1 = L_{\mathrm{loc}}^1(\mathbf{R}^n) = L_{\mathrm{loc}}^1(\mathbf{R}^n, \mathbf{R})$

$= \{\text{the set of real-valued locally integrable functions on } \mathbf{R}^n\},$

$C(\mathbf{R}^n) = C(\mathbf{R}^n, \mathbf{R})$

$= \{\text{the set of real-valued continuous functions on } \mathbf{R}^n\},$

$C^k(\mathbf{R}^n) = C^k(\mathbf{R}^n, \mathbf{R})$

$= \{\text{the set of real-valued } k\text{-times continuously}$

differentiable functions on $\mathbf{R}^n$},

$\mathcal{D} = \mathcal{D}(\mathbf{R}^n) = \mathcal{D}(\mathbf{R}^n, \mathbf{R}) = \{f \in C^\infty(\mathbf{R}^n) : \operatorname{supp} f \text{ is compact}\}$,

$\mathcal{S} = \mathcal{S}(\mathbf{R}^n) = \mathcal{S}(\mathbf{R}^n, \mathbf{R})$

$$= \left\{ f \in C^\infty(\mathbf{R}^n) : (1 + |x|)^m \sum_{\alpha:|\alpha| \le m} |D_x^\alpha f(x)| \in L^\infty \text{ for any } m \in \mathbf{N} \right\}.$$

If B is a topological linear space, if $g \in B$ and if f is a continuous linear functional on B, then we write

$$\langle g, f \rangle_B = f(g).$$

$\mathcal{D}' = \mathcal{D}(\mathbf{R}^n)' = \mathcal{D}(\mathbf{R}^n, \mathbf{R})'$
$= \{\text{the set of all continuous real-valued linear functionals on } \mathcal{D}\}$,

(as for the topology of $\mathcal{D}$, see K. Yosida [68] p. 28.)

If $\{f_n\}_{n \in \mathbf{N}} \subset \mathcal{D}'$, $f \in \mathcal{D}'$ and if $\lim_{n \to \infty} \langle \phi, f_n \rangle_\mathcal{D} = \langle \phi, f \rangle_\mathcal{D}$ for any $\phi \in \mathcal{D}$,

then we write

$$f_n \to f \ (n \to \infty) \text{ in } \mathcal{D}'.$$

$\mathcal{S}' = \mathcal{S}(\mathbf{R}^n)' = \mathcal{S}(\mathbf{R}^n, \mathbf{R})'$
$= \{\text{the set of all continuous real-valued linear functionals on } \mathcal{S}\}$

$$= \left\{ f \in \mathcal{D}' : \inf_{m \in \mathbf{N}} \sup \left\{ \frac{|\langle \phi, f \rangle_\mathcal{D}|}{\displaystyle\sum_{|\alpha| \le m} \|(1 + |x|)^m D_x^\alpha \phi(x)\|_{L^\infty}} : \phi \in \mathcal{D} \right\} < \infty \right\},$$

$$f * g(x) = \int_{\mathbf{R}^n} f(x - y) g(y) dy.$$

If $f \in \mathcal{D}'$ and $\phi \in \mathcal{D}$, then $f * \phi(x) = \langle \phi(\cdot - x), f \rangle_\mathcal{D}$.
The spaces

$$L^p(\mathbf{R}^n, \mathbf{C}), \ \mathcal{D}(\mathbf{R}^n, \mathbf{C}), \ \mathcal{S}(\mathbf{R}^n, \mathbf{C}), \ \cdots$$

are the spaces of complex-valued functions with the same conditions as the real-valued cases.

$\mathcal{D}\left(\mathbf{R}^{n},\mathbf{C}\right)'_{c}$ = {the set of all continuous complex-valued linear
functionals on $\mathcal{D}\left(\mathbf{R}^{n},\mathbf{C}\right)$, where the subindex **c**
means that we regard $\mathcal{D}\left(\mathbf{R}^{n},\mathbf{C}\right)$ as a linear topological
space over the **complex** number field}
= $\{f_{1}+if_{2}: f_{1},f_{2}\in\mathcal{D}'\}\supset\mathcal{D}'$,

$\mathcal{S}\left(\mathbf{R}^{n},\mathbf{C}\right)'_{c}$ = {the set of all continuous complex-valued linear
functionals on $\mathcal{S}\left(\mathbf{R}^{n},\mathbf{C}\right)$}
= $\{f_{1}+if_{2}: f_{1},f_{2}\in\mathcal{S}'\}\supset\mathcal{S}'$,

$$\mathcal{F}f(\xi)=\int_{\mathbf{R}^{n}}f(x)e^{-2\pi ix\cdot\xi}dx, \text{ where } x\cdot\xi=x_{1}\xi_{1}+\cdots+x_{n}\xi_{n},\ f\in L^{1}\left(\mathbf{R}^{n},\mathbf{C}\right).$$

This Fourier transform maps $\mathcal{S}\left(\mathbf{R}^{n},\mathbf{C}\right)$ onto itself and can be extended as
an operator from $\mathcal{S}\left(\mathbf{R}^{n},\mathbf{C}\right)'_{c}$ onto itself by the usual manner.

$\mathcal{S}_{0}=\mathcal{S}_{0}(\mathbf{R}^{n})=\mathcal{S}_{0}(\mathbf{R}^{n},\mathbf{R})$
$=\{f\in\mathcal{S}(\mathbf{R}^{n},\mathbf{R}): \operatorname{supp}\mathcal{F}f \text{ is compact and away from the origin}\}$,
$\mathcal{S}_{0}(\mathbf{R}^{n},\mathbf{C})=\{f\in\mathcal{S}(\mathbf{R}^{n},\mathbf{C}): \operatorname{supp}\mathcal{F}f \text{ is compact and}$
$\text{away from the origin}\}$.

In Sections 1–13, all functions and distributions considered, except Fourier
transforms, are real-valued.

After Section 16, we will treat vector-valued functions. The spaces

$$L^{p}(\mathbf{R}^{n},\mathbf{R}^{m}),\ \mathcal{D}(\mathbf{R}^{n},\mathbf{R}^{m}),\ \mathcal{S}(\mathbf{R}^{n},\mathbf{R}^{m}),\cdots$$

are spaces of $\mathbf{R}^{m}$-valued functions

$$\vec{f}(x)=(f_{1}(x),\cdots,f_{m}(x))$$

with the same definitions as the scalar-valued cases with their
absolute values $|f(x)|$ replaced by

$$\left|\vec{f}(x)\right|=\sqrt{\sum_{j=1}^{m}f_{j}(x)^{2}}\ .$$

$\mathcal{D}(\mathbf{R}^{n},\mathbf{R}^{m})'$
= {the set of real-valued continuous linear functionals on $\mathcal{D}\left(\mathbf{R}^{n},\mathbf{R}^{m}\right)$}
= $\{(f_{1},\cdots,f_{m}): f_{1},\cdots,f_{m}\in\mathcal{D}'\}$.

By identifying $\mathbf{C}^{m}$ with $\mathbf{R}^{2m}$, we define

$$L^{p}(\mathbf{R}^{n},\mathbf{C}^{m}) \text{ and some other spaces of } \mathbf{C}^{m}\text{-valued functions.}$$

When we treat $\mathbf{C}^m$-valued functions in Sections 21, 24 and 26, in some cases we disregard the complex structure of $\mathbf{C}^m$ and completely identify $\mathbf{C}^m$ with $\mathbf{R}^{2m}$. Before reading those sections, please read Remark 21.1.

Very often we will use the following abbreviations:

$$\int \cdots dx = \int_{\mathbf{R}^n} \cdots dx, \quad \iint \cdots dxdt/t = \iint_{\mathbf{R}_+^{n+1}} \cdots dxdt/t.$$

The letters C and c denote various positive constants.

Next, we sketch the backbround of real Hardy spaces.

Let

$$p \in (0,1], \quad \delta, \delta' \in (0, +\infty).$$

Let us call by the name $\mathcal{H}^p(\mathbf{R}_+^2)$ the space of analytic functions $F(x+it)$ on $\mathbf{R}_+^2 = \{(x,t) : x \in \mathbf{R}, \ t > 0\}$ such that

$$\left(\|F\|_{\mathcal{H}^p(\mathbf{R}_+^2)} = \right) \sup_{t>0} \left\{ \int_{-\infty}^{+\infty} |F(x+it)|^p dx \right\}^{1/p} < \infty.$$

For a real-valued harmonic function $u(x,t)$ defined on $\mathbf{R}_+^2$ and $x \in \mathbf{R}$ let

$$N_\delta u(x) = \sup \left\{ |u(y,t)| : (y,t) \in \mathbf{R}_+^2, \ |y-x| < \delta t \right\}.$$

D. L. Burkholder–R. F. Gundy–M. L. Silverstein [71] showed that a real-valued harmonic function $u(x,t)$ is the real part of some $F \in \mathcal{H}^p(\mathbf{R}_+^2)$ if and only if

$$\int_{-\infty}^{+\infty} N_\delta u(x)^p dx < \infty$$

and that

$$c(p,\delta)\|N_\delta u\|_{L^p(\mathbf{R}^1)} \le \|F\|_{\mathcal{H}^p(\mathbf{R}_+^2)} \le C(p,\delta)\|N_\delta u\|_{L^p(\mathbf{R}^1)}. \qquad (0.1)$$

This showed that we could remove the analyticity from the study of $\mathcal{H}^p(\mathbf{R}_+^2)$.

C. Fefferman and E. M. Stein [72] showed a lot of things. Let $n \in \mathbf{N}$. For a real-valued harmonic function $u(x,t) \ (= u(x_1, \cdots, x_n, t))$ defined on $\mathbf{R}_+^{n+1} = \{(x,t) : x = (x_1, \cdots, x_n) \in \mathbf{R}^n, \ t > 0\}$ and for $x \in \mathbf{R}^n$ let

$$N_\delta u(x) = \sup \left\{ |u(y,t)| : (y,t) \in \mathbf{R}_+^{n+1}, \ |y-x| < \delta t \right\}. \qquad (0.2)$$

For a distribution $f \in \mathcal{S}(\mathbf{R}^n)'$, for

$$\phi \in \mathcal{S}(\mathbf{R}^n) \quad \text{such that} \quad \int \phi \, dx \ne 0 \qquad (0.3)$$

and for $x \in \mathbf{R}^n$ let

$$N_{\phi,\delta} f(x) = \sup \left\{ |f * (\phi)_t(y)| : (y,t) \in \mathbf{R}_+^{n+1}, \ |y-x| < \delta t \right\}, \qquad (0.4)$$

where $*$ denotes the convolution on $\mathbf{R}^n$ and $(\phi)_t(x) = t^{-n}\phi(x/t)$.
With many other important results, C. Fefferman and E. M .Stein [72] showed
that if $u(x,t)$ is a real-valued harmonic function on $\mathbf{R}_+^{n+1}$ and if

$$N_\delta u \in L^p(\mathbf{R}^n),$$

then there exists a tempered distribution $f \in \mathcal{S}(\mathbf{R}^n)'$ such that

$$u(x,t) = \lim_{\varepsilon \to +0} f * \mathcal{F}^{-1}\left\{(1 - \eta(\xi/\varepsilon))\,e^{-2\pi t|\xi|}\right\}(x), \qquad (0.5)$$

where $\eta \in \mathcal{S}(\mathbf{R}^n)$ and $\eta(\xi) \equiv 1$ near $\xi = 0$, and

$$c(p,\delta,\delta',\phi)\,\|N_\delta u\|_{L^p(\mathbf{R}^n)} \le \|N_{\phi,\delta'}f\|_{L^p(\mathbf{R}^n)} \le C(p,\delta,\delta',\phi)\,\|N_\delta u\|_{L^p(\mathbf{R}^n)}. \qquad (0.6)$$

Conversely, if $N_{\phi,\delta'}f \in L^p(\mathbf{R}^n)$, then $u(x,t)$ in (0.5) is well defined and harmonic on $\mathbf{R}_+^{n+1}$ and (0.6) holds. (Note that $e^{-2\pi t|\xi|}$ is the Fourier transform of $P(x,t) = C(n)t(|x|^2 + t^2)^{-(n+1)/2}$ which is the Poisson kernel of $\mathbf{R}_+^{n+1}$.)

Taking these facts into account, let us define

$$H_N^p(\mathbf{R}^n) = \left\{f \in \mathcal{S}(\mathbf{R}^n)' : N_{\phi,\delta}f \in L^p(\mathbf{R}^n)\right\},$$

where ϕ is as in (0.3). (It is known that this definition is independent of the choice of ϕ and δ.) Then, combining (0.1) and (0.6), we can identify $\mathcal{H}^p(\mathbf{R}_+^2)$ with $H_N^p(\mathbf{R}^1)$. Thus, we can remove even the harmonicity from the study of $\mathcal{H}^p(\mathbf{R}_+^2)$. Namely, we can remove differential equations from the study of $\mathcal{H}^p(\mathbf{R}_+^2)$.

The above argument holds for all $p \in (0, +\infty)$. But, if $p \in (1, +\infty)$, then $H_N^p(\mathbf{R}^n)$ is nothing but $L^p(\mathbf{R}^n)$. So, our concern is the case

$$p \in (0,1], \qquad (0.7)$$

especially the case $p = 1$. The space $H_N^1(\mathbf{R}^n)$ is a proper subspace of $L^1(\mathbf{R}^n)$. And, another important result of C. Fefferman–E. M. Stein [72] (and C. Fefferman [71]) is the discovery of the dual space of $H_N^1(\mathbf{R}^n)$. They showed that the dual space of $H_N^1(\mathbf{R}^n)$ can be identified with $\mathrm{BMO}(\mathbf{R}^n)$, where for a real-valued $g \in L_{\mathrm{loc}}^1(\mathbf{R}^n)$ we define

$$\|g\|_{\mathrm{BMO}} = \sup\left\{\inf_{c \in \mathbf{R}} \frac{1}{|B|}\int_B |g(x) - c|\,dx : B \text{ is taken over all balls in } \mathbf{R}^n\right\},$$

$$\mathrm{BMO}(\mathbf{R}^n) = \left\{g \in L_{\mathrm{loc}}^1(\mathbf{R}^n) : \|g\|_{\mathrm{BMO}} < \infty\right\}.$$

(The H_N^1-BMO duality means that T is a continuous linear functional on H_N^1 if and only if there exists $g \in \mathrm{BMO}(\mathbf{R}^n)$ such that

$$T(f) = \int f(x)g(x)dx \quad \text{for any } f \in \mathcal{S}(\mathbf{R}^n) \cap H_N^1(\mathbf{R}^n)$$

and

$$c(\delta,\phi)\|T\|_{(H^1_N)'} \le \|g\|_{\text{BMO}} \le C(\delta,\phi)\|T\|_{(H^1_N)'}\, .)$$

We continue to assume (0.7). We call a function $a(x) \in L^\infty(\mathbf{R}^n)$ by the name (p,∞)-atom if there exists a ball B in $\mathbf{R}^n$ such that

$$\text{supp}\, a \subset B, \quad \|a\|_{L^\infty} \le |B|^{-1/p},$$

$$\int a(x)x^\alpha dx = 0 \text{ if } |\alpha| \le n\,(1/p - 1).$$

For $f \in \mathcal{D}(\mathbf{R}^n)'$ let

$$\|f\|_{H^p_{at}} = \inf\left\{\left(\sum_{j=1}^\infty |\lambda_j|^p\right)^{1/p} : \text{there exists a sequence of } (p,\infty)\right.$$

$$\left. \text{-atoms } \{a_j\} \text{ such that } f = \sum_{j=1}^\infty \lambda_j a_j \text{ in } \mathcal{D}(\mathbf{R}^n)'\right\},$$

$$H^p_{at}(\mathbf{R}^n) = \left\{f \in \mathcal{D}(\mathbf{R}^n)' : \|f\|_{H^p_{at}} < \infty\right\},$$

where $\inf \emptyset = \infty$. Since it is clear that

$$\|g\|_{\text{BMO}} = \sup\left\{\left|\int g(x)a(x)dx\right| : a(x) \text{ is taken over all } (1,\infty)\text{-atoms}\right\},$$

with some additional argument we can show that the dual space of $H^1_{at}(\mathbf{R}^n)$ is also $\text{BMO}(\mathbf{R}^n)$. Then this fact and the H^1_N-BMO duality, with a slight additional argument, imply

$$H^1_N(\mathbf{R}^n) = H^1_{at}(\mathbf{R}^n). \tag{0.8}$$

(The fact (0.8) seems to have been discovered by C. Fefferman.)
R. Coifman [74] extended (0.8) to all $p \in (0,1]$, namely, he showed

$$H^p_N(\mathbf{R}^n) = H^p_{at}(\mathbf{R}^n) \text{ for all } p \in (0,1]$$

and

$$c(p,\delta,\phi)\|f\|_{H^p_{at}} \le \|N_{\phi,\delta}f\|_{L^p} \le C(p,\delta,\phi)\|f\|_{H^p_{at}}. \tag{0.9}$$

(R. Coifman showed the case $n = 1$. R. Latter [78] showed the case $n \ge 2$.)

In this book, we adopt $H^p_{at}(\mathbf{R}^n)$ as the definition of our real Hardy space $H^p(\mathbf{R}^n)$. Since $H^p_{at}(\mathbf{R}^1)$ can be identified with $\mathcal{H}^p(\mathbf{R}^2_+)$ by (0.1), (0.6) and (0.9), $\mathcal{H}^p_{at}(\mathbf{R}^n)$ can be regarded as one natural generalization of $\mathcal{H}^p(\mathbf{R}^2_+)$. In Sections 1–13 we will show several characterizations of $H^p(\mathbf{R}^n)$, including (0.9).

For $f \in L^2(\mathbf{R}^1)\, (= L^2(\mathbf{R}^1,\mathbf{R}))$ let

$$Hf = \mathcal{F}^{-1}(-i\mathrm{sign}\,\xi \mathcal{F} f(\xi)).$$

This H is called the Hilbert transform. Note that if $f \in L^2(\mathbf{R}^1)$ and if

$$
\begin{aligned}
F(x+it) &= \mathcal{F}^{-1}\{e^{-2\pi t|\xi|}\mathcal{F}(f+iHf)(\xi)\}(x) \\
&\left(= 2\mathcal{F}^{-1}\{e^{-2\pi t\xi}\chi_{(0,+\infty)}(\xi)\mathcal{F}f(\xi)\}(x)\right),
\end{aligned}
$$

then $F(x+it)$ is analytic on $\mathbf{R}_+^2$ and the real part of F is $\mathcal{F}^{-1}\{e^{-2\pi t|\xi|}\mathcal{F}f(\xi)\}(x)$ $(= u(x,t)$ in (0.5)). So, we have the following characterization of $H_{at}^p(\mathbf{R}^1)$ in terms of the Hilbert transform: if $f \in H_{at}^p(\mathbf{R}^1) \cap L^2(\mathbf{R}^1)$, then

$$
\begin{aligned}
\|f\|_{H_{at}^p} &\approx \|N_{\phi,\delta}f\|_{L^p} \approx \|N_\delta u\|_{L^p} \text{ by (0.9) and (0.6)} \\
&\approx \|F\|_{\mathcal{H}^p(\mathbf{R}_+^2)} \text{ by (0.1)} \\
&= \sup_{t>0}\left\{\int_{-\infty}^{+\infty}|F(x+it)|^p dx\right\}^{1/p} \\
&= \lim_{t\to+0}\{\cdots\cdots\cdots\cdots\}^{1/p} \qquad \text{by the subharmonicity of } |F|^p \\
&= \left\{\int_{-\infty}^{+\infty}|f(x)+iHf(x)|^p dx\right\}^{1/p} \\
&\approx \|f\|_{L^p} + \|Hf\|_{L^p}. \qquad\qquad\qquad\qquad\qquad (0.10)
\end{aligned}
$$

In Sections 14–27, we will extend (0.10) to $H_{at}^p(\mathbf{R}^n)$ for $n \geq 2$.

For $f \in L^2(\mathbf{R}^n)$ let

$$R_j f = \mathcal{F}^{-1}\{-i\xi_j|\xi|^{-1}\mathcal{F}f(\xi)\}, \quad j = 1,\cdots,n.$$

These R_j's are called the Riesz transforms. For the sake of convenience let

$$R_0 f = f.$$

When we try to treat $\mathcal{H}^p(\mathbf{R}_+^2)$ from the view point of real analysis, what plays a pivotal role is the subharmonicity of $|F|^p$. (In the proof of the first inequality of (0.1), the subharmonicity of $|F|^p$ comes in.) In the 1960's, E. Stein–G. Weiss [60] and A. P. Calderón–A. Zygmund [64] obtained very important results on the subharmonicity concerning certain vector-valued harmonic functions defined on a domain in $\mathbf{R}^m$, where $m \geq 3$. As a result of their theorems we have that if $n \geq 2$, if $p \geq (n-1)/n$ and if $f \in L^2(\mathbf{R}^n) (= L^2(\mathbf{R}^n, \mathbf{R}))$, then

$$\left\{\sum_{j=0}^{n}\left(\int_{\mathbf{R}^n} P(y,t)R_j f(x-y)dy\right)^2\right\}^{p/2}$$

is subharmonic as a function of $(x,t) \in \mathbf{R}_+^{n+1}$, where $P(y,t)$ is the Poisson kernel of $\mathbf{R}_+^{n+1}$. In general, if $n \geq 2$, $k \in \mathbf{N}$, $p \geq (n-1)/(n+k-1)$ and if $f \in L^2(\mathbf{R}^n)$, then

$$\left\{\sum_{j_1=0}^{n}\cdots\sum_{j_k=0}^{n}\left(\int P(y,t)R_{j_1}\cdots R_{j_k}f(x-y)dy\right)^2\right\}^{p/2}$$

is subharmonic on $\mathbf{R}_+^{n+1}$. Using this fact, C. Fefferman–E. M. Stein [72] showed that if $k \in \mathbf{N}$, $p > (n-1)/(n+k-1)$, and if $f \in L^2(\mathbf{R}^n)$, then

$$\|N_{\phi,\delta}f\|_{L^p} \approx \sum_{j_1=0}^{n}\cdots\sum_{j_k=0}^{n}\|R_{j_1}\cdots R_{j_k}f\|_{L^p}. \tag{0.11}$$

We will explain this result in Sections 14–17.

The equivalence (0.11) for the case $p=1$ and $k=1$, combined with (0.9), turns out to be

$$\|f\|_{H_{at}^1} \approx \|f\|_{L^1} + \sum_{j=1}^{n}\|R_jf\|_{L^1}. \tag{0.12}$$

Taking the dual of (0.12), we have

$$\mathrm{BMO}(\mathbf{R}^n) = L^\infty(\mathbf{R}^n) + \sum_{j=1}^{n}R_jL^\infty(\mathbf{R}^n)$$

at least formally. This is the so-called Fefferman-Stein decomposition of $\mathrm{BMO}(\mathbf{R}^n)$. In Sections 21–28, we will discuss the Fefferman-Stein decomposition of BMO and its extensions.

Finally, we prepare some definitions and results.

Definition 0.1. For a measurable function $f(x)$ defined on $\mathbf{R}^n$, for $x \in \mathbf{R}^n$ and for $p > 0$, let

$$Mf(x) = \sup\{\mathrm{av}(|f|, B(x,t)) : t > 0\},$$
$$M_pf(x) = M(|f|^p)(x)^{1/p},$$

where $\mathrm{av}(|f|, B(x,t)) = |B(x,t)|^{-1}\int_{B(x,t)}|f(y)|dy$.

Theorem 0.1. *Let f be a measurable function defined on $\mathbf{R}^n$. Then*
(i) *if $\lambda > 0$, then*

$$\lambda|\{x \in \mathbf{R}^n : Mf(x) > \lambda\}| \leq C_{0.1}(n)\|f\|_{L^1},$$

(ii) *if $p \in (1, +\infty]$, then*

$$\|Mf\|_{L^p} \leq C_{0.2}(p,n)\|f\|_{L^p}.$$

Corollary 0.1. *Let $f \in L_{\mathrm{loc}}^1$. Then*

$$\lim_{t\to+0}\mathrm{av}(|f(\cdot)-f(x)|, B(x,t)) = 0 \quad \text{a.e. } x \in \mathbf{R}^n. \tag{0.13}$$

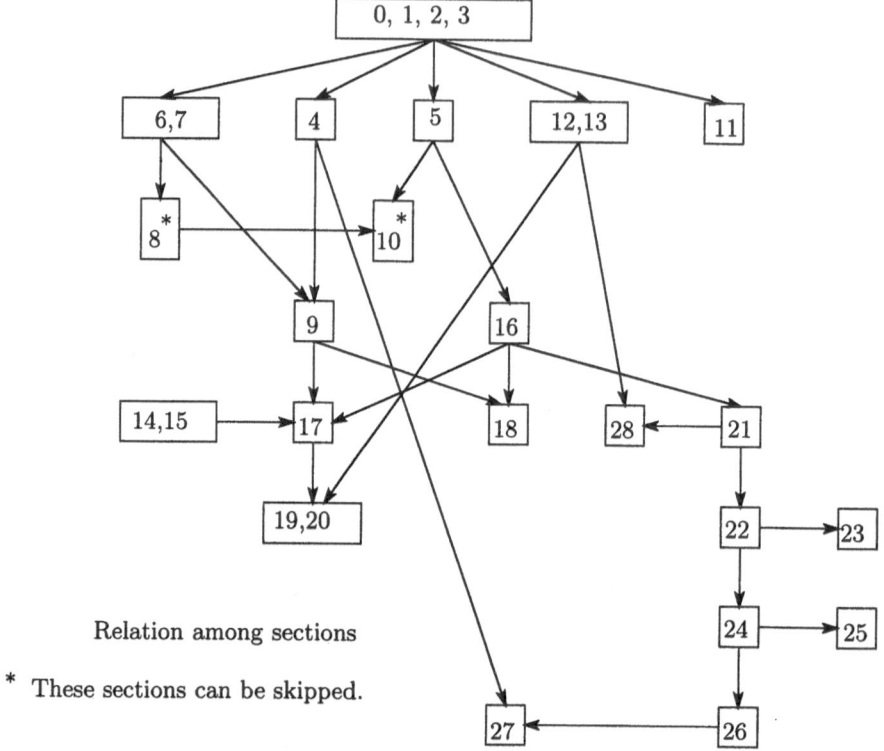

Relation among sections

* These sections can be skipped.

Corollary 0.2. *Let* $f \in L_{\mathrm{loc}}^1$, $\phi \in L^\infty$ *and let* $\operatorname{supp} \phi$ *be compact. Then*

$$f * (\phi)_t(y) \to f(x) \int_{\mathbf{R}^n} \phi(y)dy \qquad ((y,t) \in \Gamma(x,\delta), \ t \to +0)$$

for any $\delta > 0$ *for a.e.* $x \in \mathbf{R}^n$.

Corollary 0.3. *Let* $E \subset \mathbf{R}^n$ *be a measurable set. Then*

$$\lim_{t \to +0} \frac{|B(x,t) \cap E|}{|B(x,t)|} = 1 \text{ for a.e. } x \in E. \tag{0.14}$$

Corollary 0.4. *Let* $0 < q < p \le \infty$ *and* $f \in L_{\mathrm{loc}}^q$. *Then*

$$\|M_q f\|_{L^p} \le C(p,q,n)\|f\|_{L^p} .$$

For the proofs of Theorem 0.1 and Corollary 0.1 see E. M. Stein [70] pp. 4–12. Corollaries 0.2–0.4 are easy from Theorem 0.1 and Corollary 0.1. Mf is called the Hardy-Littlewood maximal function. The set of $x \in \mathbf{R}^n$ at which (0.13) holds is called the Lebesgue set of f. When x satisfies (0.14), x is called a point of density of E.

Remark 0.1. In Theorem 0.1, we can take $C_{0.2}(p,n) = 2(5^n p/(p-1))^{1/p}$. Recently, E. M. Stein [82] showed that we can take a constant $C_{0.2}$ independent of the dimension n. (See also E. M. Stein [83,76] E. M. Stein–J. O. Strömberg [83], M. Cowling–G. Mauceri [85], J. Bourgain [86a, b, c] and A. Carbery [86].)

Notes. As for the famous books that relate to our topics, there are E. M. Stein [70], E. M. Stein–G. Weiss [71], J. B. Garnett [81], García-Cuerva–Rubio de Francia [85], J. L. Journé [83] and A. Torchinsky [86]. (As for the reference of maximal theorems, there is M. de Guzman [70], too.)

As for the expository papers on $H^p(\mathbf{R}^n)$ and BMO($\mathbf{R}^n$), there are C. Fefferman [74,76], L. Carleson [81] and G. Weiss [89]. Some research problems are listed in G. Weiss [79] and V. P. Havin–S. V. Hruscev–N. K. Nikol'skii [84].

I. Lipschitz spaces and BMO

In Sections 1 and 2 we define Lipschitz spaces and atomic H^p spaces.

Definition 1.1. For $a \geq 0$ and $f \in L^1_{\text{loc}} = L^1_{\text{loc}}(\mathbf{R}^n)$ let

$$\|f\|_{\Lambda_a} = \sup_B \inf_{P:\deg P \leq a} \int_B |f(y) - P(y)| dy |B|^{-1-a/n},$$

where B is taken over all balls in $\mathbf{R}^n$ and where P is taken over all polynomials of $y_1, \cdots, y_n$ with degree $\leq a$. Let

$$\Lambda_a = \left\{ f \in L^1_{\text{loc}} : \|f\|_{\Lambda_a} < +\infty \right\},$$

$$\|f\|_{\text{BMO}} = \|f\|_{\Lambda_0} \text{ and } \text{BMO} = \Lambda_0 .$$

For an open set $\Omega \subset \mathbf{R}^n$ and $f \in L^1_{\text{loc}}(\Omega)$ let

$$\|f\|_{\Lambda_a(\Omega)} = \sup_{B:2B \subset \Omega} \inf_{P:\deg P \leq a} \int_B |f(y) - P(y)| dy |B|^{-1-a/n}$$

and

$$\|f\|_{\text{BMO}(\Omega)} = \|f\|_{\Lambda_0(\Omega)}.$$

Remark 1.1. If $f \in L^1_{\text{loc}}$ and if $E \subset \mathbf{R}^n$ is bounded and $|E| > 0$, then

$$\text{av}(|f - \text{av}(f, E)|, E) = \text{av}(|f - c - \text{av}(f - c, E)|, E) \leq 2\,\text{av}(|f - c|, E)$$

for any $c \in \mathbf{R}$. Thus, in the case $a \in [0, 1)$, we have

$$\|f\|_{\Lambda_a} \leq \sup_B \text{av}(|f - \text{av}(f, B)|, B)|B|^{-a/n} \leq 2\|f\|_{\Lambda_a},$$

where B is taken over all balls in $\mathbf{R}^n$.

Remark 1.2. The following estimate will be sometimes used: If B is a ball and if $E \subset B$ and $|E| > 0$, then

$$
\begin{aligned}
|\text{av}(f, E) - \text{av}(f, B)| &= |\text{av}(f - \text{av}(f, B), E)| \\
&\leq \text{av}(|f - \text{av}(f, B)|, E) \\
&\leq \frac{|B|}{|E|}\text{av}(|f - \text{av}(f, B)|, B) \\
&\leq \frac{|B|}{|E|} \cdot 2\|f\|_{\text{BMO}}.
\end{aligned}
$$

Remark 1.3. As a consequence of Remark 1.2, it holds that if B is a ball and if $k = -1, -2, \cdots$, then

$$\left| av(f, B) - av(f, 2^k B) \right| \leq \sum_{j=k}^{-1} \left| av(f, 2^{j+1} B) - av(f, 2^j B) \right|$$

$$\leq C(n)|k|\|f\|_{\text{BMO}}.$$

The same thing holds for $k = 1, 2, \cdots$.

Remark 1.4. Since $\left| |f(x)| - |c| \right| \leq |f(x) - c|$, we have

$$\||f|\|_{\Lambda_a} \leq \|f\|_{\Lambda_a} \qquad \text{if } a \in [0, 1).$$

Definition 1.2. Let $p \in (0, 1]$. A function $a(x) \in L^\infty$ is called a (p, ∞)-*atom* if there exists a ball B such that

$$\text{supp}\, a(x) \subset B, \tag{1.1}$$
$$\|a(x)\|_{L^\infty} \leq |B|^{-1/p}, \tag{1.2}$$
$$\int a(x) x^\alpha dx = 0 \text{ provided } |\alpha| \leq n(1/p - 1). \tag{1.3}$$

Lemma 1.1. *Let $p \in (0, 1]$ and $f \in \Lambda_{n(1/p-1)}$. Then*

$$\|f\|_{\Lambda_{n(1/p-1)}} = \sup \left\{ \left| \int f(x) a(x) dx \right| : a(x) \text{ is taken over all } (p, \infty)\text{-atoms} \right\}.$$

This is clear from the argument of dual spaces.

Lemma 1.2. *Let $a > 0$, $\phi \in \mathcal{D}$, $\int \phi(x) dx = 1$ and $f \in L^1_{\text{loc}}$. Then the followings are equivalent:*

(i) $\sup \left\{ \left| D_t^k D_x^\alpha \left(f * (\phi)_t(x) \right) \right| t^{k+|\alpha|-a} : (x, t) \in \mathbf{R}_+^{n+1} \right\} < +\infty$
      for all k, α such that $k + |\alpha| > a$,

(ii) $\|f\|_{\Lambda_a} < +\infty$,

(iii) $\sup_{B} \inf_{P: \deg P \leq a} \|f - P\|_{L^\infty(B)} |B|^{-a/n} < +\infty$,

(iv) *there exists a $C^{[a]+1}$-function $u(x, t)$ defined on $\mathbf{R}_+^{n+1}$ such that*
      $$\sup \left\{ \left| D_t^k D_x^\alpha u(x, t) \right| t^{[a]+1-a} : (x, t) \in \mathbf{R}_+^{n+1}, \ k + |\alpha| = [a] + 1 \right\}$$
      $$< +\infty, \quad (1.4)$$
      $$u(x, t) \to f(x) \ (t \to +0) \text{ a.e. } x \in \mathbf{R}^n.$$

Furthermore,

$c(\phi, a, \alpha, k)$ {*the left-hand side of* (i)}

$\leq$ {*the left-hand side of* (ii)}

$\leq$ {*the left-hand side of* (iii)}

$\leq C(a,n)$ {*the left-hand side of* (iv) (1.4)},

{*the case* $u(x,t) = f * (\phi)_t(x)$ *of the left-hand side of* (iv) (1.4)}

$$\approx \sum_{\alpha,k:|\alpha|+k=[a]+1} \{\textit{the left-hand side of } (i)\}$$

$\leq C(a,\phi)$ {*the left-hand side of* (ii)}.

The implications (i) $\rightarrow$ (iv) and (iii) $\rightarrow$ (ii) are clear.

proof of (ii) $\rightarrow$ (i). Since

$$\mathcal{F}\left(D_x^\alpha D_t^k(\phi)_t\right)(\xi) = (2\pi i \xi)^\alpha \sum_{j_1=1}^n \cdots \sum_{j_k=1}^n \xi_{j_1} \cdots \xi_{j_k} \left(D_{\xi_{j_1}} \cdots D_{\xi_{j_k}} \mathcal{F}\phi\right)(t\xi),$$

by taking the Fourier transform we can see that

$$\int D_x^\alpha D_t^k(\phi)_t(x) x^\beta dx = 0 \text{ provided } |\beta| < |\alpha| + k.$$

Thus, if $|\alpha| + k > a$, then $c(\phi, a, \alpha, k) t^{|\alpha|+k-a} D_x^\alpha D_t^k(\phi)_t(x)$ is an $(n/(n+a), \infty)$-atom. So, Lemma 1.1 implies

$$\left|D_x^\alpha D_t^k (f * (\phi)_t(x))\right| = \left|f * D_x^\alpha D_t^k(\phi)_t(x)\right| \leq C(\phi, a, \alpha, k) t^{a-|\alpha|-k} \|f\|_{\Lambda_a}.$$
$$(1.5)$$

$\square$

proof of (iv) $\rightarrow$ (iii). Take any ball $B = B(x_0, t_0)$. Let

$$p(x,t) = \sum_{|\alpha|+k\leq a} \frac{1}{\alpha! k!} D_x^\alpha D_t^k u(x_0,t_0)(x-x_0)^\alpha (t-t_0)^k.$$

Then,

$|f(x) - p(x,0)|$

$= \lim_{\varepsilon \to +0} |u(x,\varepsilon) - p(x,\varepsilon)| \quad$ a.e. $x \in \mathbf{R}^n$

$= \lim \left| \frac{(-1)^{[a]}}{[a]!} \int_{t_0}^\varepsilon (t-\varepsilon)^{[a]} \frac{d^{[a]+1}}{dt^{[a]+1}} \left\{ u\left(x_0 + \frac{t_0-t}{t_0-\varepsilon}(x-x_0), t\right) \right\} dt \right|$

by repeating integrations by parts on the line segment connecting the points (x_0, t_0) and (x, ε)

$\leq \lim C \int_\varepsilon^{t_0} (t-\varepsilon)^{[a]} \sum_{\substack{k,\alpha: \\ k+|\alpha|=[a]+1}} \left(\frac{|x-x_0|}{t_0-\varepsilon}\right)^{|\alpha|} \left|D_t^k D_x^\alpha u\left(x_0 + \frac{t_0-t}{t_0-\varepsilon}(x-x_0), t\right)\right| dt$

$$\leq C \int_0^{t_0} t^{[a]} \left(1 + \frac{|x - x_0|}{t_0}\right)^{[a]+1} \times t^{a-[a]-1} \{\text{the left-hand side of } (1.4)\} \, dt$$

$$\leq C t_0^a \left(1 + \frac{|x - x_0|}{t_0}\right)^{[a]+1} \{\text{the left-hand side of } (1.4)\}.$$

So, if we put $P(x) = p(x, 0)$, then

$$\|f - P\|_{L^\infty(B)} \leq C t_0^a \{\text{the left-hand side of } (1.4)\}.$$

$\square$

Remark 1.5. Let $k \in \mathbf{N}_0$. Let ϕ be as in Lemma 1.2. If $a \in (k, k+1)$ and if $f \in \Lambda_a$, then Lemma 1.2 (i) implies

$$\left|\nabla_{x,t} \nabla_x^k \left(f * (\phi)_t(x)\right)\right| \leq C(a, \phi, f) t^{a-k-1} \qquad \text{for } (x, t) \in \mathbf{R}_+^{n+1}.$$

Since $a - k - 1 > -1$, this estimate implies that $\nabla_x^k \left(f * (\phi)_t(x)\right)$ can be extended continuously to the closure of $\mathbf{R}_+^{n+1}$. If $f \in \Lambda_{k+1}$, then Lemma 1.2 (i) implies

$$\left|\nabla_{x,t}^2 \nabla_x^k \left(f * (\phi)_t(x)\right)\right| \leq C(k, \phi, f) t^{-1} \qquad \text{for } (x, t) \in \mathbf{R}_+^{n+1},$$

which implies

$$\left|\nabla_{x,t} \nabla_x^k \left(f * (\phi)_t(x)\right)\right| \leq C(k, r, \phi, f) \log(2+1/t) \qquad \text{for } (x, t) \in B(0, r) \times (0, r)$$

for any $r > 0$. Thus $\nabla_x^k \left(f * (\phi)_t(x)\right)$ can be extended continuously to the closure of $\mathbf{R}_+^{n+1}$. Consequently, if $a \in (k, k+1]$, then Λ_a can be regarded as a subspace of $C^k(\mathbf{R}^n)$. In particular, $\bigcup_{a>0} \Lambda_a \subset C(\mathbf{R}^n)$.

Remark 1.6. If $a \in (0, 1)$, then Lemma 1.2 (iii) implies

$$\|f\|_{\Lambda_a} \approx \sup_{\substack{x, y \in \mathbf{R}^n \\ x \neq y}} \frac{|f(x) - f(y)|}{|x - y|^a}.$$

The following is clear from the definition of Λ_1

$$\|f\|_{\Lambda_1} \approx C \sup_{\substack{x, y \in \mathbf{R}^n \\ x \neq y}} \frac{|f(x) - f(y)|}{|x - y|}.$$

Remark 1.7. If

$$f \in \Lambda_1 \text{ and } \operatorname{supp} f \subset B(0, 1),$$

then it holds that

$$C(n) \|f\|_{\Lambda_1} \geq \sup_{\substack{x, y \in \mathbf{R}^n \\ x \neq y}} \frac{|f(x) - f(y)|}{|x - y| \log\left(2 + |x - y|^{-1}\right)}.$$

We will show this. Let ϕ be as in Lemma 1.2 and let $u(x,t) = f * (\phi)_t(x)$. The existence of the polynomial P such that

$$\deg P \leq 1 \quad \text{and} \quad \|f - P\|_{L^1(B(0,2))} \leq |B(0,2)|^{1+1/n} \|f\|_{\Lambda_1}$$

implies

$$\begin{aligned}
\|f\|_{L^1} &\leq \|f - P\|_{L^1(B(0,1))} + \|P\|_{L^1(B(0,1))} \leq \cdots + C\|P\|_{L^1(B(0,2)\setminus B(0,1))} \\
&= \cdots + C\|f - P\|_{L^1(B(0,2)\setminus B(0,1))} \leq C\|f - P\|_{L^1(B(0,2))} \leq C\|f\|_{\Lambda_1}.
\end{aligned}$$

So,

$$\begin{aligned}
|\nabla_{t,x} u(x,t)| &= |f * \nabla_{t,x}(\phi)_t(x)| \leq \|f\|_{L^1} \|\nabla_{t,x}(\phi)_t\|_{L^\infty} \\
&\leq C\|f\|_{\Lambda_1} t^{-n-1} \leq C\|f\|_{\Lambda_1} \qquad \text{if } t \geq 1.
\end{aligned}$$

Since

$$|\nabla^2_{t,x} u(x,t)| \leq Ct^{-1}\|f\|_{\Lambda_1} \quad \text{for any } (x,t) \in \mathbf{R}^{n+1}_+ \text{ by Lemma 1.2 (i)},$$

we get that if $t \in (0,1]$, then

$$\begin{aligned}
|\nabla_{t,x} u(x,t)| &\leq |\nabla_{t,x} u(x,1)| + \int_t^1 |\nabla^2_{t,x}(x,t)| \, dt \\
&\leq C\|f\|_{\Lambda_1} + C\log(2 + 1/t) \cdot \|f\|_{\Lambda_1}.
\end{aligned}$$

Thus, if $B = B(x_0, t_0)$, then

$$\begin{aligned}
\text{ess} \cdot \sup_{x \in B} |f(x) - u(x_0, t_0)| \\
= \sup_{x \in B} \lim_{\varepsilon \to +0} |u(x,\varepsilon) - u(x_0, t_0)| \\
= \sup \lim \left| \int_{t_0}^{\varepsilon} \frac{d}{dt}\left\{ u\left(x_0 + \frac{t_0 - t}{t_0 - \varepsilon}(x - x_0), t\right)\right\} dt \right| \\
\leq C \sup \lim \int_{\varepsilon}^{t_0} \left| \nabla_{t,x} u\left(x_0 + \frac{t_0 - t}{t_0 - \varepsilon}(x - x_0), t\right)\right| dt \quad \text{by } \frac{|x - x_0|}{t_0} \leq 1 \\
\leq C \int_0^{t_0} \log(2 + 1/t) \|f\|_{\Lambda_1} dt \\
\leq Ct_0 \log(2 + 1/t_0) \|f\|_{\Lambda_1}.
\end{aligned}$$

This implies the desired estimate.

Lemma 1.3. *Let $a > 1$. Let $f \in C^1(\mathbf{R}^n)$. Then*

$$c(a,n)\|f\|_{\Lambda_a} \leq \sum_{j=1}^n \|D_{x_j} f\|_{\Lambda_{a-1}} \leq C(a,n)\|f\|_{\Lambda_a}. \tag{1.6}$$

Proof Let ϕ be as in Lemma 1.2. If $k + |\alpha| > a - 1$, then

$$\left| D_t^k D_x^\alpha \left((D_{x_j} f) * (\phi)_t(x) \right) \right| t^{(k+|\alpha|)-(a-1)}$$
$$= \left| D_t^k D_x^\alpha D_{x_j} \left(f * (\phi)_t(x) \right) \right| t^{(k+|\alpha|+1)-a} \le C\|f\|_{\Lambda_a} \text{ by (1.5)}.$$

So, the second inequality of (1.6) follows from applying Lemma 1.2 (i) to $D_{x_j} f$ with $a - 1$ in place of a.

Similarly, if $k + |\alpha| + 1 > a$, then

$$\left| D_t^k D_x^\alpha D_{x_j} \left(f * (\phi)_t(x) \right) \right| t^{(k+|\alpha|+1)-a}$$
$$= \left| D_t^k D_x^\alpha \left((D_{x_j} f) * (\phi)_t(x) \right) \right| t^{(k+|\alpha|)-(a-1)} \le C\|D_{x_j} f\|_{\Lambda_{a-1}}, \quad (1.7)$$

and if $k > a$, then

$$\left| D_t^k \left(f * (\phi)_t(x) \right) \right| t^{k-a} = \left| \sum_{j=1}^n D_t^{k-1} \left((D_{x_j} f) * (\phi_j)_t(x) \right) \right| t^{(k-1)-(a-1)}$$

$$\text{where } \phi_j(x) = -x_j \phi(x),$$

$$\le C \sum_{j=1}^n \|D_{x_j} f\|_{\Lambda_{a-1}},$$

where the last inequality follows from (1.5) with $D_{x_j} f$, $a-1$ and ϕ_j in places of f, a and ϕ, respectively. Thus, the first inequality of (1.6) follows from Lemma 1.2 (i). $\square$

Definition 1.3. For $a > 0$ let

$$\mathcal{B}_a = \{ f \in \Lambda_a : \operatorname{supp} f \subset B(0,1), \ \|f\|_{\Lambda_a} \le 1 \}.$$

Lemma 1.4. *Let $0 < b < a$ and $f \in \mathcal{B}_a$. Then*

$$\|f\|_{\Lambda_b} \le C(a,b,n).$$

Consequently, $\|f\|_{L^\infty} \le C(a,n)$ and

$$c(a,b,n) f \in \mathcal{B}_b \text{ for some } c(a,b,n) > 0.$$

Proof.

Case 1 : $0 < a \le 1$. This case is clear from Remarks 1.6 and 1.7.
Case 2 : $0 < b \le 1 < a \le 2$.

$$\sup_{x \ne y} \frac{|f(x) - f(y)|}{|x - y|} \le \|\nabla_x f\|_{L^\infty}$$

$$\le C\|\nabla_x f\|_{\Lambda_{a-1}} \text{ by } 0 < a - 1 \le 1 \text{ and by Case 1}$$

$$\approx \|f\|_{\Lambda_a} \text{ by Lemma 1.3}.$$

So, $\|f\|_{\Lambda_b} \le C$ follows from Remark 1.6.

Case 3 : $a - b \le 1$. Let $m = \lim_{\varepsilon \to +0} [b - \varepsilon]$. Then

$$
\begin{aligned}
\|f\|_{\Lambda_b} &\approx \|\nabla_x^m f\|_{\Lambda_{b-m}} && \text{by Lemma 1.3} \\
&\le C\|\nabla_x^m f\|_{\Lambda_{a-m}} && \text{by Cases 1 and 2} \\
& && (\text{because } 0 < b - m \le 1,\ b - m < a - m \le 2\ ) \\
&\approx \|f\|_{\Lambda_a} && \text{by Lemma 1.3.}
\end{aligned}
$$

Finally, the general case follows from the repeated use of Case 3. □

Lemma 1.5. *Let* $a > 0$, $r > 0$, $f \in L^1(B(0, 2r))$, $\phi \in \Lambda_a$ *and* $\operatorname{supp} \phi \subset B(0, r)$. *Then*

$$\|\phi f\|_{\Lambda_a} \le C(a, n)\|\phi\|_{\Lambda_a} \left\{ \operatorname{av}\left(|f|, B(0, 2r)\right) + r^a\|f\|_{\Lambda_a(B(0,2r))} \right\}, \tag{1.8}$$

$$\|\phi f\|_{\text{BMO}} \le C(a, n)r^a\|\phi\|_{\Lambda_a} \left\{ \operatorname{av}\left(|f|, B(0, 2r)\right) + \|f\|_{\text{BMO}(B(0,2r))} \right\}. \tag{1.9}$$

Proof of (1.8). Since

$$r^a\|\phi f\|_{\Lambda_a} = \|\phi(r\,\cdot)f(r\,\cdot)\|_{\Lambda_a}$$

and

$$
\begin{aligned}
r^a\|\phi\|_{\Lambda_a} &\left\{ \operatorname{av}\left(|f|, B(0, 2r)\right) + r^a\|f\|_{\Lambda_a(B(0,2r))} \right\} \\
&= \|\phi(r\,\cdot)\|_{\Lambda_a} \left\{ \operatorname{av}\left(|f(r\,\cdot)|, B(0, 2)\right) + \|f(r\,\cdot)\|_{\Lambda_a(B(0,2))} \right\},
\end{aligned}
$$

it is enough to show the case $r = 1$. So, we assume

$$r = 1,\ \operatorname{av}\left(|f|, B(0, 2)\right) + \|f\|_{\Lambda_a(B(0,2))} \le 1 \text{ and } \phi \in \mathcal{B}_a. \tag{1.10}$$

Take $\eta \in \mathcal{D}(\mathbf{R}^n)$ and $v \in C^\infty(\mathbf{R}^1)$, depending only on n, so that

$$\operatorname{supp} \eta \subset B(0, 1),\ \int \eta\, dx = 1 \text{ and}$$

$$
v(t) = \begin{cases} 1 \text{ if } t \in (-\infty, 0], \\ 0 \text{ if } t \in [1/5, +\infty). \end{cases}
$$

For $(x, t) \in B(0, 3/2) \times (0, 1/4)$ let

$$
\begin{aligned}
u_f(x, t) &= f * (\eta)_t(x), \\
u_\phi(x, t) &= \phi * (\eta)_t(x).
\end{aligned}
$$

For $(x, t) \in \mathbf{R}_+^{n+1}$ let

$$u(x, t) = u_f(x, t)u_\phi(x, t)v\left(|x| - 1\right)v(t),$$

where this is defined to be zero outside the support of $v\left(|x| - 1\right)v(t)$.

By $\operatorname{av}(|f|, B(0, 2)) \le 1$ (see (1.10)) we get that if $x \in B(0, 3/2)$, then

$$\left| \nabla^k_{t,x} u_f \left(x, 1/4 \right) \right| \;\leq\; \left\| \left\{ \nabla^k_{t,x}(\eta)_t \right\} \right|_{t=1/4} \right\|_{L^\infty} \| f \|_{L^1(B(0,2))}$$
$$\leq\; C(k,n) \, \mathrm{av}(|f|, B(0,2))$$
$$\leq\; C(k,n) \qquad \text{for any } k \in \mathbf{N}_0.$$

By $\| f \|_{\Lambda_a(B(0,2))} \leq 1$ (see (1.10)) and by localizing the argument of (ii) $\to$ (i) of Lemma 1.2, we get that if $(x,t) \in B(0,3/2) \times (0,1/4)$, then

$$\left| \nabla^{[a]+1}_{t,x} u_f(x,t) \right| \leq C(a,n) t^{a-[a]-1}.$$

Then, combining these estimates gives that if $(x,t) \in B(0,3/2) \times (0,1/4)$ then

$$
\left| \nabla^{[a]}_{t,x} u_f(x,t) \right| \;\leq\; \left| \nabla^{[a]}_{t,x} u_f \left(x, 1/4 \right) \right| + \int_t^{1/4} \left| \nabla^{[a]+1}_{t,x} u_f(x,t) \right| dt
$$
$$
\leq\; \begin{cases} C(a,n) + C(a,n) & \text{if } a \notin \mathbf{N}, \\ C(a,n) + C(a,n) \log\left(2 + 1/t \right) & \text{if } a \in \mathbf{N}. \end{cases}
$$

Repeating integrations, we get that if $k \in \{0,1,2,\cdots,[a]+1\}$, then

$$
\left| \nabla^k_{t,x} u_f(x,t) \right| \leq \begin{cases} C(a,n) t^{a-[a]-1} & \text{if } k = [a]+1, \\ C(a,n) \log\left(2 + 1/t\right) & \text{if } a \in \mathbf{N} \text{ and } k = a, \\ C(a,n) & \text{if } k < a. \end{cases} \qquad (1.11)
$$

Since the estimate similar to (1.11) holds for u_ϕ, differentiating $u(x,t)$ directly gives

$$\left| \nabla^{[a]+1}_{t,x} u(x,t) \right| \leq C(a,n) t^{a-[a]-1}$$

on $\mathbf{R}^{n+1}_+$, which combined with Lemma 1.2 (iv) implies (1.8). $\square$

Proof of (1.9). We may assume

$$r = 1, \quad \mathrm{av}\left(|f|, B(0,2)\right) + \| f \|_{\mathrm{BMO}(B(0,2))} \leq 1 \text{ and } \phi \in \mathcal{B}_a. \qquad (1.12)$$

Note that Remark 1.3 and (1.12) imply that if $B \subset B(0,3/2)$, then

$$|\mathrm{av}(f,B)| \leq C \log\left(2 + |B|^{-1}\right). \qquad (1.13)$$

Take any ball B. If $\ell(B) \geq 1/2$, then

$$\mathrm{av}\left(|\phi f - \mathrm{av}(\phi f, B)|, B\right) \leq 2\mathrm{av}\left(|\phi f|, B\right) \leq C\mathrm{av}\left(|\phi f|, B(0,2)\right) \leq C$$

by $\| \phi \|_{L^\infty} \leq C$ (see Lemma 1.4) and by $\mathrm{av}(|f|, B(0,2)) \leq 1$. If $\ell(B) < 1/2$ and $B \cap B(0,1) \neq \emptyset$, then

$$\mathrm{av}\big(|\phi f - \mathrm{av}(\phi, B)\mathrm{av}(f, B)|, B\big)$$
$$\leq \mathrm{av}\big(|\phi - \mathrm{av}(\phi, B)|, B\big)|\mathrm{av}(f, B)| + \mathrm{av}\big(|\phi||f - \mathrm{av}(f, B)|, B\big)$$
$$\leq C\|\phi\|_{\Lambda_{\min\{a,1/2\}}} \ell(B)^{\min\{a,1/2\}} \cdot C \log\big(2 + |B|^{-1}\big)$$
$$+\|\phi\|_{L^\infty} \cdot 2\|f\|_{\mathrm{BMO}(B(0,2))} \qquad \text{by Remark 1.1 and (1.13)}$$
$$\leq C \quad \text{by (1.12) and by } \|\phi\|_{L^\infty} + \|\phi\|_{\Lambda_{\min\{a,1/2\}}} \leq C \text{ (see Lemma 1.4).}$$

Thus, we get (1.9). $\qquad\qquad\qquad\qquad\qquad\qquad\qquad\qquad\qquad\qquad\square$

It is easy to see that $\|f\|_{\Lambda_a}$ is zero if and only if f is itself a polynomial of degree $\leq a$. Identifying f and g if $f - g$ is a polynomial of degree $\leq a$, Λ_a becomes a Banach space endowed with the norm $\|\cdot\|_{\Lambda_a}$. Let Λ'_a be its dual space. (To be precise, we should use the notation $(\Lambda_a/\{\text{polynomials of degree} \leq a\})'$. But for simplicity we use the notation Λ'_a.)

Lemma 1.6. *Let* $a \geq 0$. *Then*

$$\left\{ g \in L^\infty : \mathrm{supp}\, g \text{ is compact and } \int g(x)x^\alpha dx = 0 \text{ if } |\alpha| \leq a \right\} \qquad (1.14)_a$$

can be regarded as a subspace of Λ'_a *by identifying* $g \in (1.14)_a$ *with the functional* "$f \in \Lambda_a \mapsto \int gf dx$".

Lemma 1.7. *Let* $a \geq 0$ *and* $g \in (1.14)_a$. *Then,*

$$c(a, n)\|g\|_{\Lambda'_a} \leq \sup\left\{ \left|\int gf dx\right| : f \in \mathcal{D}, \|f\|_{\Lambda_a} \leq 1 \right\} \leq \|g\|_{\Lambda'_a}. \qquad (1.15)$$

Proof. The second inequality is clear.

Take $r > 0$ so that $\mathrm{supp}\, g \subset B(0, r)$. Let $f \in \Lambda_a$. Then, there exists a polynomial $P(x)$ of degree $\leq a$ such that

$$\int_{B(0,2r)} |f(x) - P(x)|\, dx\, |B(0, 2r)|^{-1-a/n} \leq \|f\|_{\Lambda_a}.$$

Take $\phi, \eta \in \mathcal{D}$, depending only on the dimension n, so that $\phi(x) = 1$ on $B(0, 1)$, $\mathrm{supp}\,\phi \subset B(0, 2)$, $\int \eta dx = 1$ and $\eta(x) \geq 0$. Let

$$\tilde{f}(x) = \phi(x/r)\,(f(x) - P(x)).$$

If $a > 0$, Then (1.8) implies

$$\left\|\tilde{f}\right\|_{\Lambda_a} \leq C\|\phi(\cdot/r)\|_{\Lambda_a} \left\{ \mathrm{av}\,(|f - P|, B(0, 2r)) + r^a\|f - P\|_{\Lambda_a} \right\}$$
$$\leq Cr^{-a}\|\phi\|_{\Lambda_a} r^a\|f\|_{\Lambda_a}$$
$$\leq C(a, n)\|f\|_{\Lambda_a}.$$

If $a = 0$, then (P is a constant and) (1.9) implies

$$\|\tilde{f}\|_{\text{BMO}} \le Cr \|\phi(\cdot/r)\|_{\Lambda_1} \{\text{av}(|f-P|, B(0,2r)) + \|f-P\|_{\text{BMO}}\} \le C(n)\|f\|_{\text{BMO}}.$$

Thus, the first inequality of (1.15) follows from

$$\int g f dx = \int g \tilde{f} dx = \lim_{t \to +0} \int g(x) \tilde{f} * (\eta)_t(x) dx,$$
$$\tilde{f} * (\eta)_t \in \mathcal{D},$$
$$\|\tilde{f} * (\eta)_t\|_{\Lambda_a} \le \|\tilde{f}\|_{\Lambda_a} \le C(a,n)\|f\|_{\Lambda_a}.$$

$\square$

Definition 1.4. For $a \ge 0$ let

$$\Lambda_a^{\prime 0} = \{\text{the closure of } (1.14)_a \text{ in } \Lambda_a' \text{ with respect to the norm } \|\cdot\|_{\Lambda_a'}\}.$$

Lemma 1.8. Let $a \ge 0$. For $h \in \Lambda_a'$ let $h|_{\mathcal{D}}$ be its restriction to $\mathcal{D}(\subset \Lambda_a)$. Then, $h|_{\mathcal{D}} \in \mathcal{S}'$ and the mapping

$$h \in \Lambda_a^{\prime 0} \mapsto h|_{\mathcal{D}} \in \mathcal{S}' \tag{1.16}$$

is injective.

Proof. Let $f \in \mathcal{D}$. Then, $\|f\|_{\Lambda_0} (= \|f\|_{\text{BMO}}) \le \|f\|_{L^\infty}$ is clear. If $a > 0$, then

$$\|f\|_{\Lambda_a} \approx \|\nabla_x^m f\|_{\Lambda_{a-m}} \text{ by Lemma 1.3, where } m = \lim_{\varepsilon \to +0}[a - \varepsilon]$$
$$\le C\left(\|\nabla_x^m f\|_{L^\infty} + \|\nabla_x^{m+1} f\|_{L^\infty}\right),$$
$$\text{by } 0 < a - m \le 1 \text{ and by Remark 1.6.}$$

So,

$$\frac{|\langle f, h|_{\mathcal{D}}\rangle_{\mathcal{D}}|}{\|f\|_{L^\infty}} \le \|h\|_{\Lambda_0'} \text{ if } a = 0,$$

$$\frac{|\langle f, h|_{\mathcal{D}}\rangle_{\mathcal{D}}|}{\|\nabla_x^m f\|_{L^\infty} + \|\nabla_x^{m+1} f\|_{L^\infty}} \le \frac{\cdots\cdots}{c\|f\|_{\Lambda_a}} \le C\|h\|_{\Lambda_a'} \text{ if } a > 0.$$

This implies $h|_{\mathcal{D}} \in \mathcal{S}'$.

On the other hand, if $\{g_j\} \subset (1.14)_a$ is a Cauchy sequence in Λ_a' and if $g_j \not\to 0 \in \Lambda_a'$, then Lemma 1.7 implies the existence of $f \in \mathcal{D}$ such that

$$\|f\|_{\Lambda_a} \le 1 \text{ and } \lim_{j \to \infty} \int g_j f dx \ge c(a,n) \lim_{j \to \infty} \|g_j\|_{\Lambda_a'}.$$

Therefore, the mapping (1.16) is injective.

$\square$

Remark 1.8. Let $p \in (0,1]$, $a = n(1/p - 1)$, $\{\lambda_j\}_{j \in \mathbb{N}} \in \ell^1$ and let $\{a_j(x)\}_{j \in \mathbb{N}}$ be (p, ∞)-atoms. Then, Lemma 1.1 implies $\|a_j\|_{\Lambda'_a} \leq 1$. So, $\left\{\sum_{j=1}^m \lambda_j a_j\right\}_{m \in \mathbb{N}}$ converges in Λ'_a and

$$\lim_{m \to \infty \text{ in } \Lambda'_a} \sum_{j=1}^m \lambda_j a_j \in \Lambda'^0_a.$$

So, Lemma 1.8 implies that if

$$\lim_{m \to \infty \text{ in } \Lambda'_a} \sum_{j=1}^m \lambda_j a_j \neq 0 \ (\in \Lambda'_a),$$

then

$$\lim_{m \to \infty \text{ in } \mathcal{D}'} \sum_{j=1}^m \lambda_j a_j = \left(\lim_{m \to \infty \text{ in } \Lambda'_a} \sum_{j=1}^m \lambda_j a_j \right)\bigg|_{\mathcal{D}} \neq 0.$$

Therefore, in the following sections we will identify

$$\lim_{m \to \infty \text{ in } \Lambda'_a} \sum_{j=1}^m \lambda_j a_j \ \left(\in \Lambda'^0_a\right) \quad \text{with} \quad \lim_{m \to \infty \text{ in } \mathcal{D}'} \sum_{j=1}^m \lambda_j a_j \ (\in \mathcal{S}').$$

Finally, we give five lemmas that will be refered to in later sections.

Remark 1.9. For any cube I, there exists a ball B such that $B \supset I$ and $|B| \leq C(n)|I|$. So, Remark 1.2 implies

$$|\mathrm{av}(f, B) - \mathrm{av}(f, I)| \leq C(n)\|f\|_{\mathrm{BMO}}.$$

So,

$$
\begin{aligned}
\mathrm{av}(|f - \mathrm{av}(f, I)|, I) &\leq \mathrm{av}(|f - \mathrm{av}(f, B)|, I) + C(n)\|f\|_{\mathrm{BMO}} \\
&\leq \frac{|B|}{|I|}\mathrm{av}(|f - \mathrm{av}(f, B)|, B) + C(n)\|f\|_{\mathrm{BMO}} \\
&\leq C(n)\|f\|_{\mathrm{BMO}}.
\end{aligned}
$$

Thus,

$$\sup_I \mathrm{av}(|f - \mathrm{av}(f, I)|, I) \leq C(n)\|f\|_{\mathrm{BMO}}$$

where I is taken over all cubes in $\mathbf{R}^n$. Changing the roles of balls and cubes, we get the opposite inequality. Therefore, we have

$$\|f\|_{\mathrm{BMO}} \approx \sup_I \mathrm{av}(|f - \mathrm{av}(f, I)|, I).$$

In the proof of Lemmma 1.9 below, take into consideration this remark.

Lemma 1.9. *Let* $\|f\|_{\mathrm{BMO}} \le 1$. *Let* $B = B(x_0, t_0)$. *Let* $\lambda > 0$. *Then*

$$|\{x \in B : |f(x) - \mathrm{av}(f, B)| > \lambda\}|/|B| \le C_{1.1}(n)e^{-\lambda/C_{1.1}(n)}. \quad (1.17)$$

Proof It suffices to prove (1.17) with the ball B replaced by a cube. Furthermore, since the BMO-norm is invariant under translations and dilations, it is enough to show (1.17) with B replaced by a dyadic cube. For $\lambda > 0$ let

$$v(\lambda) = \sup_J |\{x \in J : |f(x) - \mathrm{av}(f, J)| > \lambda\}|/|J|,$$

where J is taken over all dyadic cubes in $\mathbf{R}^n$. Since

$$\mathrm{av}\,(|f - \mathrm{av}(f, J)|, J) \le C_{1.2}(n)\|f\|_{\mathrm{BMO}} \le C_{1.2}$$

for any cube J by Remark 1.9, we have

$$v(\lambda) \le C_{1.2}/\lambda. \quad (1.18)$$

Let I be any dyadic cube. Let $\lambda > 2^{n+1}C_{1.2}$. Let $\{I_j\}$ be the family of maximal dyadic subcubes of I such that

$$|\{x \in I_j : |f(x) - \mathrm{av}(f, I)| > \lambda\}|/|I_j| > 2^{-n-1}. \quad (1.19)$$

Let $\tilde{I}_j$ be the dyadic double of I_j. Then

$$\{x \in I : |f(x) - \mathrm{av}(f, I)| > \lambda\}$$
$$= \bigcup_j \{x \in I_j : |f(x) - \mathrm{av}(f, I)| > \lambda\} \text{ a.e.} \quad (1.20)$$

by Corollary 0.3,

$$\sum_j |I_j| \le \sum_j 2^{n+1} |\{x \in I_j : |f(x) - \mathrm{av}(f, I)| > \lambda\}|$$
$$= 2^{n+1}|\{x \in I : |f(x) - \mathrm{av}(f, I)| > \lambda\}|$$
$$\le 2^{n+1}v(\lambda)|I|, \quad (1.21)$$
$$|\{x \in I_j : |f(x) - \mathrm{av}(f, I)| \le \lambda\}|/|I_j|$$
$$\ge 1 - \left|\{x \in \tilde{I}_j : |f(x) - \mathrm{av}(f, I)| > \lambda\}\right|/\left(|\tilde{I}_j|2^{-n}\right)$$
$$\ge 1 - 2^{-n-1}/2^{-n} \quad \text{because } \tilde{I}_j \text{ dose not satisfy (1.19)}$$
$$= 1/2. \quad (1.22)$$

Since

$$|\{x \in I_j : |f(x) - \mathrm{av}(f, I_j)| \le 3C_{1.2}\}|/|I_j| \ge 1 - v(3C_{1.2}) \ge 2/3$$

by (1.18), this combined with (1.22) implies

$$|\mathrm{av}(f, I) - \mathrm{av}(f, I_j)| \le \lambda + 3C_{1.2}. \quad (1.23)$$

Therefore,

$$\left|\{x \in I : |f(x) - \mathrm{av}(f,I)| > \lambda + 3C_{1.2} + 2^{n+2}C_{1.2}\}\right|$$

$$\leq \sum_j \left|\{x \in I_j : |f(x) - \mathrm{av}(f,I_j)| > 2^{n+2}C_{1.2}\}\right| \quad \text{by (1.20) and (1.23)}$$

$$\leq \sum v\left(2^{n+2}C_{1.2}\right)|I_j| \leq \sum 2^{-n-2}|I_j| \qquad \text{by (1.18)}$$

$$\leq 2^{-1}v(\lambda)|I| \qquad \text{by (1.21)}.$$

Since a dyadic cube I is arbitrary, we have

$$v\left(\lambda + 3C_{1.2} + 2^{n+2}C_{1.2}\right) \leq 2^{-1}v(\lambda), \quad \text{where } \lambda \geq 2^{n+1}C_{1.2}.$$

This implies

$$v(\lambda) \leq Ce^{-\lambda/C}$$

for some $C < \infty$. $\qquad\qquad\square$

Lemma 1.10. *Let $f \in BMO$. Then*

$$\int |f(x)|\,(1+|x|)^{-n-1}\,dx < \infty.$$

Proof Let $k \in \mathbf{N}$. Then

$$\mathrm{av}\left(|f|, B(0,2^k)\right) \quad \leq \quad \mathrm{av}\left(|f|, B(0,1)\right) + Ck\|\,\|f\|\|_{\mathrm{BMO}} \qquad \text{by Remark 1.3}$$

$$\leq \quad \mathrm{av}\left(|f|, B(0,1)\right) + Ck\|f\|_{\mathrm{BMO}} \qquad \text{by Remark 1.4}.$$

Thus

$$\int_{\mathbf{R}^n} |f(x)|\,(1+|x|)^{-n-1}\,dx$$

$$\leq \sum_{k \in \mathbf{N}} 2^{-k(N+1)} \int_{B(0,2^k)} |f|\,dx$$

$$= C \sum 2^{-k}\mathrm{av}\left(|f|, B\left(0,2^k\right)\right)$$

$$\leq C \sum 2^{-k}\left\{\mathrm{av}\left(|f|, B(0,1)\right) + Ck\|f\|_{\mathrm{BMO}}\right\} < \infty.$$

$$\square$$

Lemma 1.11. *Let $a > b \geq 0$ and $\phi \in \mathcal{B}_a$. Let μ be a signed measure on $\mathbf{R}^{n+1}_+$ such that*

$$\sup\left\{|\mu|\left(Q(B)\right)/|B|^{1+b/n} : B \text{ is taken over all balls in } \mathbf{R}^n\right\} \leq 1, \quad (1.24)$$

where $|\mu|$ is the total variation of μ. let $M \in (0,+\infty)$ and

$$f(x) = \iint_{\mathbf{R}^n \times (0,M]} (\phi)_t(x-y)\,d\mu(y,t).$$

Then,

$$\|f\|_{\Lambda_b} \leq C_{1.3}(a,b,n).$$

Proof. Take $b' \in (b, a]$ so that $[b'] = [b]$. By Lemma 1.4 $\|\phi\|_{\Lambda_{b'}} \leq C(b', a, n)$. Take any ball $B = B(x_B, r_B)$. Put

$$
\begin{aligned}
D_1 &= \left\{ (y, t) \in \mathbf{R}_+^{n+1} : t \in (0, r_B],\ t \leq M,\ |y - x_B| < t + r_B \right\}, \\
D_2 &= \left\{ (y, t) \in \mathbf{R}_+^{n+1} : t > r_B,\ t \leq M,\ |y - x_B| < t + r_B \right\}.
\end{aligned}
$$

If $x \in B$, then

$$
\begin{aligned}
f(x) &= \iint_{D_1} (\phi)_t(x - y) d\mu(y, t) + \iint_{D_2} (\phi)_t(x - y) d\mu(y, t) \\
&= f_1(x) + f_2(x), \qquad \text{say.}
\end{aligned}
$$

By (1.24)

$$
\int |f_1(x)|\, dx \leq C \iint_{D_1} d|\mu| \leq C r_B^{n+b}. \tag{1.25}
$$

On the other hand, since

$$
\begin{aligned}
\|f_2\|_{\Lambda_{b'}} &\leq \iint_{D_2} \|(\phi)_t\|_{\Lambda_{b'}}\, d|\mu| \\
&\leq \sum_{j \in \mathbf{N}} \iint_{D_2 \cap (\mathbf{R}^n \times (2^{j-1} r_B, 2^j r_B])} C t^{-n-b'} d|\mu| \\
&\leq \sum C \left(2^j r_B \right)^{-n-b'} |\mu| \left(Q \left(2^{j+1} B \right) \right) \\
&\leq \sum C \left(2^j r_B \right)^{-n-b'} \left(2^{j+1} r_B \right)^{n+b} \qquad \text{by (1.24)} \\
&\leq C r_B^{b-b'},
\end{aligned}
$$

we have

$$
\inf_{P:\deg P \leq [b']} \int_B |f_2(x) - P(x)|\, dx \leq C \|f_2\|_{\Lambda_{b'}} r_B^{n+b'} \leq C r_B^{n+b}. \tag{1.26}
$$

Combining (1.25) and (1.26) gives the desired result. $\qquad \square$

Definition 1.5. For a signed measure μ on $\mathbf{R}_+^{n+1}$, let

$$
\|\mu\|_C = \sup \left\{ |\mu| \left(Q(B) \right) / |B| : B \text{ is taken over all balls in } \mathbf{R}^n \right\}.
$$

Lemma 1.12. *Let* $\chi(x) = \chi_{B(0,1)}(x)$. *Let* $k(y, t)$ *be a measurable function defined on* $\mathbf{R}_+^{n+1}$ *such that*

$$
\|k(y, t) dy dt\|_C \leq 1, \tag{1.27}
$$

$$
0 \leq k(y, t) \leq 1/t. \tag{1.28}
$$

Let

$$
f(x) = \iint_{\mathbf{R}_+^{n+1}} (\chi)_t(x - y) k(y, t) dy dt.
$$

Then, $f(x) \equiv +\infty$ *or*

$$
\|f\|_{\mathrm{BMO}} \leq C_{1.4}(n).
$$

Proof. We may assume $f(x) \not\equiv +\infty$. Take any ball $B = B(x_B, r_B)$. Let

$$
\begin{aligned}
D_1 &= \left\{ (y,t) \in \mathbf{R}_+^{n+1} : t \in (0, r_B], \ |y - x_B| < t + r_B \right\}, \\
D_2 &= \left\{ (y,t) \in \mathbf{R}_+^{n+1} : t > r_B \right\}, \\
f_j(x) &= \iint_{D_j} (\chi)_t (x - y) k(y,t) \, dy \, dt \qquad (j = 1, 2).
\end{aligned}
$$

(It follows easily from (1.28) that if $f(x) \not\equiv +\infty$, then $f_2(x) < \infty$ for all $x \in \mathbf{R}^n$.) Then

$$
\|f_1\|_{L^1} \leq \|\chi\|_{L^1} \iint_{D_1} |k(y,t)| \, dy \, dt \leq C \iint_{Q(2B)} |k| \, dy \, dt \leq C|B| \quad \text{by (1.27)},
$$

$$
\begin{aligned}
|f_2(x) - f_2(x')| &\leq \iint_{D_2} |(\chi)_t(x - y) - (\chi)_t(x' - y)| \, |k(y,t)| \, dy \, dt \\
&\leq \int_{r_B}^{+\infty} \frac{dt}{t} \int_{\mathbf{R}^n} |(\chi)_t(x - y) - (\chi)_t(x' - y)| \, dy \quad \text{by (1.28)} \\
&\leq C \int_{r_B}^{+\infty} |x - x'| \frac{dt}{t^2} \\
&\leq C|x - x'|/r_B.
\end{aligned}
$$

Thus

$$
\begin{aligned}
|B|^{-1} \int_B |f(x) - f_2(x_B)| \, dx &= |B|^{-1} \int_B |f_1(x) + (f_2(x) - f_2(x_B))| \, dx \\
&\leq |B|^{-1} \|f_1\|_{L^1} + C \leq C.
\end{aligned}
$$

$\square$

Lemma 1.13. *Let* $\phi \in C^1(\mathbf{R}^n)$,

$$
|\phi(x)| + |\nabla \phi(x)| \leq (1 + |x|)^{-n-1}.
$$

Let μ *be a signed measure on* $\mathbf{R}_+^{n+1}$ *such taht* $\|\mu\|_C \leq 1$ *and that*

$$
\iint_{\mathbf{R}_+^{n+1} \setminus Q(B(0,1))} t^{-n} (1 + |x|/t)^{-n-1} \, d|\mu|(y,t) < \infty.
$$

Let

$$
f(x) = \iint_{\mathbf{R}_+^{n+1}} (\phi)_t(x - y) \, d\mu(y,t).
$$

Then

$$
\|f\|_{\mathrm{BMO}} \leq C_{1.5}(n).
$$

Proof. Take any ball $B = B(x_B, r_B)$. Put

$$
\begin{aligned}
f_1(x) &= \iint_{Q(2B)} (\phi)_t(x - y) d\mu(y, t), \\
f_2(x) &= \iint_{\mathbf{R}_+^{n+1} \setminus Q(2B)} (\phi)_t(x - y) d\mu(y, t).
\end{aligned}
$$

Then

$$
\|f_1\|_{L^1} \leq \|\phi\|_{L^1} |\mu| \left(Q(2B) \right) \leq \|\phi\|_{L^1} |2B|.
$$

Since $|\nabla_y(\phi)_t(y)| \leq C (t + |y|)^{-n-1}$ and since

$$
\begin{aligned}
\iint_{\mathbf{R}_+^{n+1} \setminus Q(2B)} C (t + |y|)^{-n-1} d|\mu|(y, t) &\leq C \sum_{k \in \mathbf{N}} \iint_{Q(2^k B)} (2^k r_B)^{-n-1} d|\mu| \\
&\leq C \sum (2^k r_B)^{-n-1} |2^k B| \\
&\leq C r_B^{-1},
\end{aligned}
$$

we have

$$
\begin{aligned}
\sup_{x \in B} |\nabla f_2(x)| &= \sup_{x \in B} \left| \iint_{\mathbf{R}_+^{n+1} \setminus Q(2B)} \nabla_x(\phi)_t(x - y) d\mu(y, t) \right| \\
&\leq \iint_{\mathbf{R}_+^{n+1} \setminus Q(2B)} C(t + |y|)^{-n-1} d|\mu| \\
&\leq C r_B^{-1}.
\end{aligned}
$$

Thus,

$$
|B|^{-1} \int_B |f(x) - f_2(x_B)| \, dx \leq |B|^{-1} \|f_1\|_{L^1} + r_B \sup_{x \in B} |\nabla f_2(x)| \leq C.
$$

$\square$

Notes. Lemma 1.9 is due to F. John–L. Nirenberg [61]. Their argument was extended by N. G. Meyers [64] and S. Campanato [63] and was refined by J.-O. Strömberg [79]. J. B. Garnett–P. W. Jones [78] analysed the result of John-Nirenberg in the quotient space BMO/{the closure of L^∞ in BMO}.

Definition 1.2 is due to R. Coifman [74] and R. Coifman–G. Weiss [77]. (See also C. Herz [74a].) More precise results than (1.9) are in S. Janson [76], D. A. Stegenga [76] and E. Nakai–K. Yabuta [85]. Definition 1.5 is due to L. Carleson [62]. (See also L. Hörmander [67].) The condition (1.24) is due to P. Duren [69]. (See also E. Amar–A. Bonami [79] and A. Uchiyama [80b].)

H. M. Reimann–T. Rychener [75] is a good textbook of BMO.

II. Atomic H^p spaces

Definition 2.1. For $p \in (0,1]$ and for $f \in \mathcal{D}'$ let

$$\|f\|_{H^p} = \inf \left\{ \|\{\lambda_j\}\|_{\ell^p} : \begin{array}{l} \text{there exists a sequence of } (p,\infty)\text{-atoms} \\ \{a_j(x)\}_{j\in\mathbf{N}} \text{ such that} \end{array} \right.$$

$$f = \lim_{m\to\infty \text{ in } \mathcal{D}'} \sum_{j=1}^{m} \lambda_j a_j \Big\} \tag{2.1}$$

where $\inf \emptyset = \infty$. Let

$$H^p = \{ f \in \mathcal{D}' : \|f\|_{H^p} < \infty \} .$$

As we saw in Remark 1.8, H^p is included in $\mathcal{S}'$ and can be regarded as a subspace of $\Lambda'^0_{n(1/p-1)}$ $(\subset \Lambda'_{n(1/p-1)})$ by $\ell^p \subset \ell^1$. Of course, $H^1 \subset L^1$.

Lemma 2.1. *Let $p \in (0,1]$. Then H^p is a complete metric space with respect to the metric*

$$(\text{the distance between } f \text{ and } g) = \|f - g\|_{H^p}^p .$$

This is clear from the above definition.

Lemma 2.2. *Let $p \in (0,1]$ and $f \in H^p$. Then*

$$\|f\|_{\Lambda'_{n(1/p-1)}} \le \|f\|_{H^p} .$$

This follows from Lemma 1.1 and from $\|\cdot\|_{\ell^1} \le \|\cdot\|_{\ell^p}$.

Definition 2.2. For $m \in \mathbf{N}_0$, for a nonnegative function $\psi \in L^1\backslash\{0\}$ such that $\operatorname{supp}\psi$ is compact and for $f \in \mathcal{D}'$ such that $\operatorname{supp} f$ is compact (or for $f \in L^1$ such that $\int |f(x)||x|^m dx < \infty$), let $\mathbf{P}(f,\psi,m)(x)$ be the polynomial of degree $\le m$ such that

$$\langle x^\alpha, f \rangle_{\mathcal{D}} - \int x^\alpha \mathbf{P}(f,\psi,m)(x)\psi(x)dx = 0 \text{ if } |\alpha| \le m \tag{2.2}$$

(or such that

$$\int x^\alpha \left(f(x) - \mathbf{P}(f, \psi, m)(x)\psi(x) \right) dx = 0 \quad \text{if } |\alpha| \le m, \qquad (2.2)'$$

respectively.) Let $\mathbf{P}(f, 0, m)(x) \equiv 0$.

Lemma 2.3. *Let $f \in L^1$ and $\operatorname{supp} f \subset \overline{B} = \overline{B(x_0, t_0)}$ (or $\operatorname{supp} f \subset \overline{I} = \overline{I(x_0, t_0)}$). Then,*

$$\|\mathbf{P}(f, \chi_B, m)\|_{L^\infty(B)} \le C(m, n)\operatorname{av}(|f|, B)$$

$$\left(\text{or } \|\mathbf{P}(f, \chi_I, m)\|_{L^\infty(I)} \le C(m, n)\operatorname{av}(|f|, I), \text{ respectively.} \right)$$

Proof. We may assume $B = B(0, 1)$. Let $\{\pi_j\}_{j=1,2,\cdots,k(m,n)}$ be an orthonormal basis of the polynomials of degree $\le m$ with respect to the inner product

$$(P, Q) = \int P(x)Q(x)\chi_B(x)dx.$$

Then, the desired estimate follows from the formula

$$\mathbf{P}(f, \chi_B, m)(x) = \sum_{j=1}^{k} \int f(y)\pi_j(y)dy \, \pi_j(x).$$

$\square$

Definition 2.3. For $f \in L^1_{\text{loc}}$, let

$$M_d f(x) = \sup \left\{ \operatorname{av}(|f|, I) : I \text{ is taken over all dyadic cubes such that } I \ni x \right\}.$$

($M_d f$ is called the dyadic maximal function.)

It is clear that

$$M_d f(x) \le C(n) M f(x). \qquad (2.3)$$

Lemma 2.4. *Let $p \in (0, 1]$, $q \in [1, \infty]$, $p < q$, $f \in L^q$, $\operatorname{supp} f \subset \overline{B} = \overline{B(x_0, t_0)}$ and*

$$\int f(x)x^\alpha dx = 0 \quad \text{if } |\alpha| \le n(1/p - 1). \qquad (2.4)$$

Then,

$$\|f\|_{H^p} \le C(p, q, n)|B|^{1/p - 1/q}\|f\|_{L^q}.$$

Proof. By translation, dilation and by multiplying a constant, we may assume

$$\operatorname{supp} f \subset [0, 1]^n, \qquad (2.5)$$
$$\|f\|_{L^q} = 1. \qquad (2.6)$$

Let $I_{0,1} = [0, 1]^n$ and $I_{0,2} = I_{0,3} = \cdots = \emptyset$. For $k \in \mathbf{N}$ let $\{I_{k,j}\}_{j=1,2,\dots}$ be the maximal dyadic cubes such that

$$\text{av}(|f|, I_{k,j}) \geq 2^k. \tag{2.7}$$

For $k \in \mathbf{N}_0$ let $\chi_{k,j} = \chi_{I_{k,j}}$ and

$$u_{k,j} = \mathbf{P}\left(f\chi_{k,j}, \chi_{k,j}, n(1/p - 1)\right).$$

Then,

$$\{I_{k,j}\}_{j=1,2,\dots} \text{ are mutually disjoint,} \tag{2.8}$$

for each $I_{k+1,i}$ there exists an $I_{k,j}$ such that $\tag{2.9}$

$$I_{k,j} \supset I_{k+1,i},$$

$$\text{av}\left(|f|, I_{k,j}\right) \leq 2^{k+n}, \tag{2.10}$$

$$|f(x)| \leq 2^k \text{ a.e. } x \in \left(\bigcup_j I_{k,j}\right)^c, \tag{2.11}$$

$$\int \left(f(x) - u_{k,j}(x)\right) x^\alpha \chi_{k,j}(x) dx = 0 \text{ if } |\alpha| \leq n(1/p - 1), \tag{2.12}$$

$$\|u_{k,j}\|_{L^\infty(I_{k,j})} \leq C(p,n)2^k, \tag{2.13}$$

$$u_{0,1} = 0. \tag{2.14}$$

((2.10) follows from (2.6) and from the maximality of $I_{k,j}$. (2.11) follows from Corollary 0.2. (2.13) follows from Lemma 2.3 and (2.10). (2.14) follows from (2.4).)

Let

$$g_k = f - \sum_j (f - u_{k,j}) \chi_{k,j}.$$

Then,

$$g_0 = 0 \text{ by (2.14)},$$

$$\|f - g_k\|_{L^1} \leq \int_{\bigcup_j I_{k,j}} \left(|f| + C2^k\right) dx \text{ by (2.13)}$$

$$\leq C \int_{\bigcup_j I_{k,j}} |f| dx \text{ by (2.7)}$$

$$\to 0 \ (k \to \infty)$$

by $f \in L^1$ and by $\left|\bigcup_j I_{k,j}\right| \to 0 \ (k \to \infty)$, which follows from Theorem 0.1(i). Thus

$$f = \lim_{k \to \infty} g_k$$

$$= \sum_{k=0}^{\infty} (g_{k+1} - g_k)$$

$$= \sum_k \sum_j \left\{ (f - u_{k,j}) \chi_{k,j} - \sum_{i: I_{k+1,i} \subset I_{k,j}} (f - u_{k+1,i}) \chi_{k+1,i} \right\} \quad \text{by (2.9)}$$

$$= \sum \sum b_{k,j}(x) \text{ say,}$$

$$= \sum_{k=0}^{\infty} \sum_j \left\{ f \left(1 - \sum_i \chi_{k+1,i} \right) - u_{k,j} + \sum_i u_{k+1,i} \chi_{k+1,i} \right\} \chi_{k,j},$$

where equalities are in the sense of L^1. The last formula, (2.11) and (2.13) imply $\|b_{k,j}\|_{L^\infty} \leq C2^{k+1}$. So, (2.12) implies that $c(p,n)2^{-k} |I_{k,j}|^{-1/p} b_{k,j}$ ($= a_{k,j}$, say) is a (p,∞)-atom. Then, since

$$f = C \sum_k \sum_j 2^k |I_{k,j}|^{1/p} a_{k,j},$$

we have

$$c\|f\|_{H^p}^p \leq \sum_{k \in \mathbb{N}_0} \sum_j \left(2^k |I_{k,j}|^{1/p} \right)^p \leq 1 + \sum_{k \in \mathbb{N}} 2^{kp} \left| \{ M_d f(x) \geq 2^k \} \right|$$

$$\leq 1 + C \int_1^\infty \lambda^{p-1} \left| \{ M_d f(x) \geq \lambda \} \right| d\lambda$$

$$\leq 1 + C \int_1^\infty \lambda^{p-1} \lambda^{-q} d\lambda \qquad \text{by (2.3), (2.6) and Theorem 0.1}$$

$$= C(p,q,n) \qquad \text{by } q > p.$$

$\square$

Theorem 2.1. *Let $p \in (0,1]$. Let T be a linear functional defined on H^p such that*

$$\|T\| = \sup \left\{ |T(f)| / \|f\|_{H^p} : f \in H^p \setminus \{0\} \right\} < \infty. \tag{2.15}$$

Then, there exists $h \in \Lambda_{n(1/p-1)}$ satisfying the following:

$$T(f) = \langle h, f \rangle_{\Lambda_{n(1/p-1)}} \qquad \text{for any } f \in H^p, \tag{2.16}$$

$$\|T\| = \|h\|_{\Lambda_{n(1/p-1)}}. \tag{2.17}$$

Proof. Let $k \in \mathbb{N}$. Then,

$$\sup \left\{ |T(f)| / \|f\|_{L^2} : f \in (2.18)_k \right\} < \infty$$

by (2.15) and by Lemma 2.4, where

$$(2.18)_k = \{ f \in L^2 : \quad \operatorname{supp} f \subset B(0,k),$$

$$\int f(x) x^\alpha dx = 0 \text{ if } |\alpha| \leq n(1/p-1) \}.$$

So, there exists $h_k \in L^2 (B(0,k))$ such that

$$T(f) = \int fh_k dx \text{ for any } f \in (2.18)_k.$$

Since it is clear that $h_{k+1} - h_k$ is a polynomial of degree $\leq n(1/p - 1)$ on $B(0, k)$, there exists $h \in L^2_{loc}$ such that $h - h_k$ is a polynomial of degree $\leq n(1/p - 1)$ on $B(0, k)$ for any $k \in \mathbf{N}$. So,

$$T(f) = \int fh \, dx \text{ for any } f \in \bigcup_k (2.18)_k. \tag{2.19}$$

Then, (2.17) follows from

$$\|h\|_{\Lambda_{n(1/p-1)}} = \sup\left\{ \left| \int fh dx \right| : f \text{ is taken over all } (p, \infty)\text{-atoms} \right\}$$

$$\text{by Lemma 1.1}$$

$$= \sup\{|T(f)| : \cdots\cdots\cdots\cdots\cdots\cdots\cdots\} \text{ by } (2.19)$$

$$= \|T\|.$$

Take any $f \in H^p$. Let $\{\lambda_j\} \in \ell^p$ and let $\{a_j(x)\}$ be a sequence of (p, ∞)-atoms such that

$$\sum_{j=1}^{m} \lambda_j a_j \to f \quad \text{in } H^p \ (m \to \infty). \tag{2.20}$$

Then, (2.16) follows from

$$T(f) = \lim_{m\to\infty} T\left(\sum_{j=1}^{m} \lambda_j a_j\right) \quad \text{by } (2.15) \text{ and } (2.20)$$

$$= \lim_{m\to\infty} \left\langle h, \sum_{j=1}^{m} \lambda_j a_j \right\rangle_{\Lambda_{n(1/p-1)}} \quad \text{by } (2.19)$$

$$= \langle h, f \rangle_{\Lambda_{n(1/p-1)}} \quad \text{by } \left\| \sum_{j=1}^{m} \lambda_j a_j - f \right\|_{\Lambda'_{n(1/p-1)}} \leq \| \cdots \|_{H^p} \to 0.$$

$\square$

Lemma 2.1 implies that the space H^1 is a Banach space. Theorem 2.1 implies that the dual space of the Banach space H^1 can be regarded as BMO. But, the function $f(x)h(x)$, where $f \in H^1$ and $h \in$ BMO, is not necessarily integrable. But, we have the following.

Theorem 2.2. Let $f \in H^1$ and $h \in$ BMO. For $r > 0$ let

$$\text{tr}(h, r)(x) = h(x)/\max\{1, |h(x)|/r\}. \tag{2.21}$$

Then

$$\|\mathrm{tr}(h,r)\|_{L^\infty} \le r, \tag{2.22}$$

$$\|\mathrm{tr}(h,r)\|_{\mathrm{BMO}} \le \|h\|_{\mathrm{BMO}}, \tag{2.23}$$

$$\langle h, f\rangle_{\mathrm{BMO}} = \lim_{r \to +\infty} \int f(x)\mathrm{tr}(h,r)(x)dx. \tag{2.24}$$

In particular, if $f \in H^1$, $h \in BMO$ and if $fh \in L^1$, then

$$\langle h, f\rangle_{\mathrm{BMO}} = \int f(x)h(x)dx.$$

Proof. (2.22) is clear. (2.23) is easy from

$$|\mathrm{tr}(h,r)(x) - c/\max\{1, |c|/r\}| \le |h(x) - c| \qquad \text{for any } c \in \mathbf{R}.$$

Take $\{\lambda_j\}_j \in \ell^1$ and a sequence of $(1,\infty)$-atoms $\{a_j\}_j$ so that

$$\sum_{j=1}^m \lambda_j a_j \to f \text{ in } H^1 \ (m \to \infty). \tag{2.25}$$

Then,

$$\left| \int a_j \mathrm{tr}(h,r)dx \right| \le \|\mathrm{tr}(h,r)\|_{\mathrm{BMO}} \qquad \text{by Lemma 1.1}$$

$$\le \|h\|_{\mathrm{BMO}} \qquad \text{by (2.23),} \tag{2.26}$$

$$\int a_j \mathrm{tr}(h,r)dx \to \int a_j h dx = \langle h, a_j\rangle_{\mathrm{BMO}} \qquad (r \to \infty). \tag{2.27}$$

Therefore,

$$\int f\mathrm{tr}(h,r)dx$$

$$= \sum_{j=1}^\infty \lambda_j \int a_j \mathrm{tr}(h,r)dx \text{ by (2.22) and } \sum_{j=1}^m \lambda_j a_j \to f \text{ in } L^1 \ (m \to \infty)$$

$$\to \sum_{j=1}^\infty \lambda_j \langle h, a_j\rangle_{\mathrm{BMO}} \qquad (r \to \infty)$$

by (2.26)–(2.27), $\sum |\lambda_j| < \infty$ and the dominated
convergence theorem

$$= \langle h, f\rangle_{\mathrm{BMO}} \text{ by } \left\|\sum_{j=1}^m \lambda_j a_j - f\right\|_{\mathrm{BMO}'} = \|\cdot\cdot\|_{H^1} \to 0.$$

□

As a byproduct of Lemma 2.4 we get the following Theorem 2.3.

Definition 2.4. Let $p \in (0,1]$, $q \in [1,\infty]$ and $p < q$. A function $a(x) \in L^q$ is called a (p,q)-*atom* if there exists a ball B such that (1.1), (1.2)′ and (1.3) hold, where

$$\|a\|_{L^q} \leq |B|^{-1/p+1/q}. \tag{1.2}'$$

A (p,∞)-atom is a (p,q)-atom. And, Lemma 2.4 implies that if $a(x)$ is a (p,q)-atom, then $\|a\|_{H^p} \leq C(p,q,n)$. Therefore, we get

Theorem 2.3. Let $p \in (0,1]$, $q \in [1,\infty]$ and $p < q$. Then,

$$c(p,q,n)\|f\|_{H^p} \leq \inf\{\|\{\lambda_j\}\|_{\ell^p} : \text{there exists a sequence of } (p,q)\text{-atoms}$$
$$\{a_j(x)\} \text{ such that (2.1) holds}\}$$
$$\leq \|f\|_{H^p}.$$

Finally, we add the following.

Theorem 2.4. Let $p \in (0,1]$ and $f \in H^p$. Then, $\mathcal{F}f \in C(\mathbf{R}^n, \mathbf{C})$ and

$$|\mathcal{F}f(\xi)| \leq C(p,n)\|f\|_{H^p}|\xi|^{n(1/p-1)}.$$

Proof. We may assume that f is a (p,∞)-atom $a(x)$. Then $\mathcal{F}a \in C(\mathbf{R}^n)$ and

$$\begin{aligned}
|\mathcal{F}a(\xi)| &= \left|\int e^{-2\pi i\xi \cdot x} a(x)dx\right| \\
&\leq \|e^{-2\pi i\xi \cdots}\|_{\Lambda_{n(1/p-1)}} \quad \text{by Lemma 1.1} \\
&\leq C|\xi|^{n(1/p-1)}.
\end{aligned}$$

□

Theorem 2.5. Let $p \in (0,1]$ and $f \in L^1_{\text{loc}}$. Then

$$\|f\|_{L^p} \leq \|f\|_{H^p}.$$

Proof. (The case $p = 1$ is clear.) We may assume $f \in H^p$. Let $\{\lambda_j\} \in \ell^p$, $\lambda_j > 0$ and let $\{a_j\}_j$ be a sequence of (p,∞)-atoms such that

$$\text{supp}\, a_j \subset B_j, \quad \|a_j\|_{L^\infty} \leq |B_j|^{-1/p}, \quad \text{where } B_j\text{'s are balls,}$$

and such that

$$\sum_{j=1}^m \lambda_j a_j \to f \text{ in } \mathcal{D}' \ (m \to \infty). \tag{2.28}$$

Let $\varepsilon > 0$. It is enough to show

$$|f(x)| \leq \sum_1^\infty \lambda_j |B_j|^{-1/p}\chi_{(1+\varepsilon)B_j}(x) \quad \text{a.e. } x. \tag{2.29}$$

First note that since

$$\left\| \sum_{j:\ell(B_j)<t} \lambda_j^p |B_j|^{-1} \chi_{B_j} \right\|_{L^1} \to 0 \qquad (t \to +0),$$

Theorem 0.1 (i) implies

$$M\left(\sum_{j:\ell(B_j)<t} \lambda_j^p |B_j|^{-1} \chi_{B_j} \right)(x) \to 0 \qquad (t \to +0) \text{ a.e. } x.$$

So, in particular,

$$\lim_{t \to +0} \left\{ \sum_{j:B_j \subset B(x,t)} \lambda_j^p \right\} / t^n \to 0 \qquad (t \to +0) \text{ a.e. } x. \tag{2.30}$$

Let $\phi \in \mathcal{D}$, $\phi(x) \geq 0$, $\operatorname{supp}\phi \subset B(0,1)$ and $\int \phi \, dx = 1$. Then

$$
\begin{aligned}
|f(x)| &= \lim_{t \to +0} |f * (\phi)_t(x)| \qquad \text{a.e. } x \text{ by Corollary 0.2} \\
&= \lim_{t \to +0} \lim_{m \to \infty} \left| \left(\sum_{j=1}^m \lambda_j a_j \right) * (\phi)_t(x) \right| \qquad \text{by (2.28)} \\
&\leq \limsup_{t \to +0} \sum_{j=1}^\infty \lambda_j |a_j * (\phi)_t(x)| \\
&\leq \limsup_{t \to +0} \sum_{j \in (2.31)_{t,x}} \lambda_j |a_j * (\phi)_t(x)| \\
&\quad + \limsup_{t \to +0} \sum_{j \in (2.32)_{t,x}} \lambda_j |a_j * (\phi)_t(x)| \\
&= (2.33)_x + (2.34)_x, \quad \text{say,}
\end{aligned}
$$

where

$$
\begin{aligned}
(2.31)_{t,x} &= \{ j \in \mathbf{N} : \ell(B_j) > t/\varepsilon, B_j \text{ intersects } B(x,t) \}, \\
(2.32)_{t,x} &= \{ j \in \mathbf{N} : \ell(B_j) \leq t/\varepsilon, B_j \text{ intersects } B(x,t) \}.
\end{aligned}
$$

Then

$$
\begin{aligned}
(2.33)_x &\leq \limsup_{t \to +0} \sum_{j \in (2.31)_{t,x}} \lambda_j |B_j|^{-1/p} \chi_{(1+\varepsilon)B_j}(x) \\
&\leq \sum_{j \in \mathbf{N}} \lambda_j |B_j|^{-1/p} \chi_{(1+\varepsilon)B_j}(x), \\
(2.34)_x &\leq \limsup_{t \to +0} \sum_{j \in (2.32)_{t,x}} \lambda_j \|(\phi)_t\|_{\Lambda_{n(1/p-1)}} \qquad \text{by Lemma 1.1}
\end{aligned}
$$

$$\leq \limsup_{t \to +0} \sum_{j \in (2.32)_{t,x}} \lambda_j t^{-n/p} \|\phi\|_{\Lambda_{n(1/p-1)}}$$

$$\leq \left\{ C \limsup_{t \to +0} \left(\sum_{j \in (2.32)_{t,x}} \lambda_j^p \right) t^{-n} \right\}^{1/p} \qquad \text{by } p \leq 1$$

$$\leq \left\{ C \limsup_{t \to +0} \left(\sum_{j: B_j \subset B(x,(1+2/\varepsilon)t)} \lambda_j^p \right) t^{-n} \right\}^{1/p}$$

$$\to 0 \qquad \text{a.e. } x \text{ by } (2.30).$$

Thus, we get (2.29). $\qquad\qquad\qquad\qquad\qquad\qquad\qquad\qquad\qquad\qquad\qquad$ $\square$

Remark 2.1. Taking the dual of Lemma 2.4 with $p < 1$ and $q = 1$ gives another proof of the implication (ii) $\to$ (iii) of Lemma 1.2.

Remark 2.2. By the same argument as Lemma 2.4 we can show that if $f \in L^1$, $\operatorname{supp} f \subset B = B(x_0, t_o)$,

$$\int |f(x)| \log^+ (|B| \, |f(x)|) \, dx \leq 1,$$

$$\int f(x) dx = 0,$$

then

$$\|f\|_{H^1} \leq C(n).$$

Taking the dual of this fact and using the theory of Orlicz spaces, we can get another proof of Lemma 1.9. We omit the details.

Notes. Definition 2.1 is due to R. Coifman [74] and R. Coifman–G. Weiss [77]. (See also C. Herz [74a].) R. Coifman–G. Weiss [77] showed Lemma 2.4, Theorem 2.1, Theorem 2.3 and Remark 2.1 very generally on spaces of homogeneous type. The discovery of the dual space of analytic H^p ($p < 1$) on the unit disc in $\mathbf{C}^1$ is due to P. Duren–B. Romberg–A. Shields [69]. (See also A. P. Frazier [72], T. Walsh [73] and C. Fefferman–E. M. Stein [72].) The discovery of the dual space of H^1, which is defined in terms of the generalized Cauchy-Riemann equations, is due to C. Fefferman [71]. We learned the argument of the proof of Lemma 2.3 from C. Fefferman–N. Riviere–Y. Sagher [74]. Theorem 2.4 is in C. Fefferman–E. M. Stein [72]. Remark 2.2 is suggested in C. Herz [74b]. A. Garsia [73] is also important.

R. Coifman–G. Weiss [77] and M. Taibleson–G. Weiss [80] introduced the notion of "molecule" that includes "atom". A. Baernstein II–E. T. Sawyer [85] developed it. Though "molecule" is very important, we do not treat it in this book.

As for the analytic H^p on the unit disc of $\mathbf{C}^1$, P. Duren [70] is a popular textbook.

III. Operators on H^p

We introduce several important operators on H^p and give easy estimates to them. As for the definitions of M, χ_E and $\Gamma(x, \delta)$ recall Section 0. Recall that {the volume of the unit ball in $\mathbf{R}^n$} $= \pi^{n/2}/\Gamma((n+2)/2)$.

Lemma 3.1. *Let $\Omega \subset \mathbf{R}^n$ be measurable. Let $0 < \delta < \delta'$ and*

$$\Omega' = \left\{ x \in \mathbf{R}^n : M\chi_\Omega(x) > (\delta/(\delta + \delta'))^n /2 \right\}. \tag{3.1}$$

Then,

$$|\Omega'| \leq C(n)(\delta'/\delta)^n|\Omega|, \tag{3.2}$$

$$\bigcup_{x \in \Omega'^c} \Gamma(x, \delta') \subset \left\{ (y, t) \in \mathbf{R}_+^{n+1} : \right.$$

$$\left. \Gamma((n+2)/2)\,\pi^{-n/2}(\delta t)^{-n} \int_{\Omega^c} \chi_{\Gamma(x,\delta)}(y,t)dx > 1/2 \right\}, \tag{3.3}$$

$$\bigcup_{x \in \Omega'^c} \Gamma(x, \delta') \subset \bigcup_{x \in \Omega^c} \Gamma(x, \delta). \tag{3.4}$$

Proof. (3.2) is clear from Theorem 0.1.
Let $x \in \Omega'^c$ and $(y, t) \in \Gamma(x, \delta')$. Since

$$B(y, \delta t) \subset B\left(x, |x - y| + \delta t\right) \subset B\left(x, (\delta + \delta')t\right)$$

and since

$$|\Omega \cap B\left(x, (\delta + \delta')t\right)| / |B\left(x, (\delta + \delta')t\right)| \leq (\delta/(\delta + \delta'))^n /2 \text{ by } x \in \Omega'^c,$$

we get

$$|\Omega^c \cap B(y + \delta t)| / |B(y, \delta t)|$$
$$= 1 - |\Omega \cap B(y, \delta t)| / |B(y, \delta t)|$$
$$\geq 1 - |\Omega \cap B\left(x, (\delta + \delta')t\right)| / |B(y, \delta t)|$$
$$= 1 - \left(|\Omega \cap B\left(x, (\delta + \delta')t\right)| / |B\left(x, (\delta + \delta')t\right)|\right)\left((\delta + \delta')/\delta\right)^n > 1/2.$$

This means $(y, t) \in$ {the right-hand side of (3.3)}. Thus, we get (3.3).
(3.4) is clear from (3.3). $\qquad\square$

Definition 3.1. For $u \in C(\mathbf{R}_+^{n+1})$, $x \in \mathbf{R}^n$ and $\delta > 0$ let

$$N_\delta u(x) = \sup\{|u(y,t)| : (y,t) \in \Gamma(x,\delta)\},$$
$$N_0 u(x) = \sup\{|u(x,t)| : t > 0\},$$

$$S_\delta u(x) = \left\{ \Gamma((n+2)/2)\,\pi^{-n/2} \iint_{\Gamma(x,\delta)} |u(y,t)|^2 (\delta t)^{-n} dy\,dt/t \right\}^{1/2},$$

$$S_0 u(x) = \left\{ \int_0^{+\infty} |u(x,t)|^2\, dt/t \right\}^{1/2}.$$

Lemma 3.2. *Let $u \in C(\mathbf{R}_+^{n+1})$ and $0 < \delta < \delta'$. Then*

$$N_\delta u(x) \le N_{\delta'} u(x), \tag{3.5}$$
$$S_\delta u(x) \le (\delta'/\delta)^{n/2} S_{\delta'} u(x), \tag{3.6}$$
$$\|N_{\delta'} u\|_{L^p} \le C(p,n)(\delta'/\delta)^{n/p} \|N_\delta u\|_{L^p} \quad \text{if } p \in (0,\infty], \tag{3.7}$$
$$\|S_{\delta'} u\|_{L^p} \le C(p,n)(\delta'/\delta)^{n/p-n/2} \|S_\delta u\|_{L^p} \quad \text{if } p \in (0,2]. \tag{3.8}$$

(3.5) and (3.6) are clear.

Proof of (3.7). Let $\lambda > 0$ and

$$\Omega_\lambda = \{x \in \mathbf{R}^n : N_\delta u(x) > \lambda\}.$$

Let Ω' be as in (3.1) with $\Omega = \Omega_\lambda$. Then, (3.4) implies that

$$\bigcup_{x \in \Omega'^c} \Gamma(x,\delta') \subset \bigcup_{x \in \Omega_\lambda^c} \Gamma(x,\delta) \subset \{(y,t) : |u(y,t)| \le \lambda\}.$$

Thus, $N_{\delta'} u(x) \le \lambda$ if $x \in \Omega'^c$. So,

$$|\{x \in \mathbf{R}^n : N_{\delta'} u(x) > \lambda\}| \le |\Omega'| \le C(\delta'/\delta)^n |\Omega_\lambda| \quad \text{by (3.2).}$$

Operating $\displaystyle\int_0^{+\infty} \lambda^{p-1} \cdot d\lambda$ to the both sides of the above gives (3.7). $\qquad\square$

Proof of (3.8). It is clear that $\|S_{\delta'} u\|_{L^2} = \|S_\delta u\|_{L^2}$. So, we may assume that $p \in (0,2)$. Let $\lambda > 0$ and

$$\Omega_\lambda = \{x \in \mathbf{R}^n : S_\delta u(x) > \lambda\}.$$

Let Ω' be as in (3.1) with $\Omega = \Omega_\lambda$. Then

$$\int_{\Omega'^c} S_{\delta'} u(x)^2 dx \le \iint u(y,t)^2 \chi_{\bigcup_{x \in \Omega'^c} \Gamma(x,\delta')}(y,t)\,dy\,dt/t$$

$$\le 2 \iint u(y,t)^2 \left\{ \frac{\Gamma((n+2)/2)}{\pi^{n/2}} (\delta t)^{-n} \int_{\Omega_\lambda^c} \chi_{\Gamma(x,\delta)}(y,t)\,dx \right\} dy\,dt/t$$

by (3.3)

$$= 2 \int_{\Omega_\lambda^c} S_\delta u(x)^2 dx. \tag{3.9}$$

Let $q > 0$. Then

$$|\{x \in \mathbf{R}^n : S_{\delta'}u(x) > q\lambda\}|$$
$$\leq |\Omega'| + |\{x \in \Omega'^c : S_{\delta'}u(x) > q\lambda\}|$$
$$\leq C(\delta'/\delta)^n|\Omega_\lambda| + (q\lambda)^{-2}2\int_{\Omega_\lambda^c} S_\delta u(x)^2 dx \qquad (3.10)$$

by (3.2) and (3.9).

Thus,

$$\|S_{\delta'}u\|_{L^p}^p = pq^p\int_0^{+\infty} \lambda^{p-1}\,|\{S_{\delta'}u(x) > q\lambda\}|\,d\lambda$$

$$\leq Cq^p\int_0^{+\infty} \lambda^{p-1}\Big\{(\delta'/\delta)^n\,|\{S_\delta u(x) > \lambda\}|$$

$$+ (q\lambda)^{-2}\int_{\{S_\delta u(x)\leq\lambda\}} S_\delta u(x)^2 dx\Big\}d\lambda \ \text{ by (3.10)}$$

$$\leq C\Big\{q^p(\delta'/\delta)^n\int_0^{+\infty} \lambda^{p-1}\,|\{S_\delta u(x) > \lambda\}|\,d\lambda$$

$$+ q^{p-2}\int_{\mathbf{R}^n} S_\delta u(x)^2 dx\int_{S_\delta u(x)}^{+\infty} \lambda^{p-3}d\lambda\Big\}$$

$$\leq C\left(q^p(\delta'/\delta)^n + q^{p-2}\right)\|S_\delta u\|_{L^p}^p.$$

Thus, putting $q = (\delta'/\delta)^{-n/2}$ we get (3.8). $\qquad\qquad\square$

Definition 3.2. When the convolution $f * (\phi)_t$ can be defined in some sense, let

$$N_{\phi,\delta}f(x) = N_\delta u(x), \ \ S_{\phi,\delta}f(x) = S_\delta u(x),$$

with

$$u(x,t) = f * (\phi)_t(x). \qquad (3.11)$$

Definition 3.3. For $a > 0$, $f \in \mathcal{D}'$ and $x \in \mathbf{R}^n$ let

$$G_a f(x) = \sup\{|\langle(\phi)_t(\cdot - x), f\rangle_{\mathcal{D}}| : t > 0, \phi \in \mathcal{B}_a \cap \mathcal{D}\}.$$

Remark 3.1. Let $f \in \mathcal{D}'$ and $\psi \in \mathcal{D}$. Let $y \in \mathbf{R}^n$, $r > 0$ and

$$\text{supp}\,\psi \subset B(y,r).$$

Then

$$\text{supp}\,(\psi(\cdot + y))_{1/r} \subset B(0,1).$$

So,

$$\begin{aligned}|\langle \psi, f\rangle_{\mathcal{D}}| &= \left|\left\langle \left((\psi(\cdot + y))_{1/r}\right)_r (\cdot - y), f\right\rangle_{\mathcal{D}}\right| \\ &\leq \|(\psi)_{1/r}\|_{\Lambda_a} G_a f(y) \\ &= r^{n+a}\|\psi\|_{\Lambda_a} G_a f(y).\end{aligned}$$

Remark 3.2. Let $f \in \mathcal{D}'$, $\psi \in \mathcal{D}$ and

$$\operatorname{supp}\psi \subset B(0, r).$$

Then, for any $x \in \mathbf{R}^n$

$$\operatorname{supp}\psi \subset B(x, r + |x|).$$

So, Remark 3.1 implies

$$G_a f(x) \geq \|\psi\|_{\Lambda_a}^{-1} (r + |x|)^{-n-a} |\langle\psi, f\rangle_{\mathcal{D}}|.$$

Thus, if $f \in \mathcal{D}'\backslash\{0\}$, $p \in (0, 1)$ and if $0 < a \leq n(1/p - 1)$, then

$$\int G_a f(x)^p dx = \infty.$$

Remark 3.3. Let $f \in \mathcal{D}'$, $a > 0$ and $\phi \in \mathcal{B}_a \cap \mathcal{D}$. Let $x, y \in \mathbf{R}^n$, $t > 0$ and $|x - y| \leq \delta t$. Then

$$\operatorname{supp}(\phi)_t(y - \cdot) \subset B(x, (1 + \delta)t).$$

Thus, Remark 3.1 implies

$$|f * (\phi)_t(y)| \leq ((1 + \delta)t)^{n+a} \|(\phi)_t\|_{\Lambda_a} G_a f(x) = (1 + \delta)^{n+a} G_a f(x).$$

Therefore

$$N_{\phi, \delta} f(x) \leq (1 + \delta)^{n+a} G_a f(x).$$

Theorem 3.1. Let $p \in (0, 1]$, $a > n(1/p - 1)$ and $f \in H^p$. Then,

$$\|G_a f\|_{L^p} \leq C(a, p, n)\|f\|_{H^p}.$$

It is very easy to show this when f is a (p, ∞)-atom. (See the proof of Lemma 3.4 below.) We omit the proof of Theorem 3.1.

Definition 3.4. Let Ψ_0, $\Psi_1 \in \mathcal{D}$ be such that

$$\operatorname{supp}\Psi_0 \subset B(0, 1/2), \quad \operatorname{supp}\Psi_1 \subset B(0, 1/2)\backslash B(0, 1/8),$$
$$\Psi_0(x) + \sum_{k \in \mathbf{N}} \Psi_1(2^{-k}x) \equiv 1, \quad \Psi_0(x) \geq 0, \quad \Psi_1(x) \geq 0. \tag{3.12}$$

(We freeze these Ψ_0 and Ψ_1.) For m, $a > 0$ and $\phi \in \Lambda_a$ let

$$\|\phi\|_{\mathcal{B}, a, m} = \|\phi\Psi_0\|_{\Lambda_a} + \sup_{k \in \mathbf{N}} 2^{k(n+m)} \|\phi(2^k \cdot)\Psi_1(\cdot)\|_{\Lambda_a}.$$

Theorem 3.2. *Let $p \in (0,1]$, $a > n(1/p - 1)$, $m > 0$, $\delta \geq 0$ and $f \in H^p$. Let $\phi \in \Lambda_a$ and $\|\phi\|_{\mathcal{B},a,m} \leq 1$. Then,*

$$\|N_{\phi,\delta}f\|_{L^p} \leq C(a,p,m,n)(1+\delta)^{n/p}\|f\|_{H^p}.$$

Proof. It is clear that $\|\phi\|_{\Lambda_{n(1/p-1)}} \leq C(a,p,m,n)\|\phi\|_{\mathcal{B},a,m}$ by $a > n(1/p - 1)$. So, $f * (\phi)_t(x)$ is well defined. Let

$$\phi_0(x) = \phi(x)\Psi_0(x), \quad \phi_k(x) = 2^{k(n+m)}\phi(2^k x)\Psi_1(x) \quad (k \in \mathbf{N}).$$

Then

$$\phi_k \in \mathcal{B}_a \quad \text{and} \quad \phi = \sum_{k \in \mathbf{N}_0} 2^{-km}(\phi_k)_{2^k}.$$

So,

$$f * (\phi)_t(y) = \sum_{k \in \mathbf{N}_0} 2^{-km} f * (\phi_k)_{2^k t}(y)$$

and

$$
\begin{aligned}
N_{\phi,1}f(x) &\leq \sum 2^{-km} N_{(\phi_k)_{2^k},1}f(x) \\
&= \sum 2^{-km} N_{\phi_k,2^{-k}}f(x) \\
&\leq C\sum 2^{-km} G_a f(x) \qquad \text{by Remark 3.3} \\
&\leq C G_a f(x).
\end{aligned}
$$

Thus, the desired result follows from Theorem 3.1 and (3.7). $\qquad \square$

Definition 3.5. For $a > 0$ and $m \geq 0$ let

$$\mathcal{B}_a^m = \left\{ \phi \in \mathcal{B}_a : \int \phi(x)x^\alpha dx = 0 \text{ if } |\alpha| \leq m \right\}.$$

Lemma 3.3. *Let $a > 0$, $\phi \in \mathcal{B}_a^0$ and $f \in L^2$. Then,*

$$\iint |f * (\phi)_t(x)|^2 \, dxdt/t \leq C(a,n)\|f\|_{L^2}^2.$$

This is clear from Plancherel's theorem and from

$$|\mathcal{F}\phi(\xi)| \leq C(a,n) \min\left\{ |\xi|, |\xi|^{-a} \right\}.$$

Lemma 3.4. *Let $p \in (0,1]$, $a > n(1/p - 1)$, $\delta \in [0,1]$, $f \in H^p$ and $\phi \in \mathcal{B}_a^0$. Then,*

$$\|S_{\phi,\delta}f\|_{L^p} \leq C(a,p,n)\|f\|_{H^p}.$$

Proof. We give a proof only for the case $\delta = 1$. The case $\delta \in [0, 1)$ follows from the same argument. By $S_{\phi,\delta}\{f(t_0 \cdot + x_0)\}(x) = (S_{\phi,\delta}f)(t_0 x + x_0)$, by translation and dilation, we may assume that f is a (p, ∞)-atom such that

$$\operatorname{supp} f \subset B(0, 1),$$

$$\int f(x)x^\alpha dx = 0 \text{ if } |\alpha| \le n(1/p - 1),$$

$$\|f\|_{L^\infty} \le |B(0, 1)|^{-1/p}.$$

Let

$$a' = \min\{a, (n(1/p - 1) + [n(1/p - 1)] + 1)/2\}.$$

Then, since $|B(0, 1)|^{-(1+a'/n)+1/p}f$ is a $((1 + a'/n)^{-1}, \infty)$-atom and since $\|\phi\|_{\Lambda_{a'}} \le C\|\phi\|_{\Lambda_a}$ by $a' \le a$, we have

$$|f * (\phi)_t(x)| \le \begin{cases} 0 \text{ if } |x| > 1 + t, \\ C\|(\phi)_t\|_{\Lambda_{a'}} \le Ct^{-n-a'} \text{ if } |x| \le 1 + t. \end{cases}$$

Thus, if $|x| > 2$, then

$$\begin{aligned} S_{\phi,1}f(x)^2 &\le \iint_{\Gamma(x,1) \cap (\mathbf{R}^n \times ((|x|-1)/2, +\infty))} Ct^{-2(n+a')-n} dydt/t \\ &\le C|x|^{-2(n+a')}, \end{aligned}$$

which combined with the condition $a' > n(1/p - 1)$ implies

$$\int_{|x| \ge 2} S_{\phi,1}f(x)^p dx \le C.$$

On the other hand, Lemma 3.3 implies

$$\begin{aligned} \int_{|x|<2} S_{\phi,1}f(x)^p dx &\le C\left(\int_{|x|<2} S_{\phi,1}f(x)^2 dx\right)^{p/2} \\ &\le C\left(\iint |f * (\phi)_t(y)|^2 dydt/t\right)^{p/2} \\ &\le C\|f\|_{L^2}^p \le C. \end{aligned}$$

$\square$

Lemma 3.5. *Let $0 \le m' < m$, $a > 0$ and $\phi \in \Lambda_a$. Let*

$$\|\phi\|_{\mathcal{B},a,m} \le 1, \tag{3.13}$$

$$\int \phi(x)x^\alpha dx = 0 \text{ if } |\alpha| \le m'. \tag{3.14}$$

Then, there exists a sequence of functions $\{\phi_j\}_{j\in\mathbf{N}_0} \subset \Lambda_a$ such that

$$\phi(x) = \sum_{j \in \mathbf{N}_0} 2^{-jm} (\phi_j)_{2^j}(x), \tag{3.15}$$

$$c(a, m, n)\phi_j \in \mathcal{B}_a^{m'}. \tag{3.16}$$

Proof. First, note that (3.13) implies

$$|\phi(x)| \le C(1 + |x|)^{-n-m}. \tag{3.17}$$

Let Ψ_0 and Ψ_1 be as in Definition 3.4. Let $\{\pi_i\}_{i=1}^s$ be an orthonormal basis of the polynomials of degree $\le m'$ with respect to the inner product $(P, Q) = \int P(x)Q(x)\Psi_1(x)dx$. Let

$$
\begin{aligned}
u_j(x) &= \mathbf{P}\left(\phi(\cdot)\sum_{k=j}^{\infty} \Psi_1(2^{-k}\cdot), \Psi_1(2^{-j}\cdot), m'\right)\Psi_1(2^{-j}x) \\
&= \left\{2^{-jn}\sum_{i=1}^{S}\int \phi(y)\sum_{k=j}^{\infty}\Psi_1(2^{-k}y)\pi_i(2^{-j}y)dy\,\pi_i(2^{-j}x)\right\}\Psi_1(2^{-j}x) \\
&= \{(3.18)_j\} \times \Psi_1(2^{-j}x), \quad \text{say.}
\end{aligned}
$$

Then, since (3.17) implies

$$\left(c2^{ja}\|(3.18)_j\|_{\Lambda_a(B(0,2^j))} \le\right) \; |(3.18)_j\|_{L^\infty(B(0,2^j))} \le C2^{-j(n+m)},$$

(1.8) implies

$$\|u_j\|_{\Lambda_a} \le C2^{-j(n+m+a)} \quad \text{and} \quad \operatorname{supp} u_j \subset B(0, 2^{j-1}). \tag{3.19}$$

Note that

$$
\begin{aligned}
\phi(x) = \sum_{j \in \mathbf{N}}\Bigg\{&\left(\sum_{k \ge j}\Psi_1(2^{-k}x)\phi(x) - u_j(x)\right) \\
&- \left(\sum_{k \ge j+1}\Psi_1(2^{-k}x)\phi(x) - u_{j+1}(x)\right)\Bigg\} \\
&+ \{\Psi_0(x)\phi(x) + u_1(x)\} \\
= \sum_{j \in \mathbf{N}}&\{(3.20)_j\} + \{(3.20)_0\}, \quad \text{say.}
\end{aligned}
$$

Then

$$\|(3.20)_0\|_{\Lambda_a} \le C \quad \text{and} \quad \operatorname{supp}(3.20)_0 \subset B(0,1)$$

are clear. Since

$$(3.20)_j = \Psi_1(2^{-j}x)\phi(x) - u_j(x) + u_{j+1}(x), \quad (j \in \mathbf{N}),$$

(3.12), (3.13) and (3.19) imply

$$\|(3.20)_j\|_{\Lambda_a} \le C2^{-j(n+m+a)}, \quad \operatorname{supp}(3.20)_j \subset B(0, 2^j).$$

On the other hand, since

$$\int \left\{ \sum_{k \geq j} \Psi_1(2^{-k}x)\phi(x) - u_j(x) \right\} x^\alpha dx = 0, \quad |\alpha| \leq m'$$

by the definition of u_j, it follows that

$$\int (3.20)_j x^\alpha dx = 0 \text{ if } |\alpha| \leq m' \text{ and } j \in \mathbf{N}.$$

So,

$$\begin{aligned} \int (3.20)_0 x^\alpha dx &= \int \phi(x)x^\alpha dx - \sum_{j \in \mathbf{N}} \int (3.20)_j x^\alpha dx \\ &= \int \phi x^\alpha dx - \sum 0 \\ &= 0 \text{ by (3.14) if } |\alpha| \leq m'. \end{aligned}$$

Define $\phi_j(x)$ by

$$(3.20)_j = 2^{-jm}(\phi_j)_{2^j} \quad j \in \mathbf{N}_0.$$

Then, (3.15) is clear. (3.16) follows from the above properties of $(3.20)_j$. $\square$

Remark 3.4. Let $p \in (0,1]$. Let $m' = n(1/p - 1) < m$, $\phi \in L^\infty$ and

$$|\phi(x)| \leq (1 + |x|)^{-n-m},$$
$$\int \phi(x)x^\alpha dx = 0 \text{ if } |\alpha| \leq m'.$$

Then, applying the same procedure with the proof of Lemma 3.5 to this ϕ gives the decomposition

$$\phi = \sum_{j \in \mathbf{N}_0} (3.20)_j.$$

This time, $\{(3.20)_j\}$ satisfy

$$\|(3.20)_j\|_{L^\infty} \leq C2^{-j(n+m)}, \quad \text{supp} (3.20)_j \subset B(0, 2^j),$$
$$\int (3.20)_j x^\alpha dx = 0 \text{ if } |\alpha| \leq m' = n(1/p - 1).$$

Therefore,

$$c(m, p, n)2^{jm - jn(1/p-1)}(3.20)_j$$

are (p, ∞)-atoms. So,

$$\|\phi\|_{H^p}^p \leq C \sum_{j \in \mathbf{N}_0} \left(2^{-jm + jn(1/p-1)} \right)^p \leq C.$$

(This is a special case of the so called molecule.)

Theorem 3.3. *Let* $p \in (0,1]$, $a > n(1/p - 1)$, $m > 0$, $\delta \geq 0$ *and* $f \in H^p$. *Let* $\phi \in \Lambda_a$, $\|\phi\|_{\mathcal{B},a,m} \leq 1$ *and* $\int \phi \, dx = 0$. *Then,*

$$\|S_{\phi,\delta}f\|_{L^p} \leq C(a,p,m,n)(1+\delta)^{n/p-n/2}\|f\|_{H^p}.$$

Proof. By (3.8) it is enough to show only the case

$$\delta \in [0,1]. \tag{3.21}$$

Applying Lemma 3.5 with $m' = 0$, ϕ can be decomposed into the form (3.15) with

$$c(a,m,n)\phi_j \in \mathcal{B}_a^0.$$

Then,

$$
\begin{aligned}
\|S_{\phi,\delta}f\|_{L^p}^p &\leq \left\| \sum_{j \in \mathbb{N}_0} 2^{-jm} S_{(\phi_j)_{2^j},\delta}f \right\|_{L^p}^p = \left\| \sum 2^{-jm} S_{\phi_j, 2^{-j}\delta}f \right\|_{L^p}^p \\
&\leq \sum 2^{-jmp} \left\| S_{\phi_j, 2^{-j}\delta}f \right\|_{L^p}^p \leq C\|f\|_{H^p}^p \qquad \text{by Lemma 3.4.}
\end{aligned}
$$

$\square$

Definition 3.6. When $\delta \geq 0$, $m \in \mathbb{R}$ and when the convolution $f * (\phi)_t(x)$ is defined in some sense, let

$$S_{\phi,\delta,m}f(x) = S_\delta u(x)$$

with

$$u(x,t) = t^m f * (\phi)_t(x).$$

Theorem 3.4. *Let* $p \in (0,1]$, $m > 0$, $\delta \geq 0$, $f \in H^p$ *and* $\phi \in \mathcal{S}$. *Then,*

$$\left\| S_{\phi,\delta,m} \mathcal{F}^{-1}\left(|\xi|^m \mathcal{F}f(\xi)\right) \right\|_{L^p} \leq C(p,m,\phi)(1+\delta)^{n/p-n/2}\|f\|_{H^p}. \tag{3.22}$$

(By Theorem 2.4 $\mathcal{F}^{-1}\left(|\xi|^m \mathcal{F}f(\xi)\right)$ can be defined as a tempered distribution.)

Proof. By (3.8) it is enough to show the case (3.21). We assume (3.21). Let Ψ_1 be as in Definition 3.4. For $j \in \mathbb{N}$ define ϕ_j by

$$\mathcal{F}^{-1}\left(|\xi|^m \Psi_1(2^j \xi) \mathcal{F}\phi(\xi)\right) = (\phi_j)_{2^j}.$$

Then, since

$$\mathcal{F}\phi_j(\xi) = 2^{-jm}|\xi|^m \Psi_1(\xi)\mathcal{F}\phi(2^{-j}\xi)$$

and since $\phi \in \mathcal{S}$, we have

$$\|D_\xi^\alpha \mathcal{F}\phi_j\|_{L^\infty} \leq C(\alpha,\phi,\Psi_1,m)2^{-jm} \quad \text{for any} \quad \alpha$$

and

$$\operatorname{supp}\mathcal{F}\phi_j \subset \operatorname{supp}\Psi_1 \subset B(0,1/2)\backslash B(0,1/8).$$

So,

$$|D_x^\alpha \phi_j(x)| \le C(\alpha, \phi, \Psi_1, m, k)2^{-jm}(1+|x|)^{-k}$$

for any α and k. Thus,

$$\|\phi_j\|_{B,a,1} \le C(a, \phi, \Psi_1, m)2^{-jm} \qquad \text{for any } a \qquad (3.23)$$

and

$$\int \phi_j dx = 0. \qquad (3.24)$$

Let

$$\phi_0 = \mathcal{F}^{-1}\left(|\xi|^m \sum_{j=-\infty}^{0} \Psi_1(2^j\xi)\mathcal{F}\phi(\xi)\right).$$

Then, since $\phi \in \mathcal{S}$, (3.23) and (3.24) hold for $j = 0$, too. Thus,

$$t^m(\phi)_t * \mathcal{F}^{-1}(|\xi|^m \mathcal{F}f(\xi))(x)$$
$$= \sum_{j\in\mathbf{Z}} \mathcal{F}^{-1}\left(\Psi_1(2^j t\xi)t^m\mathcal{F}\phi(t\xi)|\xi|^m\mathcal{F}f(\xi)\right)(x)$$
$$= \sum_{j\in\mathbf{N}_0} f * (\phi_j)_{2^j t}(x) \qquad \text{for any } (x,t) \in \mathbf{R}_+^{n+1}.$$

So,

$$\{\text{the left-hand side of (3.22)}\}^p \le \sum_{j\in\mathbf{N}_0} \left\|S_{(\phi_j)_{2^j},\delta}f\right\|_{L^p}^p$$
$$= \sum_{j\in\mathbf{N}_0} \left\|S_{\phi_j,2^{-j}\delta}f\right\|_{L^p}^p \le C\sum 2^{-jmp}\|f\|_{H^p}^p$$
$$\qquad \text{by (3.23)–(3.24) and by Theorem 3.3}$$
$$\le C\|f\|_{H^p}^p.$$

$\square$

Finally, we add small lemmas that will be referred to in later sections.

Lemma 3.6. *Let* $u \in C(\mathbf{R}_+^{n+1})$, $\delta > 0$, $\varepsilon > 0$, $p \in (0,1]$ *and* $S_\delta u \in L^p$. *Then*

$$\left\|S_\delta\left(u\chi_{\mathbf{R}^n \times (\varepsilon,+\infty)}\right)\right\|_{L^\infty} \le C(p,n)(\delta\varepsilon)^{-n/p}\|S_\delta u\|_{L^p}.$$

Proof. Take any $x \in \mathbf{R}^n$. Then, there exist balls $B_1, \cdots, B_{k(n)} \subset \mathbf{R}^n$ with radius $\delta\varepsilon/2$ such that

$$B(x, \delta\varepsilon) \subset \bigcup_{j=1}^{k} B_j.$$

So, if $z_j \in B_j$ $(j = 1, \cdots, k)$, then

$$\Gamma(x,\delta) \cap (\mathbf{R}^n \times (\varepsilon, +\infty)) \subset \bigcup_{j=1}^{k} \Gamma(z_j, \delta).$$

Thus,

$$\Gamma((n+2)/2)\,\pi^{-n/2} \iint_{\Gamma(x,\delta)} u^2 \chi_{\mathbf{R}^n \times (\varepsilon, +\infty)}(\delta t)^{-n} dy dt/t$$

$$\leq \sum_{j=1}^{k} \inf \left\{ S_\delta u(z)^2 : z \in B_j \right\}$$

$$\leq C(p,n)(\delta\varepsilon/2)^{-2n/p} \| S_\delta u \|_{L^p}^2.$$

$\square$

Lemma 3.7. *Assume all the conditions in Lemma 3.6. Then*

$$\iint_{\mathbf{R}^n \times (\varepsilon, 2\varepsilon)} |u(y,t)|\, dy dt/t \leq C(p,n)(\delta\varepsilon)^{n(1-1/p)} \| S_\delta u \|_{L^p}. \qquad (3.25)$$

Proof. { the left-hand side of (3.25) }

$$= C \int_{\mathbf{R}^n} \left\{ \iint_{\Gamma(x,\delta) \cap (\mathbf{R}^n \times (\varepsilon, 2\varepsilon))} |u|(t\delta)^{-n} dy dt/t \right\} dx$$

$$\leq C \int_{\mathbf{R}^n} \left\{ \iint_{\Gamma(x,\delta) \cap (\mathbf{R}^n \times (\varepsilon, 2\varepsilon))} u^2 (t\delta)^{-n} dy dt/t \right\}^{1/2} dx$$

$$\leq C \int_{\mathbf{R}^n} \min \left\{ S_\delta u(x), (\delta\varepsilon)^{-n/p} \| S_\delta u \|_{L^p} \right\} dx \qquad \text{by Lemma 3.6}$$

$$\leq C \int_{\mathbf{R}^n} S_\delta u(x)^p \left\{ (\delta\varepsilon)^{-n/p} \| S_\delta u \|_{L^p} \right\}^{1-p} dx$$

$$= \{\text{the right-hand side of } (3.25)\}.$$

$\square$

Lemma 3.8. *Let $u \in C(\mathbf{R}_+^{n+1})$, $\varepsilon > 0$ and $p \in (0,1]$. Then*

$$\iint |u(x,t)|^p\, t^\varepsilon \, (1+|x|+t)^{-2\varepsilon}\, dx dt/t \leq C(p,\varepsilon,n) \| S_1 u \|_{L^p}^p. \qquad (3.26)$$

Proof. { the left-hand side of (3.26) }

$$\leq \sum_{j \in \mathbf{N}} 2^{-\varepsilon j} \iint_{Q(B(0,2^j))} |u(x,t)|^p\, (t/2^j)^\varepsilon dx dt/t$$

$$\leq C \sum_{j} 2^{-\varepsilon j} \int_{B(0,2^j)} \left\{ \iint_{\Gamma(x,1,2^j)} |u(y,t)|^p\, (t/2^j)^\varepsilon t^{-n} dy dt/t \right\} dx$$

$$\leq C \sum 2^{-\epsilon j} \int_{B(0,2^j)} \left\{ \iint_{\Gamma(x,1,2^j)} |u(y,t)|^2 \, (t/2^j)^\epsilon t^{-n} dy dt/t \right\}^{p/2} dx$$

$$\leq C \sum 2^{-\epsilon j} \int_{B(0,2^j)} S_1 u(x)^p dx$$

$$\leq \{\text{the right-hand side of (3.26)}\}.$$

$\square$

Notes. $N_{\phi,\delta} f$, $N_{\phi,0} f$ and $G_a f$ are called "nontangential maximal function", "radial maximal function" and "grand maximal function", respectively. These were investigated by C. Fefferman–E. M. Stein [72]. (3.7) is in it. (Our notation is defferent from theirs. In their paper, a grand maximal function was introduced in a little bit different from.) $S_{\phi,\delta} f$ and $S_{\phi,0} f$ are called "area integral" and "g-function", respectively. $S_{\phi,\delta} f$ was investigated by A. P. Calderón–A. Torchinsky [75]. (3.8) is in it.

IV. Atomic decomposition from grand maximal functions

In this section we prove the following:

Theorem 4.1. *Let $p \in (0,1]$, $a > 0$, and $f \in \mathcal{D}'$. Then*

$$\|f\|_{H^p} \leq C(p,a,n)\|G_a f\|_{L^p}.$$

The following Lemmas 4.1 and 4.2 are very easy. Lemma 4.3 is clear from Lemma 4.2.

Lemma 4.1. *Let $a > 0$ and $f \in \mathcal{D}'$. Then $G_a f(x)$ is lower semicontinuous.*

Lemma 4.2. *Let $a > 0$, $t_1, t_2 > 0$ and $\eta_1, \eta_2 \in \mathcal{B}_a$. Then,*

$$c(a,n)\left((\eta_1)_{t_1} * (\eta_2)_{t_2}\right)_{1/(t_1+t_2)} \in \mathcal{B}_a.$$

Lemma 4.3. *Let $a > 0$, $t > 0$, $f \in \mathcal{D}'$ and $\eta \in \mathcal{B}_a \cap \mathcal{D}$. Then,*

$$G_a\left(f * (\eta)_t\right)(x) \leq C(a,n)G_a f(x).$$

Lemma 4.4. *Let $a > 0$, $t > 0$, $f \in \mathcal{D}'$, $\phi \in \mathcal{D}$, $\operatorname{supp}\phi \subset B = B(x_B, r)$ and $\psi \in C^\infty$. Then,*

$$
\begin{aligned}
r^{-n}\left|\langle\psi, \phi f\rangle_{\mathcal{D}}\right| &\leq C(a,n)(t+1)^{n+a}\inf\left\{G_a f(y) : y \in tB\right\}r^a\|\phi\|_{\Lambda_a} \\
&\quad \times \left(r^{-n}\|\psi\|_{L^1(2B)} + r^a\|\psi\|_{\Lambda_a(2B)}\right).
\end{aligned}
$$

Proof Let $y \in tB$. Then

$$\operatorname{supp}\phi \subset B\left(y, (t+1)r\right).$$

So,

$$
\begin{aligned}
\left|\langle\psi, \phi f\rangle_{\mathcal{D}}\right| &= \left|\langle\psi\phi, f\rangle_{\mathcal{D}}\right| \\
&\leq \left((t+1)r\right)^{n+a}\|\psi\phi\|_{\Lambda_a}G_a f(y) \qquad \text{by Remark 3.1} \\
&\leq C(t+1)^{n+a}r^{n+a}\|\phi\|_{\Lambda_a}\left\{r^{-n}\|\psi\|_{L^1(2B)} + r^a\|\psi\|_{\Lambda_a(2B)}\right\}G_a f(y) \\
&\qquad \text{by (1.8).}
\end{aligned}
$$

$\square$

Lemma 4.5. *Let* $a > 0$, $s > 1$, $f \in \mathcal{D}'$, *supp* f *be compact and*

$$r^{-n} |\langle \psi, f \rangle_{\mathcal{D}}| \leq r^{-n} \|\psi\|_{L^1(sB)} + r^a \|\psi\|_{\Lambda_a(2sB)} \tag{4.1}$$

for any $\psi \in C^\infty$. *Then,*

$$\|\mathbf{P}(f, \chi_B, a) \chi_B\|_{L^\infty} \leq C(a, s, n) \tag{4.2}$$

and

$$r^{-n} |\langle \psi, f - \mathbf{P}(f, \chi_B, a) \chi_B \rangle_{\mathcal{D}}| \leq C(a, s, n) r^a \|\psi\|_{\Lambda_a(2sB)} \tag{4.3}$$

for any $\psi \in C^\infty$.

Proof. It is clear that we may assume

$$x_B = 0.$$

Let $f_r \in \mathcal{D}'$ be such that

$$\langle \psi, f_r \rangle_{\mathcal{D}} = \langle (\psi)_r, f \rangle_{\mathcal{D}} \qquad \text{for any } \psi \in \mathcal{D}.$$

Then

$$r^{-n} \langle \psi, f \rangle_{\mathcal{D}} = \langle \psi(r \cdot), f_r \rangle_{\mathcal{D}},$$
$$r^{-n} \|\psi\|_{L^1(B(0,sr))} = \|\psi(r \cdot)\|_{L^1(B(0,s))},$$
$$r^a \|\psi\|_{\Lambda_a(B(0,2sr))} = \|\psi(r \cdot)\|_{\Lambda_a(B(0,2s))},$$
$$\mathbf{P}(f, \chi_{B(0,r)}, a)(rx) = \mathbf{P}(f_r, \chi_{B(0,1)}, a)(x),$$
$$r^{-n} \langle \psi, f - \mathbf{P}(f, \chi_{B(0,r)}, a) \chi_{B(0,r)} \rangle_{\mathcal{D}}$$
$$= \langle \psi(r \cdot), f_r - \mathbf{P}(f_r, \chi_{B(0,1)}, a) \chi_{B(0,1)} \rangle_{\mathcal{D}}.$$

By substituting these formulae into (4.1)–(4.3), we can reduce the proof of our Lemma to the special case $B = B(0, 1)$.

So, in the following part of this proof, we assume

$$B = B(0,1), \quad r = 1. \tag{4.4}$$

(So, "(4.1)" will mean "(4.1) with (4.4)". Similar for (4.2) and (4.3).)
Let $\{\pi_j(x)\}_{j=1,2,\cdots,k(a,n)}$ be as in the proof of Lemma 2.3 with $m = a$. Then

$$\mathbf{P}(f, \chi_B, a)(x) = \sum_{j=1}^{k} \langle \pi_j, f \rangle_{\mathcal{D}} \pi_j(x). \tag{4.5}$$

So, (4.2) follows from (4.1), (4.5) and

$$\|\pi_j\|_{L^1(sB)} + \|\pi_j\|_{\Lambda_a(2sB)} \leq C(a, s, n).$$

Take any $\psi \in C^\infty$. Take a polynomial $P_0(x)$ of degree $\leq a$ such that

$$\|\psi - P_0\|_{L^1(sB)} \le C(a,s,n)\|\psi\|_{\Lambda_a(2sB)}. \tag{4.6}$$

Then, (4.3) follows from

$$
\begin{aligned}
|\langle \psi, f - \mathbf{P}(f, \chi_B, a)\chi_B\rangle_{\mathcal{D}}| &= |\langle \psi - P_0, f - \mathbf{P}(f, \chi_B, a)\chi_B\rangle_{\mathcal{D}}| \\
&\le |\langle \psi - P_0, f\rangle_{\mathcal{D}}| + |\langle \psi - P_0, \mathbf{P}\chi_B\rangle_{\mathcal{D}}| \\
&\le \left(\|\psi - P_0\|_{L^1(sB)} + \|\psi - P_0\|_{\Lambda_a(2sB)}\right) + \|\psi - P_0\|_{L^1(B)}\|\mathbf{P}\chi_B\|_{L^\infty} \text{ by (4.1)} \\
&\le C(a,s,n)\|\psi\|_{\Lambda_a(2sB)} \quad \text{by (4.6) and (4.2).}
\end{aligned}
$$

$\square$

Lemma 4.6. *Let $\Omega \subset \mathbf{R}^n$ be open. Then, there exists $\{x_j\}_{j\in\mathbf{N}} \subset \Omega$ such that*

$$B(x_i, r_i) \cap B(x_j, r_j) = \emptyset \quad \text{if } i \ne j, \tag{4.7}$$

$$\bigcup_j B(x_j, 3r_j) = \Omega, \tag{4.8}$$

where

$$r_j = 100^{-1}\text{dist}(x_j, \Omega^c). \tag{4.9}$$

Proof. Consider the family $\mathcal{A}$ of all the subsets A of Ω satisfying

$$B\left(x, 100^{-1}\text{dist}(x, \Omega^c)\right) \quad \cap \quad B\left(y, 100^{-1}\text{dist}(y, \Omega^c)\right) = \emptyset$$
$$\text{if } x, y \in A \text{ and if } x \ne y.$$

Take a maximal element A_0 in $\mathcal{A}$ with respect to the inclusion relation. Then, this A_0 is the desired $\{x_j\}$. We will show it. The countability of A_0 and (4.7) are clear. Let $y \in \Omega$. Then, the maximality of A_0 implies the existence of $x_{j_0} \in \{x_j\} = A_0$ such that

$$B\left(y, 100^{-1}\text{dist}(y, \Omega^c)\right) \cap B(x_{j_0}, r_{j_0}) \ne \emptyset.$$

Then, Remark 4.1 below implies

$$\text{dist}(y, \Omega^c) \le (101/99)^2\text{dist}(x_{j_0}, \Omega^c).$$

So,

$$y \in B\left(x_{j_0}, \left(1 + (101/99)^2\right) r_{j_0}\right) \subset B(x_{j_0}, 3r_{j_0}).$$

$\square$

Remark 4.1. Let Ω be open. Let $x \in \Omega$ and $s \in (0,1)$. Then

$$\sup\left\{\text{dist}(y, \Omega^c)/\text{dist}(z, \Omega^c) : y, z \in B\left(x, s \times \text{dist}(x, \Omega^c)\right)\right\} \le (1+s)/(1-s).$$

Remark 4.2. Let Ω, $\{x_j\}$ and $\{r_j\}$ be as in Lemma 4.6. Let $x \in \Omega$ and

$$r = 100^{-1}\mathrm{dist}(x, \Omega^c).$$

Then, the following are easy:

(i) If $B(x_i, 4r_i)$ intersects $B(x, 4r)$, then

$$\left(\frac{24}{26}\right)^2 r \le r_i \le \left(\frac{26}{24}\right)^2 r, \ \ B(x_i, r_i) \subset B(x, 10r), \ \ B(x, r) \subset B(x_i, 10r_i)$$

(ii) $\sharp\{i : B(x_i, 4r_i) \text{ intersects } B(x, 4r)\}$

$$\le \sharp\{i : B(x_i, r_i) \subset B(x, 10r), r_i \ge (24/26)^2 r\} \le 10^n/(24/26)^{2n}.$$

Lemma 4.7. *Let Ω, $\{x_j\}$ and $\{r_j\}$ be as in Lemma 4.6. Let $a > 0$. Let $\phi \in \mathcal{D}$ be such that*

$$\phi(x) = 1 \text{ on } B(0, 3), \ \ \phi(x) \ge 0 \text{ and } \mathrm{supp}\,\phi \subset B(0, 4).$$

(We freeze this ϕ depending only on the dimension n.) Put

$$\phi_j(x) = \phi\left((x - x_j)/r_j\right) \Big/ \left(\sum_{i \in \mathbf{N}} \phi\left((x - x_i)/r_i\right)\right). \tag{4.10}$$

Then,

$$\sum_{j \in \mathbf{N}} \phi_j(x) = 1 \ \ on \ \Omega, \tag{4.11}$$

$$\mathrm{supp}\,\phi_j \subset B(x_j, 4r_j), \tag{4.12}$$

$$\|\phi_j\|_{\Lambda_a} \le C(a, n) r_j^{-a}. \tag{4.13}$$

Proof. (4.11)–(4.12) are clear. (4.13) follows from the fact that the denominator on the right-hand side of (4.10) is ≥ 1 on Ω (by (4.8)) and from Remark 4.2 with $x = x_j$. □

We begin the proof of Theorem 4.1. We may assume $G_a f \in L^p$. Then by Remark 3.2 we may assume

$$G_a f \in L^p \text{ and } a > n(1/p - 1). \tag{4.14}$$

For $k \in \mathbf{Z}$ let

$$\Omega_k = \left\{x \in \mathbf{R}^n : G_a f(x) > 2^k\right\}.$$

Let $\{x_{k,j}\}_{j \in \mathbf{N}}$ be $\{x_j\}$ in Lemma 4.6 with $\Omega = \Omega_k$. Let

$$r_{k,j} = 100^{-1}\mathrm{dist}(x_{k,j}, \Omega_k^c). \tag{4.9}^*$$

Let $\{\phi_{k,j}\}_{j \in \mathbf{N}}$ be $\{\phi_j\}$ in Lemma 4.7 with $\Omega = \Omega_k$ and with $\{x_j\} = \{x_{k,j}\}_j$. Let

$$B_{k,j} = B(x_{k,j}, r_{k,j}), \ \chi_{k,j}(x) = \chi_{B_{k,j}}(x),$$
$$u_{k,j}(x) = \mathbf{P}(\phi_{k,j} f, \chi_{k,j}, a)(x)\chi_{k,j}(x),$$
$$u_{k+1,i,j}(x) = \mathbf{P}\left(\phi_{k,j}(\phi_{k+1,i} f - u_{k+1,i}), \chi_{k+1,i}, a\right)(x)\chi_{k+1,i}(x).$$

Claim 1. If $\psi \in C^{\infty}$, then

$$|\langle \psi, \phi_{k,j} f \rangle_{\mathcal{D}}| \le C(a,n) 2^k \left(\|\psi\|_{L^1(8B_{k,j})} + r_{k,j}^{n+a} \|\psi\|_{\Lambda_a(8B_{k,j})} \right).$$

Proof. Note that

$$101 B_{k,j} \cap \Omega_k^c \ne \emptyset \quad \text{by } (4.9)^*,$$
$$\operatorname{supp} \phi_{k,j} \subset 4B_{k,j} \quad \text{by } (4.12),$$
$$\|\phi_{k,j}\|_{\Lambda_a} \le C(a,n) r_{k,j}^{-a} \quad \text{by } (4.13).$$

Then, Claim 1 follows from Lemma 4.4 (with $t = 101/4$, $\phi = \phi_{k,j}$, $B = 4B_{k,j}$) and from

$$\inf \{G_a f(y) : y \in 101 B_{k,j}\} \le 2^k \quad \text{by } 101 B_{k,j} \cap \Omega_k^c \ne \emptyset.$$

$\square$

Claim 2.
$$|u_{k,j}(x)| \le C(a,n) 2^k \chi_{k,j}(x).$$

Claim 3. If $\psi \in C^{\infty}$, then

$$|\langle \psi, \phi_{k,j} f - u_{k,j} \rangle_{\mathcal{D}}| \le C(a,n) 2^k r_{k,j}^{n+a} \|\psi\|_{\Lambda_a}.$$

These are clear from Claim 1 and Lemma 4.5 (with $\phi_{k,j} f$, $B_{k,j}$ and 8 in places of f, B and s, respectively).

Claim 4. $\sum_{j \in \mathbf{N}} \phi_{k,j} f$ converges in $\mathcal{D}'$ independently of the order of summation.

This follows from Claim 1 and from that if $\psi \in \mathcal{D}$, then

$$\sum_j \left(\|\psi\|_{L^1(8B_{k,j})} + r_{k,j}^{n+a} \|\psi\|_{\Lambda_a} \right)$$
$$\le C \sum_j \left(r_{k,j}^n \|\psi\|_{L^\infty} + r_{k,j}^{n+a} \|\psi\|_{\Lambda_a} \right)$$
$$\le C \left\{ |\Omega_k| \|\psi\|_{L^\infty} + |\Omega_k|^{1+a/n} \|\psi\|_{\Lambda_a} \right\} \quad \text{by the disjointness of } \{B_{k,j}\}_j$$
$$< \infty \quad \text{by } G_a f \in L^p. \tag{4.15}$$

Claim 5. $f - \sum_{j \in \mathbf{N}} \phi_{k,j} f \in L^\infty$ and its L^∞-norm is less than $C(a,n) 2^k$. **Proof.**

Let $t > 0$, $\eta \in \mathcal{D}$ and $\int \eta \, dx = 1$. Then,

$$f * (\eta)_t = f * (\eta)_t \chi_{\Omega_k^c} + \sum_{j \in \mathbf{N}} \phi_{k,j} f * (\eta)_t = (4.16)_t + \sum_j (4.17)_{j,t}, \text{ say.}$$

Then

$$\|(4.16)_t\|_{L^\infty} \le C2^k. \tag{4.18}$$

By Lemma 4.3 and by the same reason as Claim 1 we have

$$|\langle \psi, (4.17)_{j,t} \rangle_{\mathcal{D}}| \le C2^k \left(\|\psi\|_{L^1(8B_{k,j})} + r_{k,j}^{n+a} \|\psi\|_{\Lambda_a} \right) \text{ for any } \psi \in \mathcal{D}.$$

So, by (4.15) and by the dominated convergence theorem we have

$$\lim_{t \to +0} \sum_j \langle \psi, (4.17)_{j,t} \rangle_{\mathcal{D}} = \sum_j \lim_{t \to +0} \langle \psi, (4.17)_{j,t} \rangle_{\mathcal{D}} \left(= \sum_j \langle \psi, \phi_{k,j} f \rangle_{\mathcal{D}} \right)$$

for any $\psi \in \mathcal{D}$, namely

$$\sum_j (4.17)_{j,t} \to \sum_j \phi_{k,j} f \text{ in } \mathcal{D}' \ (t \to +0).$$

Thus,

$$(4.16)_t \to f - \sum_j \phi_{k,j} f \text{ in } \mathcal{D}' \ (t \to +0),$$

which combined with (4.18) implies the desired result.   □

Define g_k by

$$f = g_k + \sum_{j \in \mathbf{N}} (\phi_{k,j} f - u_{k,j}).$$

Claim 6.
$$\|g_k\|_{L^\infty} \le C(a,n) 2^k.$$

This is easy from Claims 5, 2 and the disjointness of $\{\operatorname{supp} u_{k,j}\}_j$.

Claim 7.
$$g_k \to f \qquad \text{in } \mathcal{D}' \ (k \to \infty).$$

Proof. Let $\psi \in \mathcal{D}$. Then,

$$\left| \left\langle \psi, \sum_j (\phi_{k,j} f - u_{k,j}) \right\rangle_{\mathcal{D}} \right| = \left| \sum_j \langle \psi, \phi_{k,j} f - u_{k,j} \rangle_{\mathcal{D}} \right|$$

$$\le \sum_j C(a,n) 2^k r_{k,j}^{n+a} \|\psi\|_{\Lambda_a} \text{ by Claim 3}$$

$$\le C(a,n) 2^k |\Omega_k|^{1+a/n} \|\psi\|_{\Lambda_a} \text{ by the disjointness of } \{B_{k,j}\}_j$$

$$\to 0 \qquad (k \to \infty)$$

by (4.14). So

$$\sum_j (\phi_{k,j} f - u_{k,j}) \to 0 \qquad \text{in } \mathcal{D}' \; (k \to \infty),$$

which implies the desired result. □

Claim 8. If $4B_{k,j} \cap 4B_{k+1,i} = \emptyset$, then $u_{k+1,i,j} = 0$.

Proof. In this case $\phi_{k,j}(\phi_{k+1,i}f - u_{k+1,i}) = 0$. □

Claim 9. If $4B_{k,j} \cap 4B_{k+1,i} \neq \emptyset$, then

$$r_{k,j} \geq (24/26)^2 r_{k+1,i}, \quad B_{k+1,i} \subset 10B_{k,j}.$$

Claim 10. $\sharp\{j \in \mathbf{N} : 4B_{k,j} \text{ intersects } 4B_{k+1,i}\} \leq 10^n/(24/26)^{2n}$.

Proof. Apply Remark 4.2 to $\Omega = \Omega_k$, $\{x_j\} = \{x_{k,j}\}_j$, $x = x_{k+1,i}$ and $r = 100^{-1}\text{dist}(x_{k+1,i}, \Omega_k^c)$. Then, since

$$r \geq r_{k+1,i} \qquad (\text{because } \Omega_k \supset \Omega_{k+1}),$$

Claims 9 and 10 follow from the fact that

if $4B_{k,j}$ intersects $4B_{k+1,i}$, then $4B_{k,j}$ intersects $B(x_{k+1,i}, 4r)$.

□

Claim 11. If $\psi \in C^\infty$, then

$$|\langle \psi, \phi_{k+1,i}\phi_{k,j}f\rangle_{\mathcal{D}}| \leq C(a,n)2^k \left(\|\psi\|_{L^1(8B_{k+1,i})} + r_{k+1,i}^{n+a}\|\psi\|_{\Lambda_a(8B_{k+1,i})} \right).$$

Proof. Since $\text{supp}\,\phi_{k+1,i}\phi_{k,j} \subset 4B_{k+1,i}$ and since

$$\|\phi_{k+1,i}\phi_{k,j}\|_{\Lambda_a} \leq C(a,n)r_{k+1,i}^{-a}$$

by (1.8), (4.13) and Claim 9, we get the desired estimate by the same argument as Claim 1. □

Claim 12.
$$|u_{k+1,i,j}(x)| \leq C(a,n)2^k \chi_{k+1,i}(x).$$

Proof. By
$$|u_{k+1,i}(x)| \leq C2^k \chi_{k+1,i}(x),$$

$\phi_{k+1,i}\phi_{k,j}f - \phi_{k,j}u_{k+1,i}$ satisfies the same estimate as Claim 11. So, applying Lemma 4.5 (4.2) (with $B_{k+1,i}$ and 8 in places of B and s, respectively) to this distribution gives the desired result. □

Claim 13.
$$\sum_i (\phi_{k+1,i}f - u_{k+1,i}) = \sum_j \sum_i (\phi_{k,j}(\phi_{k+1,i}f - u_{k+1,i}) - u_{k+1,i,j}) \quad \text{in } \mathcal{D}'$$

and the convergences of both sides in $\mathcal{D}'$ are independent of the orders of summations.

Proof. By Claims 10–11, $\sum_i \sum_j \phi_{k,j}\phi_{k+1,i}f$ converges in $\mathcal{D}'$ independently of the order of summation. By Claims 2, 8, 10 and 12, $\sum_i \sum_j |\phi_{k,j}u_{k+1,i}|$ and $\sum_i \sum_j |u_{k+1,i,j}|$ converge in L^1. Thus, the desired result follows from

$$\sum_j \phi_{k,j} \equiv 1 \text{ on } \Omega_{k+1}\left(\supset \bigcup_i \text{supp}(\phi_{k+1,i}f - u_{k+1,i})\right)$$

and from

$$
\begin{aligned}
\sum_j u_{k+1,i,j} &= \sum_j \mathbf{P}\big(\phi_{k,j}(\phi_{k+1,i}f - u_{k+1,i}), \chi_{k+1,i}, a\big)\chi_{k+1,i} \\
&= \mathbf{P}\left(\sum_j \phi_{k,j}(\phi_{k+1,i}f - u_{k+1,i}), \chi_{k+1,i}, a\right)\chi_{k+1,i} \\
&= \mathbf{P}(\phi_{k+1,i}f - u_{k+1,i}, \chi_{k+1,i}, a)\chi_{k+1,i} \\
&= 0.
\end{aligned}
$$

$\square$

Now, we enter the final step of the proof. We continue to assume (4.14). Then

$$
\begin{aligned}
f &= \lim_{k\to\infty}(g_k - g_{-k}) \qquad \text{by Claims 6 and 7} \\
&= \sum_{k=-\infty}^{\infty}(g_{k+1} - g_k) \\
&= \sum_k\left\{\sum_j(\phi_{k,j}f - u_{k,j}) - \sum_i(\phi_{k+1,i}f - u_{k+1,i})\right\} \\
&= \sum_k \sum_j\left\{\phi_{k,j}f - u_{k,j} - \sum_i(\phi_{k,j}(\phi_{k+1,i}f - u_{k+1,i}) - u_{k+1,i,j})\right\} \\
&\qquad\qquad\qquad\qquad\qquad\qquad \text{by Claim 13} \\
&= \sum_k \sum_j b_{k,j}, \text{ say,}
\end{aligned}
$$

where all the equalities hold in $\mathcal{D}'$. By the definition of the operator $\mathbf{P}$ we have

$$\langle x^\alpha, b_{k,j}\rangle = 0 \qquad \text{if } |\alpha| \le a.$$

Since

$$b_{k,j} = \phi_{k,j}\left(f - \sum_i \phi_{k+1,i}f\right) - u_{k,j} + \phi_{k,j}\sum_i u_{k+1,i} + \sum_i u_{k+1,i,j},$$

Claims 2, 5, 8, 9, 12 and the disjointness of $\{B_{k+1,i}\}_i$ imply that

$$|b_{k,j}(x)| \le C(a,n)2^k \chi_{10B_{k,j}}.$$

So, $c2^{-k}|B_{k,j}|^{-1/p}b_{k,j}(x)$ $(= a_{k,j}(x)$, say $)$ is a (p,∞)-atom,

$$f = \frac{1}{c}\sum_k \sum_j 2^k |B_{k,j}|^{1/p} a_{k,j}$$

and

$$\sum_k \sum_j \left(2^k |B_{k,j}|^{1/p}\right)^p \le \sum_k 2^{kp} \sum_j |B_{k,j}| \le \sum_k 2^{kp}|\Omega_k| \le \int G_a f(x)^p dx.$$

$\square$

Notes. Theorem 4.1 for the case $n = 1$ is due to R. Coifman [74]. This theorem for $n \ge 2$ is due to R. Latter [78]. C. Fefferman–N. Riviere–Y. Sagher [74] played a very important role in the argument of Coifman and Latter. For extensions of the argument of Coifman and Latter, see A. P. Calderón [77], J. B. Garnett–R. Latter [78], R. Macías–C. Segovia [79], R. Latter–A. Uchiyama [79], J. M. Wilson [82,85] and G. B. Folland–E. M. Stein [82]. We learned the ideas of Lemmas 4.6 and 4.7 from A. P. Calderón–A. Torchinsky [75].

V. Atomic decomposition from S-functions

In this section we prove the following:

Theorem 5.1. *Let*

$$\{\phi_1, \phi_2, \cdots, \phi_J\} \subset \mathcal{S} \tag{5.1}$$

and

$$\sup\left\{\sum_{i=1}^{J} |\mathcal{F}\phi_i(t\xi)| : t > 0\right\} > 0 \text{ for any } \xi \in \mathbf{R}^n\backslash\{0\}. \tag{5.2}$$

Let $p \in (0,1]$, $\delta > 0$, $f \in \mathcal{S}'$ and

$$\sum_{i=1}^{J} \|S_{\phi_i,\delta}f\|_{L^p} < +\infty. \tag{5.3}$$

Then, there exists a polynomial $P(x)$ such that

$$\|f - P\|_{H^p} \leq C\left(\{\phi_1, \cdots, \phi_J\}, \delta, p\right) \sum_{i=1}^{J} \|S_{\phi_i,\delta}f\|_{L^p}.$$

The following is immediate from taking the dual of Lemma 3.3.

Lemma 5.1. *Let $a > 0$, $\delta > 0$, $\eta \in \mathcal{B}_a^0$, $u \in C(\mathbf{R}_+^{n+1})$ and let $E \subset \mathbf{R}_+^{n+1}$ be compact. Let*

$$f(x) = \iint_E (\eta)_{\delta t}(x - y)u(y, t)dydt/t. \tag{5.4}$$

Then,

$$\|f\|_{L^2} \leq C(a, n)\|u\|_{L^2(\mathbf{R}_+^{n+1}, dxdt/t)}.$$

Lemma 5.2. *Let $p \in (0,1]$, $a > 0$ and*

$$\eta \in \mathcal{B}_a^{n(1/p-1)}. \tag{5.5}$$

Let $u \in C(\mathbf{R}_+^{n+1})$, $B \subset \mathbf{R}^n$ be a ball and

$$\operatorname{supp} u \subset Q(B) \text{ and } \iint u(y, t)^2 dydt/t \leq |B|. \tag{5.6}$$

Let $\delta \geq 1$, $E \subset \mathbf{R}_+^{n+1}$ be compact and let f be defined by the formula (5.4). Then

$$\|f\|_{H^p} \leq \begin{cases} C(a,p,n)\delta^{n(1/p-1)}|B|^{1/p} & \text{if } p \in (0,1), \\ C(a,n)\log(1+\delta)|B| & \text{if } p = 1. \end{cases}$$

Proof. By translation and by dilation we may assume

$$B = B(0,1). \tag{5.7}$$

We may assume $\delta = 2^k$ with $k \in \mathbf{N}_0$. Then,

$$\begin{aligned} f(x) &= \iint_{(B \times (0,2^{-k}]) \cap E} (\eta)_{2^k t}(x-y)u(y,t)\,dy\,dt/t \\ &\quad + \sum_{j=1}^{k} \iint_{(B \times (2^{j-k-1}, 2^{j-k}]) \cap E} (\eta)_{2^k t}(x-y)u(y,t)\,dy\,dt/t \\ &= (5.8)_0 + \sum_{j=1}^{k} (5.8)_j, \text{ say.} \end{aligned}$$

By (5.5)–(5.7) and by Lemma 5.1, we have

$$\operatorname{supp}(5.8)_j \subset B(0, 2^{j+1}) \qquad (0 \leq j \leq k),$$

$$\begin{aligned} \|(5.8)_j\|_{L^\infty} &\leq \sup_{t \in (2^{j-k-1}, 2^{j-k}]} \|(\eta)_{2^k t}\|_{L^\infty} \iint_{B \times (2^{j-k-1}, 2^{j-k}]} |u|\,dy\,dt/t \\ &\leq C(a,n)2^{-jn} \qquad (1 \leq j \leq k), \end{aligned}$$

$$\|(5.8)_0\|_{L^2} \leq C(a,n),$$

$$\int x^\alpha (5.8)_j\,dx = 0 \text{ if } |\alpha| \leq n(1/p-1), \qquad (0 \leq j \leq k).$$

So, $c(5.8)_0$ is a $(p,2)$-atom and $\left\{c2^{nj(1-1/p)}(5.8)_j\right\}_{j=1,\cdots,k}$ are (p,∞)-atoms. Thus

$$\|f\|_{H^p}^p \leq \sum_{j=0}^{k} C2^{-nj(1-1/p)p},$$

which implies the desired result. $\square$

Lemma 5.3. *Let $\eta \in S$ and*

$$\int \eta(x)x^\alpha\,dx = 0 \text{ if } |\alpha| \leq n(1/p-1).$$

Let p, B, u and E be as in Lemma 5.2. Let f be defined by the formula (5.4) with $\delta = 1$. Then

$$\|f\|_{H^p} \leq C(\eta,p,n)|B|^{1/p}.$$

Proof. By Lemma 3.5 η can be written in the form

$$\eta = \sum_{j\in\mathbf{N}_0} 2^{-j(n/p-n+1)}(v_j)_{2^j},$$

where

$$c(\eta,p)v_j \in \mathcal{B}_1^{n(1/p-1)}.$$

Then,

$$
\begin{aligned}
\|f\|_{H^p}^p &= \left\|\sum_{j\in\mathbf{N}_0} 2^{-j(n/p-n+1)}\iint_E (v_j)_{2^jt}(\cdot - y)u(y,t)dydt/t\right\|_{H^p}^p \\
&\leq \sum_{j\in\mathbf{N}_0} 2^{-j(n-np+p)}C\max\left\{2^{jn(1-p)}, j+1\right\}|B| \quad \text{by Lemma 5.2} \\
&\leq C|B|.
\end{aligned}
$$

$\square$

Lemma 5.4. *Let $u \in C(\mathbf{R}_+^{n+1})$, $0 < \delta' < \delta$, $\lambda > 0$,*

$$
\begin{aligned}
\Omega &= \{x \in \mathbf{R}^n : S_\delta u(x) > \lambda\}, \\
W &= \bigcup_{x\in\Omega^c} \Gamma(x,\delta').
\end{aligned}
$$

Then, Ω is open and

$$\left\|u(y,t)^2\chi_W(y,t)dydt/t\right\|_c \leq C(\delta,\delta',n)\lambda^2.$$

Proof. We omit the proof of the openness of Ω.
We may assume $\Omega^c \neq \emptyset$. First, we show that

$$\Gamma\left(\frac{n+2}{2}\right)\pi^{-n/2}\iint_{\Gamma(x,(\delta-\delta')/2)} u(y,t)^2\chi_W(y,t)(\delta t)^{-n}\frac{dydt}{t} \leq \lambda^2 \quad (5.9)$$

for all $x \in \mathbf{R}^n$.
If $x \in \Omega^c$, then (5.9) is clear because $\Gamma(x,(\delta-\delta')/2) \subset \Gamma(x,\delta)$.
Let $x \in \Omega$. Then, there exists $x^* \in \Omega^c$ such that

$$|x - x^*| = \text{dist}(x,\Omega^c),$$

because Ω^c is nonempty and closed. A geometric observation gives

$$\Gamma(x,(\delta-\delta')/2)\cap W \subset \Gamma(x^*,\delta),$$

(see Fig. 5.1), which implies

$$\text{(the left–hand of (5.9))} \leq S_\delta u(x^*)^2 \leq \lambda^2.$$

Thus, we get (5.9).

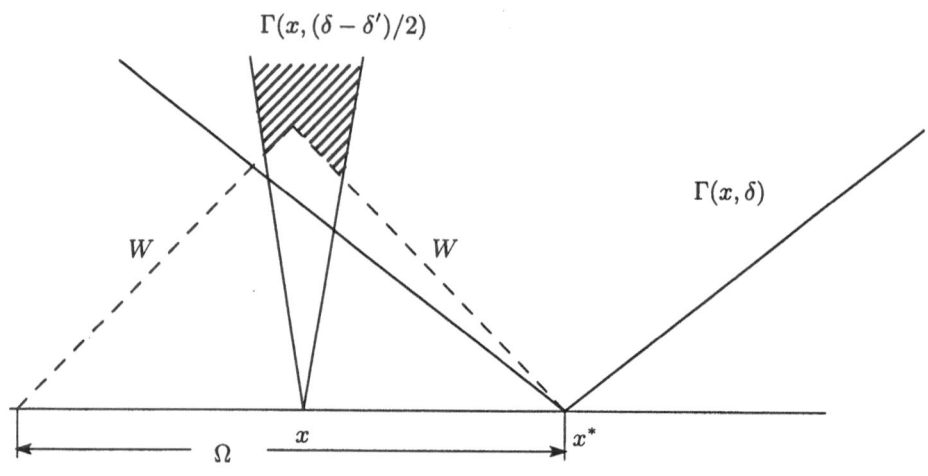

Fig. 5.1: $\Gamma(x, (\delta - \delta')/2)$ and $\Gamma(x^*, \delta)$

Take any ball $B \subset \mathbf{R}^n$. Then, by (5.9)

$$\iint_{Q(B)} u^2 \chi_W \, dy dt/t \leq C \int_B dx \iint_{\Gamma(x,(\delta-\delta')/2)} u^2 \chi_W \cdot (\delta t)^{-n} dy dt/t \leq C\lambda^2 |B|.$$

$\square$

Lemma 5.5. *Let $u \in C(\mathbf{R}_+^{n+1})$, $\delta > 0$, $p \in (0,1]$ and $S_\delta u \in L^p$. Let η be as in Lemma 5.3. Let $\varepsilon \in (0,1)$ and*

$$f_\varepsilon(x) = \iint_{\mathbf{R}^n \times (\varepsilon, 1/\varepsilon)} (\eta)_t(x-y) u(y,t) dy dt/t. \qquad (5.10)$$

(By Lemma 3.7 this integral converges for any $x \in \mathbf{R}^n$.) Then

$$\|f_\varepsilon\|_{H^p} \leq C(\eta, \delta, p) \|S_\delta u\|_{L^p}.$$

Proof. (See Fig. 5.2.) For $k \in \mathbf{Z}$ let

$$\Omega_k = \{x \in \mathbf{R}^n : S_\delta u(x) > 2^k\},$$
$$W_k = \bigcup_{x \in \Omega_k^c} \Gamma(x, \delta/2).$$

Applying Lemma 4.6 to $\Omega = \Omega_k$ gives $\{x_{k,j}\}_{j \in \mathbf{N}}$ that satisfies (4.7)–(4.8). Let

$$Q_{k,j} = \left(\left(B(x_{k,j}, 3r_{k,j}) \backslash \bigcup_{i=1}^{j-1} B(x_{k,i}, 3r_{k,i}) \right) \times (0, +\infty) \right) \cap (W_{k+1} \backslash W_k),$$

where
$$r_{k,j} = 100^{-1}\mathrm{dist}(x_{k,j}, \Omega_k^c).$$
Then, $\{Q_{k,j}\}_{k,j}$ are mutually disjoint,
$$\bigcup_{k\in\mathbf{Z}}\bigcup_j Q_{k,j} \supset \{(x,t)\in\mathbf{R}_+^{n+1} : u(x,t)\neq 0\},$$
$$Q_{k,j} \subset B(x_{k,j}, 3r_{k,j}) \times (0, (2/\delta)\cdot 103 r_{k,j}].$$

So,
$$
\begin{aligned}
f_\varepsilon(x) &= \sum_k \sum_j \iint_{Q_{k,j}\cap(\mathbf{R}^n\times(\varepsilon,1/\varepsilon))} (\eta)_t(x-y)u(y,t)dydt/t \\
&= \sum_k \sum_j (5.11)_{k,j}, \text{ say.}
\end{aligned}
$$

Since
$$
\begin{aligned}
\iint_{Q_{k,j}} u^2 dydt/t &\leq \|u^2\chi_{W_{k+1}}dydt/t\|_C\, |B(x_{k,j}, \max\{3, 206/\delta\}r_{k,j})| \\
&\leq C2^{2k}|B(x_{k,j}, r_{k,j})| \qquad \text{by Lemma 5.4,}
\end{aligned}
$$

Lemma 5.3 implies
$$\|(5.11)_{k,j}\|_{H^p} \leq C2^k |B(x_{k,j}, r_{k,j})|^{1/p}.$$

Thus,
$$\|f\|_{H^p}^p \leq \sum_k \sum_j C2^{kp}|B(x_{k,j}, r_{k,j})| \leq C\sum_k 2^{kp}|\Omega_k| \leq C\int S_\delta u(x)^p dx.$$

$\square$

Lemma 5.6. *Assume all the conditions in Lemma 5.5. Let f_ε be as in Lemma 5.5. Then, there exists $\tilde{f}\in H^p$ such that*
$$\|f_\varepsilon - \tilde{f}\|_{H^p} \to 0 \quad (\varepsilon\to +0),$$
$$\|\tilde{f}\|_{H^p} \leq C(\eta,\delta,p)\|S_\delta u\|_{L^p}.$$

Proof. Let $0 < \varepsilon_1 < \varepsilon_2 \leq 1$. Then, since
$$f_{\varepsilon_1}(x) - f_{\varepsilon_2}(x) = \iint (\eta)_t(x-y)u(y,t)\chi_{\mathbf{R}^n\times((\varepsilon_1,\varepsilon_2)\cup(1/\varepsilon_2,1/\varepsilon_1))}(y,t)dydt/t$$
and since
$$\left\|S_\delta\left(u\chi_{\mathbf{R}^n\times((\varepsilon_1,\varepsilon_2)\cup(1/\varepsilon_2,1/\varepsilon_1))}\right)\right\|_{L^p} \to 0 \quad (\varepsilon_1,\varepsilon_2\to +0),$$
Lemma 5.5 implies
$$\|f_{\varepsilon_1} - f_{\varepsilon_2}\|_{H^p} \to 0 \quad (\varepsilon_1,\varepsilon_2\to +0).$$

Thus, Lemma 2.1 implies that there exists $\tilde{f}\in H^p$ such that $f_\varepsilon \to \tilde{f}$ $(\varepsilon\to +0)$ in H^p.

$\square$

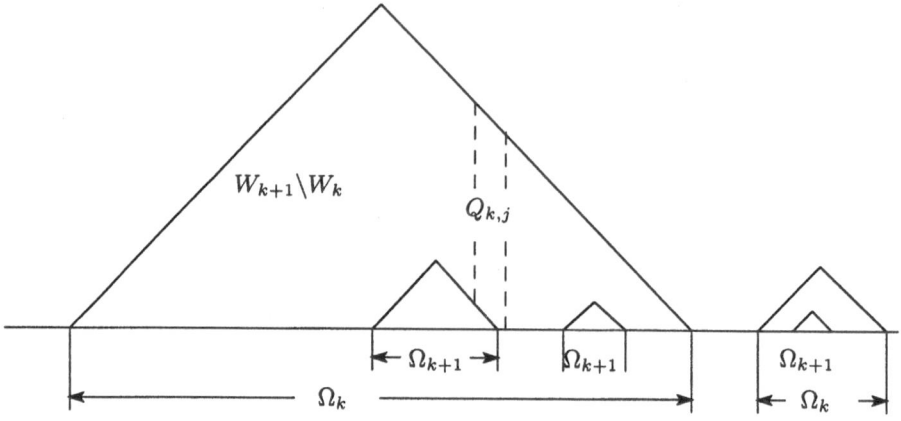

Fig. 5.2: Ω_k and W_k

Lemma 5.7. *Let $\{\psi_1, \cdots, \psi_J\} \subset C(\mathbf{R}^n \backslash \{0\}, \mathbf{C})$, $\varepsilon \in (0, 1)$ and*

$$\sup\left\{\sum_{i=1}^{J} \left|\psi_i\left(t|\xi|^{-1}\xi\right)\right| : t \in (\varepsilon, 1/\varepsilon)\right\} > 0 \qquad (5.12)$$

for any $\xi \in \mathbf{R}^n \backslash \{0\}$. Let $\theta \in C^\infty((0, +\infty), \mathbf{R})$ be nonnegative

$$\operatorname{supp}\theta \subset (\varepsilon/2, 2/\varepsilon) \quad \text{and} \quad \operatorname{supp}(1 - \theta) \subset (\varepsilon, 1/\varepsilon)^c.$$

Let

$$\Psi_i(\xi) = \begin{cases} \theta\left(|\xi|\right)\overline{\psi_i(\xi)}\Big/\left(\displaystyle\int_0^{+\infty} \theta\left(t|\xi|\right) \sum_{j=1}^{J} |\psi_j(t\xi)|^2 \dfrac{dt}{t}\right), & \xi \in \mathbf{R}^n \backslash \{0\}, \\ 0, & \xi = 0, \end{cases} \qquad (5.13)$$

$i = 1, \cdots, J$. Then,

$$\{\Psi_1, \cdots, \Psi_J\} \subset C(\mathbf{R}^n, \mathbf{C}),$$

$\operatorname{supp}\Psi_i$ are compact and away from the origin, $\qquad (5.14)$

$$\int_0^{+\infty} \sum_{i=1}^{J} \psi_i(t\xi)\Psi_i(t\xi)dt/t = 1 \quad \text{for any } \xi \in \mathbf{R}^n \backslash \{0\}. \qquad (5.15)$$

This is obvious.

Proof of Theorem 5.1. Applying Lemma 5.7 to $\{\psi_1, \cdots, \psi_J\} = \{\mathcal{F}\phi_1, \cdots, \mathcal{F}\phi_J\}$ with a sufficiently small $\varepsilon > 0$ gives $\{\eta_1, \cdots, \eta_J\} \subset \mathcal{S}_0$ such that

$$\int_0^{+\infty} \sum_{i=1}^{J} \mathcal{F}\phi_i(t\xi)\mathcal{F}\eta_i(t\xi)dt/t = 1 \quad \text{for any } \xi \in \mathbf{R}^n \backslash \{0\}. \qquad (5.16)$$

Let $\psi \in \mathcal{S}_0$. Take $\varepsilon' > 0$ so that

$$\text{supp}\mathcal{F}\psi \cap \left(\bigcup_{i=1}^{J} \text{supp}\mathcal{F}\eta_i(-t \cdot) \right) = \emptyset \text{ if } t \in (0, \varepsilon') \cup (1/\varepsilon', +\infty). \quad (5.17)$$

If $\varepsilon'' \in (0, \varepsilon']$, then

$$\mathcal{F}\psi(\xi) = \sum_{i=0}^{J} \int_{\varepsilon''}^{1/\varepsilon''} \mathcal{F}\phi_i(-t\xi)\mathcal{F}\eta_i(-t\xi)\mathcal{F}\psi(\xi)dt/t.$$

So, if $\varepsilon'' \in (0, \varepsilon']$ and if

$$f_{\varepsilon''}(x) = \sum_{i=1}^{J} \iint_{\mathbf{R}^n \times (\varepsilon'', 1/\varepsilon'')} (\eta_i)_t(x-y)f * (\phi_i)_t(y)dydt/t, \quad (5.18)$$

then

$$\langle \psi, f \rangle_\mathcal{S} = \int \psi(x)f_{\varepsilon''}(x)dx. \quad (5.19)$$

On the other hand, Lemma 5.6 and (5.3) imply the existence of $\tilde{f} \in H^p$ such that

$$\|f_{\varepsilon''} - \tilde{f}\|_{H^p} \to 0 \quad (\varepsilon'' \to +0) \quad (5.20)$$

$$\|\tilde{f}\|_{H^p} \le C\left(\{\eta_1, \cdots, \eta_J\}, \delta, p\right) \sum_{i=1}^{J} \|S_{\phi_i, \delta}f\|_{L^p}. \quad (5.21)$$

Since $\psi \in \mathcal{S}_0$ is arbitrary, (5.19) and (5.20) imply that $f - \tilde{f}$ is a polynomial. (5.21) implies the desired estimate. $\qquad \square$

Remark 5.1. If in Theorem 5.1 we impose the condition that $\mathcal{F}f \in L^1_{\text{loc}}(\mathbf{R}^n, \mathbf{C})$ or that $f \in \bigcup_{q \in [1, +\infty)} L^q$, then

$$\tilde{f} \text{ in } (5.20) = f.$$

If $f \in L^\infty$, then $P(x)$ in Theorem 5.1 is a constant.

Remark 5.2. Remark 3.4 implies

$$\mathcal{S}_0 \subset H^p.$$

On the other hand, if $f \in H^p$, if $\{\phi_1, \cdots, \phi_J\}$ is as in Theorem 5.1 and if $\int \phi_i dx = 0 \ (i = 1, \cdots, J)$, then

$$S_{\phi_i, \delta}f \in L^p$$

by Theorem 3.3 and

$$\tilde{f} \text{ in } (5.20) \ = f$$

by Remark 5.1 and Theorem 2.4. Since

$$f_{\varepsilon''} \text{ in } (5.18) \ \subset S_0,$$

combining the above gives that

"S_0 is a dense subspace of H^p."

Theorem 5.2. *Let $\phi \in S$, $\int \phi \, dx \neq 0$, $p \in (0,1]$, $\delta > 0$, $m \in \mathbf{R}$, $f \in S'$ and*

$$\|S_{\phi,\delta,m}f\|_{L^p} < +\infty. \tag{5.22}$$

Then, if $m < n(1 - 1/p)$, then

$$f = 0 \in S'. \tag{5.23}$$

If $m \geq n(1 - 1/p)$, then $\mathcal{F}f \in C(\mathbf{R}^n, \mathbf{C})$ and

$$|\mathcal{F}f(\xi)| \leq C(\phi, \delta, m, p)\|S_{\phi,\delta,m}f\|_{L^p}|\xi|^{m+n(1/p-1)} \tag{5.24}$$

and

$$\left\|\mathcal{F}^{-1}\left(|\xi|^{-m}\mathcal{F}f(\xi)\right)\right\|_{H^p} \leq C(\phi, \delta, m, p)\|S_{\phi,\delta,m}f\|_{L^p}. \tag{5.25}$$

Proof. Lemma 3.7 and (5.22) imply

$$\varepsilon^m \inf_{t \in (\varepsilon, 2\varepsilon)} \|f * (\phi)_t\|_{L^1} \leq C(\delta\varepsilon)^{n(1-1/p)}\|S_{\phi,\delta,m}f\|_{L^p}.$$

Since $\mathcal{F}\phi(0) \neq 0$, this implies $\mathcal{F}f \in C(\mathbf{R}^n, \mathbf{C})$ and

$$\inf_{t \in (\varepsilon, 2\varepsilon)} |\mathcal{F}f(\xi)\mathcal{F}\phi(t\xi)| \ \leq \ C\delta^{n(1-1/p)}\varepsilon^{n(1-1/p)-m}\|S_{\phi,\delta,m}f\|_{L^p}$$

$$\text{for all } \xi \in \mathbf{R}^n \text{ and all } \varepsilon > 0.$$

Taking $\varepsilon < c/|\xi|$ gives (5.23)–(5.24).

Applying Lemma 5.7 with a sufficiently small $\varepsilon > 0$ to $\{\psi_1\} = \{|\xi|^m \mathcal{F}\phi(\xi)\}$ gives $\{\eta\} \subset S_0$ such that

$$\int_0^\infty \mathcal{F}\eta(t\xi)|t\xi|^m \mathcal{F}\phi(t\xi)dt/t = 1 \text{ i.e. } \int_0^\infty \mathcal{F}\eta(t\xi)t^m \mathcal{F}\phi(t\xi)dt/t = |\xi|^{-m}$$

for any $\xi \in \mathbf{R}^n \backslash \{0\}$. This combined with $|\xi|^{-m} |\mathcal{F}f(\xi)| \leq C|\xi|^{n(1/p-1)}$ implies

$$\mathcal{F}^{-1}\left(|\xi|^{-m}\mathcal{F}f(\xi)\right)$$

$$= \lim_{\substack{\text{in } S' \\ \varepsilon' \to +0}} \int_{\varepsilon'}^{1/\varepsilon'} \mathcal{F}^{-1}\left(\mathcal{F}\eta(t\xi)t^m \mathcal{F}\phi(t\xi)\mathcal{F}f(\xi)\right) dt/t$$

$$= \lim_{\substack{\text{in } S' \\ \varepsilon' \to +0}} \int_{\varepsilon'}^{1/\varepsilon'} dt/t \int (\eta)_t(\cdot - y)t^m f * (\phi)_t(y)dy. \tag{5.26}$$

Applying Lemma 5.6 to the right-hand side of (5.26) gives (5.25).   □

Remark 5.3. If

$$\psi \in \mathcal{S}, \quad \int \psi \, dx \neq 0 \text{ and } m \in \mathbf{N}, \tag{5.27}$$

then

$$\{\phi_1, \cdots, \phi_J\} = \{D_x^\alpha \psi\}_{\alpha : |\alpha| = m} \tag{5.28}$$

satisfies the hypothesis of Theorem 5.1.

Remark 5.4. Assume (5.27). If $P(x)$ is a polynomial,

$$p \in (0,1], \ f \in \mathcal{S}', \ \sum_{|\alpha|=m} \left\| S_{D_x^\alpha \psi, 1} f \right\|_{L^p} < \infty, \text{ and } f - P \in H^p,$$

then

$$\sum_{|\alpha|=m} \left\| S_{\psi,1,m}(D_x^\alpha P) \right\|_{L^p} = \sum_{|\alpha|=m} \left\| S_{D_x^\alpha \psi, 1} P \right\|_{L^p}$$

$$\leq C \sum_{|\alpha|=m} \left\{ \left\| S_{D_x^\alpha \psi, 1}(f - P) \right\|_{L^p} + \left\| S_{D_x^\alpha \psi, 1} f \right\|_{L^p} \right\}$$

$$\leq C \left\{ \|f - P\|_{H^p} + \sum_{|\alpha|=m} \left\| S_{D_x^\alpha \psi, 1} f \right\|_{L^p} \right\} \quad \text{by Theorem 3.3}$$

$$< \infty,$$

which combined with Theorem 5.2 implies

$$\mathcal{F} D_x^\alpha P \in C(\mathbf{R}^n, \mathbf{C}) \ (|\alpha| = m).$$

Therefore,

$$D_x^\alpha P = 0 \ (|\alpha| = m).$$

Thus,

$$\deg P < m.$$

So, in the case (5.28) (with (5.27)), the polynomial $P(x)$ in Theorem 5.1 is of degree $< m$. In particular, if

$$\{\phi_1, \cdots, \phi_J\} = \{D_{x_j} \psi\}_{j=1,\cdots,n},$$

then

$$P(x) \equiv \text{constant}.$$

Notes. The argument we have presented in this section is essentially due to A. Chang–R. Fefferman [80] and A. Bernard [79]. As for Lemmas 5.3 and 5.5 see also R. Coifman–Y. Meyer–E. M. Stein [85, 82]. In the proof of Lemma 5.4 we borrowed an idea from R. Fefferman–R. F. Gundy–M. L. Silverstein–E. M. Stein [82]. We learned Lemma 5.7 from A. P. Calderón–A. Torchinsky [75]. A. P. Calderón–A. Torchinsky [75] obtained a result of type of Theorem 5.1 with $\|f - P\|_{H^p}$ replaced by $\|N_{\phi,1}(f - P)\|_{L^p}$, where $\phi \in \mathcal{S}$. Their result combined with R. Coifman [74] and R. Latter [78] implies Theorem 5.1. A. P. Calderón [77] is also an important paper.

VI. Hardy-Littlewood-Fefferman-Stein type inequalities, 1

Theorem 6.A . Let $m \in \{2, 3, 4, \cdots\}$ and $q > 0$. Let $u(z)$ be a harmonic function defined on the unit ball of $\mathbf{R}^m$. Then

$$|u(0)|^q \le C(q, m) \int_{\{z \in \mathbf{R}^m : |z| < 1\}} |u(z)|^q \, dz.$$

This is called the Hardy-Littlewood-Fefferman-Stein inequality. As a consequence of this we have that if $u(x, t)$ is harmonic on $\mathbf{R}_+^{n+1}$, then

$$|u(x, t)|^q \le C(q, n) t^{-n-1} \iint_{B(x, t) \times (t/2, 3t/2)} |u(y, s)|^q \, dy ds. \tag{6.1}$$

The inequality (6.1) has several important applications to the theory of $H^p(\mathbf{R}^n)$. In sections 6–8 we will give weak extensions of the inequality (6.1) to the functions of the form $f * (\phi)_t(x)$. (For the conditions on f and ϕ see Theorems 6.1, 7.1 and 8.1.) In sections 9 and 10 we will explain their applications to $H^p(\mathbf{R}^n)$. Since we do not use Theorem 6.A (or (6.1)), we do not give its proof. For its proof see Appendix.

For $\varepsilon > 0$ let

$$Q(\varepsilon) = B(0, 1 + \varepsilon) \times (0, \varepsilon]. \tag{6.2}$$

In this section we will prove the following.

Theorem 6.1. *Let* $f \in \mathcal{D}'$, $\phi \in \mathcal{D}$, $\psi \in \mathcal{D}$,

$$0 < b < a, \tag{6.3}$$

$$q \in (0, 1], \tag{6.4}$$

$$\varepsilon \in (0, +\infty), \tag{6.5}$$

$$\mathrm{supp}\, \phi \subset B(0, 1), \quad \int \phi \, dx = 1, \tag{6.6}$$

$$\mathrm{supp}\, \psi \subset B(0, 1). \tag{6.7}$$

Then,

$$|\langle \psi, f \rangle_{\mathcal{D}}|^q \leq C_{6.1}\,(a, b, q, \varepsilon, \|\phi\|_{\Lambda_a}, n)\,\|\psi\|_{\Lambda_b}^q$$

$$\times \iint_{Q(\varepsilon)} |f * (\phi)_t(x)|^q\, t^{q(n+b)-n}\, dx\, dt/t. \qquad (6.8)$$

Lemma 6.1. *Let* $k \in \mathbf{N}$, $\phi \in L^1$, $|x|^k \phi \in L^1$, $\int \phi\, dx = 1$ *and*

$$\eta(x) = \left(\prod_{i=1}^{k} (1 - 2^i) \right)^{-1}$$

$$\times \sum_{j_1=0}^{1} \cdots \sum_{j_k=0}^{1} \left(\prod_{i=1}^{k} (-2^i)^{j_i} \right) (\phi)_{2^{-j_1-j_2-\cdots-j_k}}(x). \qquad (6.9)$$

Then

$$\int \eta(x)\, dx = 1, \qquad (6.10)$$

$$\int \eta(x)\, x^\alpha\, dx = 0 \ \text{ if } \ 1 \leq |\alpha| \leq k. \qquad (6.11)$$

Proof. (6.10) is easy. (6.11) follows from

$$\int \eta(x)\, x^\alpha\, dx$$

$$= \left(\prod_{i=1}^{k} (1 - 2^i) \right)^{-1} \sum_{j_1=0}^{1} \cdots \sum_{j_k=0}^{1} \left(\prod_{i=1}^{k} (-2^i)^{j_i} \right) \int (\phi)_{2^{-j_1-\cdots-j_k}}(x)\, x^\alpha\, dx$$

$$= (\cdots)^{-1} \sum \cdots \sum \left(\prod_{i=1}^{k} (-2^i)^{j_i} \right) (2^{-j_1-\cdots-j_k})^{|\alpha|} \int \phi(x)\, x^\alpha\, dx$$

$$= (\cdots)^{-1} \sum \cdots \sum \left(\prod_{i=1}^{k} (-2^{i-|\alpha|})^{j_i} \right) \int \phi(x)\, x^\alpha\, dx$$

$$= (\cdots)^{-1} \prod_{i=1}^{k} (1 - 2^{i-|\alpha|}) \int \phi(x)\, x^\alpha\, dx$$

$$= 0 \quad \text{if } 1 \leq |\alpha| \leq k.$$

$\square$

Lemma 6.2. *Let*

$$0 < b \leq b'. \qquad (6.12)$$

Let $\phi \in L^1$ *and* $\psi \in \Lambda_{b'}$. *Assume* (6.5)–(6.7). *Then, there exists a measurable function* $k(y, t)$ *defined on* $\mathbf{R}_+^{n+1}$ *such that*

$$k(y,t) = 0 \quad on \quad \mathbf{R}^{n+1}_+ \backslash Q(\varepsilon), \tag{6.13}$$

$$|k(y,t)| \le C_{6.2}(b,b',\varepsilon,n)\|\phi\|_{L^1} \min\left\{\|\psi\|_{\Lambda_b} t^b, \|\psi\|_{\Lambda_{b'}} t^{b'}\right\}, \tag{6.14}$$

$$\psi(x) = \iint k(y,t)(\phi)_t(x-y)dydt/t. \tag{6.15}$$

Proof. If $b' < 1$, then put $\eta(x) = \phi(x)$. If $b' \ge 1$, then define η by (6.9) with $k = [b']$. Then,

$$\operatorname{supp}\eta \subset B(0,1) \quad \text{by (6.6)}, \tag{6.16}$$
$$\operatorname{supp}((\eta)_{1/e} - (\eta)_e) \subset B(0,e) \quad \text{by (6.6)}, \tag{6.17}$$
$$\int ((\eta)_{1/e}(x) - (\eta)_e(x))x^\alpha dx = 0 \quad \text{if } |\alpha| \le b' \tag{6.18}$$

by (6.11) and

$$
\begin{aligned}
\psi(x) &= \lim_{\delta \to +0} \int_\delta^{e\delta} \psi * (\eta)_{t/e} * (\eta)_t(x)dt/t \quad \text{by (6.10)} \\
&= \lim_{\delta \to +0} \iint_{\mathbf{R}^n \times (\delta,+\infty)} h(y,t)(\eta)_t(x-y)dydt/t, \tag{6.19}
\end{aligned}
$$

where

$$
h(y,t) = \begin{cases} \psi * ((\eta)_{t/e} - (\eta)_{et})(y) & \text{if } t \in (0,\varepsilon/e], \\ \psi * (\eta)_{t/e}(y) & \text{if } t \in (\varepsilon/e,\varepsilon], \\ 0 & \text{if } t > \varepsilon. \end{cases} \tag{6.20}
$$

If $t \in (0,\varepsilon/e]$, then by (6.17), (6.18) and Lemma 1.2(iii)

$$|h(y,t)| \le C(b,b',n)\left\|(\eta)_{1/e} - (\eta)_e\right\|_{L^1} \min\left\{t^b\|\psi\|_{\Lambda_b}, t^{b'}\|\psi\|_{\Lambda_{b'}}\right\}. \tag{6.21}$$

If $t \in (\varepsilon/e,\varepsilon]$, then

$$|h(y,t)| \le \|\eta\|_{L^1}\|\psi\|_{L^\infty}. \tag{6.22}$$

By (6.7), (6.16) and (6.17)

$$\operatorname{supp} h \subset Q(\varepsilon). \tag{6.23}$$

If $[b'] = 0$, then put

$$k(y,t) = h(y,t). \tag{6.24}$$

If $[b'] \ge 1$, then put $k = [b']$ and

$$
\begin{aligned}
k(y,t) &= \left(\prod_{i=1}^k (1-2^i)\right)^{-1} \\
&\quad \times \sum_{j_1=0}^1 \cdots \sum_{j_k=0}^1 \left(\prod_{i=1}^k (-2^i)^{j_i}\right) h\left(y, 2^{j_1+j_2+\cdots+j_k}t\right). \tag{6.25}
\end{aligned}
$$

Then (6.13)–(6.14) follow from (6.21)–(6.23). (6.15) follows from (6.9), (6.19) and from

$$\iint |h(y,st)| \, |(\phi)_t(x-y)| \, dydt/t < \infty \text{ for any } x \in \mathbf{R}^n \text{ and } s > 0$$

(by (6.21)–(6.22)).   □

Applying Lemma 6.2 to $\psi = \phi$ and $\varepsilon = 1/2$ gives

Lemma 6.3. *Let $b'' > 0$ and $\phi \in \Lambda_{b''}$. Assume (6.6). Then, there exists a measurable function $\tilde{k}(y,t)$ defined on $\mathbf{R}_+^{n+1}$ such that*

$$\tilde{k}(y,t) = 0 \text{ on } \mathbf{R}_+^{n+1} \setminus (B(0,2) \times (0,1/2)), \tag{6.26}$$

$$\left| \tilde{k}(y,t) \right| \leq C_{6.3}(b'',n) \|\phi\|_{L^1} \|\phi\|_{\Lambda_{b''}} t^{b''}, \tag{6.27}$$

$$\phi(x) = \iint \tilde{k}(y,t)(\phi)_t(x-y)dydt/t, \tag{6.28}$$

where

$$C_{6.3} = C_{6.2}(b'',b'',1/2,n).$$

Definition 6.1. For a measurable set $E \subset \mathbf{R}_+^{n+1}$ let

$$\text{dens}(E) = \sup \{|E \cap Q(B)| / |Q(B)| : B \text{ is taken over all balls in } \mathbf{R}^n \}.$$

Lemma 6.4. *Let $b'' > b' \geq b > 0$, $\phi \in \Lambda_{b''}$ and $\psi \in \Lambda_{b'}$. Assume (6.5)–(6.7). Let $E \subset \mathbf{R}_+^{n+1}$ be a measurable set such that*

$$\text{dens}(E) < C_{6.4} \left(b, b', b'', \|\phi\|_{\Lambda_{b''}}, n \right). \tag{6.29}$$

Then, there exists a measurable function $k(y,t)$ defined on $\mathbf{R}_+^{n+1}$ such that

$$k(y,t) = 0 \text{ on } E, \tag{6.30}$$

$$k(y,t) = 0 \text{ on } \mathbf{R}_+^{n+1} \setminus Q(\varepsilon), \tag{6.13}'$$

$$|k(y,t)| \leq C_{6.5}(b,b',\varepsilon,n)\|\phi\|_{L^1} \min \left\{ \|\psi\|_{\Lambda_b} t^b, \|\psi\|_{\Lambda_{b'}} t^{b'} \right\}, \tag{6.14}'$$

$$\psi(x) = \iint k(y,t)(\phi)_t(x-y)dydt/t. \tag{6.15}'$$

Proof. Applying Lemma 6.2 gives $k_1(y,t)$ such that

$$k_1(y,t) = 0 \quad \text{on} \quad \mathbf{R}_+^{n+1} \backslash Q(\varepsilon), \tag{6.13}_1$$

$$|k_1(y,t)| \le C_{6.2}(b,b',\varepsilon,n) \|\phi\|_{L^1} \min \left\{ \|\psi\|_{\Lambda_b} t^b, \|\psi\|_{\Lambda_{b'}} t^{b'} \right\}, \tag{6.14}_1$$

$$\psi(x) = \iint k_1(y,t)(\phi)_t(x-y) dy dt/t. \tag{6.15}_1$$

Applying Lemma 6.3 gives $\tilde{k}(y,t)$ that satisfies (6.26)–(6.28). Applying translation and dilation to (6.28) gives

$$(\phi)_t(x-y) = \iint t^{-n} \tilde{k}((z-y)/t, s/t)\,(\phi)_s(x-z) dz ds/s. \tag{6.28'}$$

Put

$$I_1 \;=\; \iint k_1(y,t)\chi_{E^c}(y,t)(\phi)_t(x-y) dy dt/t, \tag{6.31}$$

$$J_1 \;=\; \iint k_1(y,t)\chi_E(y,t)(\phi)_t(x-y) dy dt/t. \tag{6.32}$$

Substituting (6.28)′ into J_1 and interchanging the order of integrations we get

$$
\begin{aligned}
J_1 \;&=\; \iint (\phi)_s(x-z) \\
&\quad \times \left\{ \iint k_1(y,t)\chi_E(y,t) t^{-n}\tilde{k}\left((z-y)/t, s/t\right) dy dt/t \right\} dz ds/s \\
&=\; \iint (\phi)_S(x-z) k_2(z,s) dz ds/s, \quad \text{say.}
\end{aligned}
$$

By (6.26) and (6.13)$_1$

$$
k_2(z,s) \;=\; \iint_{\{(y,t)\in\Gamma(z,2):t>2s\}\cap Q(\varepsilon)} k_1(y,t)\chi_E(y,t) \\
\times t^{-n}\tilde{k}\left((z-y)/t, s/t\right) dy dt/t.
$$

Thus,

$$k_2(z,s) = 0 \quad \text{if} \quad s > \varepsilon/2 \quad \text{or} \quad |z| > 1+\varepsilon+2\varepsilon, \tag{6.13}_2$$

because in this case the domain of the above integration is empty. For general $(z,s) \in \mathbf{R}_+^{n+1}$ we have

$$|k_2(z,s)| \leq \sum_{j \in \mathbf{N}} \iint_{\{(y,t)\in\Gamma(z,2):t\in(2^j s, 2^{j+1}s]\}} C_{6.2}\|\phi\|_{L^1}$$

$$\times \min\left\{\|\psi\|_{\Lambda_b}t^b, \|\psi\|_{\Lambda_{b'}}t^{b'}\right\} \chi_E(y,t)t^{-n}$$

$$\times C_{6.3}\|\phi\|_{L^1}\|\phi\|_{\Lambda_{b''}}(s/t)^{b''} dydt/t \qquad \text{by } (6.14)_1 \text{ and } (6.27)$$

$$\leq \sum C_{6.2}\|\phi\|_{L^1}C(b,b',b'',n)\min\left\{\|\psi\|_{\Lambda_b}(2^j s)^b, \|\psi\|_{\Lambda_{b'}}(2^j s)^{b'}\right\}\text{dens}(E)$$

$$\times(2^j s)^{-n}C_{6.3}\|\phi\|_{L^1}\|\phi\|_{\Lambda_{b''}}2^{-jb''}(2^j s)^n$$

$$\leq \sum C_{6.2}\|\phi\|_{L^1}C(b,b',b'',n)C_{6.3}\|\phi\|_{L^1}\|\phi\|_{\Lambda_{b''}}$$

$$\times \min\left\{\|\psi\|_{\Lambda_b}s^b, \|\psi\|_{\Lambda_{b'}}s^{b'}\right\}\text{dens}(E).$$

So, by taking $C_{6.4}$ so small that

$$C_{6.4} < \left(2C(b,b',b'',n)C_{6.3}\|\phi\|_{L^1}\|\phi\|_{\Lambda_{b''}}\right)^{-1},$$

we get

$$|k_2(z,s)| \leq 2^{-1}C_{6.2}\|\phi\|_{L^1}\min\left\{\|\psi\|_{\Lambda_b}s^b, \|\psi\|_{\Lambda_{b'}}s^{b'}\right\}. \qquad (6.14)_2$$

Therefore,

$$\psi(x) = I_1 + J_1 = I_1 + \iint k_2(y,t)(\phi)_t(x-y)dydt/t, \qquad (6.15)_2$$

where k_2 satisfies $(6.13)_2$ and $(6.14)_2$.

Next, we apply the same argument to $k_2(y,t)$. Put

$$I_2 = \iint k_2(y,t)\chi_{E^c}(y,t)(\phi)_t(x-y)dydt/t,$$

$$J_2 = \iint k_2(y,t)\chi_E(y,t)(\phi)_t(x-y)dydt/t.$$

Substituting $(6.28)'$ into J_2 and repeating all of the above argument, we get $k_3(y,t)$ such that

$$k_3(y,t) = 0 \text{ if } t < \varepsilon/4 \text{ or } |y| > 1 + \varepsilon + 2(\varepsilon + \varepsilon/2), \qquad (6.13)_3$$

$$|k_3(y,t)| \leq 2^{-2}C_{6.2}\|\phi\|_{L^1}\min\left\{\|\psi\|_{\Lambda_b}t^b, \|\psi\|_{\Lambda_{b'}}t^{b'}\right\}, \qquad (6.14)_3$$

$$\psi(x) = I_1 + I_2 + \iint k_3(y,t)(\phi)_t(x-y)dydt/t. \qquad (6.15)_3$$

Repeating this argument infinitely, we get $k_1(y,t), k_2(y,t),\cdots,k_j(y,t),\cdots$ such that

$$k_j(y,t) = 0 \text{ if } t > \varepsilon/2^{j-1} \text{ or } |y| > 1 + \varepsilon + 2\varepsilon \sum_{i=0}^{j-2} 2^{-i}, \qquad (6.13)_j$$

$$|k_j(y,t)| \le 2^{-j+1} C_{6.2} \|\phi\|_{L^1} \min \left\{ \|\psi\|_{\Lambda_b} t^b, \|\psi\|_{\Lambda_{b'}} t^{b'} \right\}, \tag{6.14}_j$$

$$\psi(x) = I_1 + I_2 + \cdots + I_{j-1} + \iint k_j(y,t)(\phi)_t(x-y)\,dy\,dt/t, \tag{6.15}_j$$

where

$$I_i = \iint k_i(y,t)\chi_{E^c}(y,t)(\phi)_t(x-y)\,dy\,dt/t.$$

Put

$$k(y,t) = \sum_{i\in\mathbf{N}} k_i(y,t)\chi_{E^c}(y,t).$$

Then, (6.30) is clear. $(6.14)_j$ implies $(6.14)'$. Letting $j \to \infty$ in $(6.15)_j$ and using $(6.13)_j$–$(6.14)_j$ implies $(6.15)'$. $(6.13)_j$ implies

$$k(y,t) = 0 \text{ if } t > \varepsilon \text{ or } |y| > 1 + 5\varepsilon.$$

So, applying the above procedure with ε replaced by $\varepsilon/5$ gives $(6.13)'$. $\square$

Lemma 6.5. *Let $f \in \mathcal{D}'$ and $\phi, \psi \in \mathcal{D}$. Assume (6.4)–(6.7). Let $b > 0$, $b_0 > 0$,*

$$b'' > \max\{b, b_0\}, \tag{6.33}$$

$$\sup \left\{ |\langle \eta, f \rangle_{\mathcal{D}}| \, / \|\eta\|_{\Lambda_{b_0}} : \eta \in \mathcal{D}, \text{ supp}\, \eta \subset B(0, 1 + 2\varepsilon) \right\} < +\infty. \tag{6.34}$$

Then,

$$|\langle \psi, f \rangle_{\mathcal{D}}|^q \le C_{6.6} \left(b_0, b, b'', \varepsilon, q, \|\phi\|_{\Lambda_{b''}}, n \right) \|\psi\|_{\Lambda_b}^q$$

$$\times \iint_{Q((1+2\sqrt{n})\varepsilon)} |f * (\phi)_t(x)|^q \, t^{q(n+b)-n} \, dx\,dt/t. \tag{6.35}$$

Proof. Let

$$b' = (b_0 + b'')/2. \tag{6.36}$$

Let $\{I_j\}_{j\in\mathbf{N}}$ be the family of all dyadic cubes in $\mathbf{R}^n$ and let

$$T_j = I_j \times (\ell(I_j)/2, \ell(I_j)]. \tag{6.37}$$

Let

$$E = \bigcup_j \{(y,t) \in T_j : C_{6.7}(n)C_{6.4} |f * (\phi)_t(y)|^q$$

$$\ge \iint_{T_j} |f * (\phi)_s(z)|^q \, dz\,ds/|T_j| \}. \tag{6.38}$$

Note that E satisfies the condition (6.29) if we take $C_{6.7}$ (> 0) sufficiently small. Applying Lemma 6.4 with $\phi(x)$ replaced by $\phi(-x)$ and with the above E gives $k(y, t)$ that satisfies (6.30), (6.13)′, (6.14)′ and

$$\psi(x) = \iint k(y,t)(\phi)_t(y-x)dydt/t. \qquad (6.15)''$$

By (6.14)′ and by $b' > b_0$ we have

$$\sup_{y \in \mathbf{R}^n} |k(y,t)| /t^{b_0} \to 0 \ (t \to +0).$$

Let

$$\psi_\delta(x) = \iint_{\mathbf{R}^n \times (\delta,+\infty)} k(y,t)(\phi)_t(y-x)dydt/t.$$

Then, Lemma 1.11 and the above estimate on $|k(y,t)|$ imply

$$\|\psi - \psi_\delta\|_{\Lambda_{b_0}} \to 0 \ (\delta \to +0). \qquad (6.39)$$

Since

$$\operatorname{supp} \psi_\delta \subset D(0, 1 + 2\varepsilon)$$

by (6.13)′ and by (6.6), we have

$$
\begin{aligned}
\langle \psi, f \rangle_D &= \lim_{\delta \to +0} \langle \psi_\delta, f \rangle_D \qquad \text{by (6.34) and (6.39)} \\
&= \lim \iint_{\mathbf{R}^n \times (\delta,+\infty)} k(y,t) f * (\phi)_t(y) dydt/t. \qquad (6.40)
\end{aligned}
$$

Thus,

$$|\langle \psi, f \rangle_D|^q \leq \left\{ \iint_{Q(\varepsilon)} |k(y,t)| \, |f * (\phi)_t(y)| \, dydt/t \right\}^q \text{ by (6.40) and (6.13)}'$$

$$\leq \sum_{j:T_j \cap Q(\varepsilon) \neq \emptyset} \left\{ \iint_{T_j} \cdots dydt/t \right\}^q \qquad \text{by (6.4)}$$

$$\leq \sum \left\{ \iint_{T_j} |k(y,t)| \left((C_{6.7}C_{6.4})^{-1} \iint_{T_j} |f * (\phi)_s(z)|^q \, dzds/|T_j| \right)^{\frac{1}{q}} \frac{dydt}{t} \right\}^q$$

$$\text{by (6.30) with (6.38)}$$

$$\leq \sum (C_{6.7}C_{6.4})^{-1} \iint_{T_j} |f * (\phi)_s(z)|^q \, dzds/|T_j|^{-1}$$

$$\times \left\{ \iint_{T_j} C_{6.5} \|\phi\|_{L^1} \|\psi\|_{\Lambda_b} t^b dydt/t \right\}^q \qquad \text{by (6.14)}'$$

$$\leq \sum (C_{6.7}C_{6.4})^{-1} \iint_{T_j} |f * (\phi)_s(z)|^q \, dzds t_j^{-n-1}$$

$$\times (C_{6.5} \|\phi\|_{L^1} \|\psi\|_{\Lambda_b})^q \, t_j^{(n+b)q} \text{ where } t_j = \text{(the side length of } T_j)$$

$$\leq \sum (C_{6.7}C_{6.4})^{-1}C_{6.5}^q \|\phi\|_{L^1}^q \|\psi\|_{\Lambda_b}^q C(q,b,n)$$
$$\times \iint_{T_j} |f * (\phi)_s(z)|^q s^{q(n+b)-n} dz\,ds/s. \tag{6.41}$$

Since

$$\|\phi\|_{L^1} \leq C(b'',n)\|\phi\|_{\Lambda_{b''}} \qquad \text{by } \operatorname{supp}\phi \subset B(0,1),$$

(6.35) follows from (6.41). $\qquad\qquad\qquad\qquad\qquad\qquad\qquad\qquad\qquad\square$

Proof of Theorem 6.1. Since $f \in \mathcal{D}'$, there exists $b_0 > 0$ such that (6.34) holds. Take b'' so that (6.33) holds. Then, Lemma 6.5 implies

$$\sup\{|\langle\psi,f\rangle_{\mathcal{D}}|/\|\psi\|_{\Lambda_b} : \psi \in \mathcal{D},\ \operatorname{supp}\psi \subset B(0,1)\}$$
$$\leq C_{6.6}\left(b_0,b,b'',\varepsilon,q,\|\phi\|_{\Lambda_{b''}},n\right)^{1/q} I, \tag{6.42}$$

where

$$I^q = \iint_{Q((1+2\sqrt{n})\varepsilon)} |f * (\phi)_t(x)|^q t^{q(n+b)-n} dx\,dt/t. \tag{6.43}$$

Define $\tilde{f} \in \mathcal{D}'$ by

$$\langle\eta,\tilde{f}\rangle_{\mathcal{D}} = \langle(\eta)_{1/(1+2\varepsilon)}, f\rangle_{\mathcal{D}} \qquad \text{for any } \eta \in \mathcal{D}.$$

If

$$I < +\infty, \tag{6.44}$$

then (6.42) implies

$$\sup\left\{\left|\langle\eta,\tilde{f}\rangle_{\mathcal{D}}\right|/\|\eta\|_{\Lambda_b} : \eta \in \mathcal{D},\ \operatorname{supp}\eta \subset B(0,1+2\varepsilon)\right\} < +\infty. \tag{6.45}$$

Thus, applying Lemma 6.5 to $\tilde{f}$ with

$$b_0 = b,\ b'' = a$$

and with (6.34) replaced by (6.45) gives that

$$|\langle\psi,f\rangle_{\mathcal{D}}|^q = \left|\left\langle(\psi)_{1+2\varepsilon},\tilde{f}\right\rangle_{\mathcal{D}}\right|^q$$
$$\leq C_{6.6}\left(b,b,a,\varepsilon,q,\|\phi\|_{\Lambda_a},n\right)\|(\psi)_{1+2\varepsilon}\|_{\Lambda_b}^q$$
$$\times \iint_{Q((1+2\sqrt{n})\varepsilon)} \left|\tilde{f} * (\phi)_t(x)\right|^q t^{q(n+b)-n} dx\,dt/t$$
$$\leq C_{6.8}\left(b,a,\varepsilon,q,\|\phi\|_{\Lambda_a},n\right)\|\psi\|_{\Lambda_b}^q I^q \tag{6.46}$$

for any $\psi \in \mathcal{D}$ with $\operatorname{supp}(\psi)_{1+2\varepsilon} \subset B(0,1)$, namely for any

$$\psi \in \mathcal{D} \text{ with } \operatorname{supp}\psi \subset B\left(0,1/(1+2\varepsilon)\right). \tag{6.7}'$$

So far, we have shown that if (6.3)–(6.6) and (6.7)′ hold, then

$$|\langle \psi, f \rangle_{\mathcal{D}}| \le C_{6.8} \|\psi\|_{\Lambda_b} I \tag{6.47}$$

(We have shown (6.47) under the assumption (6.44). But, if (6.44) does not hold, then (6.47) is clear.) Since

$$\frac{1 + (1 + 2\sqrt{n})\varepsilon}{1/(1 + 2\varepsilon)} \to 1 \qquad (\varepsilon \to +0),$$

by choosing "our ε" so small that the above ratio is less than $1 +$ "given ε in (6.5)" we get Theorem 6.1. $\qquad\square$

Remark 6.1. As a consequence of Theorem 6.1, we get that if

$$f \in \mathcal{D}', \ \phi \in \mathcal{D}, \ \int \phi dx \ne 0, \ m \in (0, +\infty), \ m' \in (0, +\infty), \ q \in (0, +\infty),$$

and if

$$\iint |f * (\phi)_t(x)|^q \, t^m \, (1 + |x| + t)^{-m'} \, dx dt/t < \infty, \tag{6.48}$$

then

$$f \in \mathcal{S}'. \tag{6.49}$$

We will prove this. We may assume $m' > m + n$ and $q \in (0, 1]$. Let $b > 0$ be such that

$$q(n + b) - n = m.$$

Let

$$r \in [1, +\infty).$$

Then

$$\iint_{Q(B(0,2r))} |f * (\phi)_t(x)|^q \, t^{q(n+b)-n} dx dt/t$$
$$\le Cr^{m'} \{\text{the left-hand side of (6.48)}\}.$$

Thus, Theorem 6.1 with dilation implies that if

$$\psi \in \mathcal{D} \ \text{and} \ \operatorname{supp} \psi \subset B(0, r),$$

then

$$|\langle \psi, f \rangle_{\mathcal{D}}|^q \ \le \ C_{6.1} \left(b + 1, b, q, 1, \|\phi\|_{\Lambda_{b+1}}, n \right) \|\psi\|_{\Lambda_b}^q$$
$$\times \iint_{Q(B(0,2r))} |f * (\phi)_t(x)|^q \, t^{q(n+b)-n} dx dt/t$$
$$\le \ C\|\psi\|_{\Lambda_b}^q r^{m'} \{\text{the left-hand side of (6.48)}\}.$$

This implies (6.49).

Remark 6.2. Let

$$f \in \mathcal{D}', \ \phi \in \mathcal{D}, \ \int \phi \, dx \neq 0, \ q \in (0,1], \ m \in \mathbf{R}$$

and

$$\|S_{\phi,1,m}f\|_{L^q} < \infty. \tag{6.50}$$

Then

$$f \in \mathcal{S}'. \tag{6.51}$$

We will prove this. Take $\varepsilon > \max\{0, -mq\}$. Then, Lemma 3.8 and (6.50) imply

$$\iint |f * (\phi)_t(x)|^q \, t^{mq+\varepsilon} \, (1 + |x| + t)^{-2\varepsilon} \, dxdt/t < \infty.$$

So, Remark 6.1 implies (6.51).

Notes. The ideas in Sections 6–8 are in A. Uchiyama [85b, 86]. For the proof of Theorem 6.A see C. Fefferman–E. M. Stein [72]. (See also G. H. Hardy–J. E. Littlewood [32] and H-Q. Bui [83].)

VII. Hardy-Littlewood-Fefferman-Stein type inequalities, 2

In this section we will prove the following Theorem 7.1.

Definition 7.1. For $a > 0$ and $m > 0$ let

$$h_{a,m}(x,t) = t^a \left(1 + |x| + t\right)^{-n-a-m}. \tag{7.1}$$

Theorem 7.1. *Let* $f \in \mathcal{S}'$, $\phi \in \mathcal{S}$, $\psi \in \mathcal{S}$,

$$0 < b < a, \quad 0 < m, \quad 0 < \varepsilon, \tag{7.2}$$

$$q \in (0,1], \tag{7.3}$$

$$\int \phi \, dx = 1. \tag{7.4}$$

Then,

$$
\begin{aligned}
|\langle \psi, f \rangle_s|^q \;\leq\;& C_{7.1}\left(a, b, m, \varepsilon, q, \|\phi\|_{\mathcal{B},a,m+\varepsilon}, n\right) \|\psi\|_{\mathcal{B},b,m}^q \\
&\times \iint_{\mathbf{R}_+^{n+1}} |f * (\phi)_t(x)|^q \, h_{b,m}(x,t)^q t^{n(q-1)} \, dx \, dt/t.
\end{aligned}
\tag{7.5}
$$

Lemma 7.1. *Let*

$$0 < b < b'', \quad 0 < m < m''. \tag{7.6}$$

Let $E \subset \mathbf{R}_+^{n+1}$ *be a measurable set and* $(y,t) \in \mathbf{R}_+^{n+1}$. *Then,*

$$
\begin{aligned}
\iint_E h_{b,m}(x,s) s^{-n} h_{b'',m''}&\left((y-x)/s, t/s\right) dx \, ds/s \\
&\leq C_{7.2}(b, b'', m, m'', n) \mathrm{dens}(E) h_{b,m}(y,t).
\end{aligned}
\tag{7.7}
$$

Proof. Let T_j be as in (6.37). Note that

$$
\sup \left\{ \frac{h_{b,m}(x,s)}{h_{b,m}(x',s')}, \, \frac{h_{b,m}\left((y-x)/s, t/s\right)}{h_{b,m}\left((y-x')/s', t/s'\right)} : (x,s), \, (x',s') \in T_j \right\} \leq C(b,m,n).
$$

Thus,

the left-hand side of (7.7)

$$= \sum_j \iint_{T_j \cap E} h_{b,m}(x,s)s^{-n} h_{b'',m''}\left((y-x)/s, t/s\right) dx ds/s$$

$$\le C(b,b'',m,m'',n)\mathrm{dens}(E) \sum_j \iint_{T_j} \cdots dx ds/s$$

$$= C\mathrm{dens}(E) \iint_{R_+^{n+1}} h_{b,m}(x,s)s^{-n} h_{b'',m''}\left((y-x)/s, t/s\right) dx ds/s$$

$$= C\mathrm{dens}(E)u(y,t), \quad \text{say}.$$

If $|y| \le 1$ and $t \le 1$, then

$$
\begin{aligned}
u(y,t) \;\le\; & t^{b''} \iint_{Q(B(0,2))} s^{b+m''} \left(s + |x-y| + t\right)^{-n-b''-m''} dx ds/s \\
& + Ct^{b''} \iint_{Q(B(0,2))^c} s^{b+m''} \left(|x| + s\right)^{-2n-b-b''-m-m''} dx ds/s \\
\le\; & Ct^{b''} \int_0^2 s^{b+m''} (t+s)^{-b''-m''} ds/s \\
& + Ct^{b''} \int_0^{+\infty} s^{b+m''} (1+s)^{-n-b-b''-m-m''} ds/s \\
\le\; & Ct^b \qquad \text{by } b < b''.
\end{aligned}
$$

If $\max\{1, |y|\} \le t$, then

$$
\begin{aligned}
u(y,t) \;\le\; & t^{b''} \iint_{Q(B(0,2t))} s^{b+m''} \left(1 + |x| + s\right)^{-n-b-m} t^{-n-b''-m''} dx ds/s \\
& + Ct^{b''} \iint_{Q(B(0,2t))^c} s^{b+m''} \left(|x| + s\right)^{-2n-b-b''-m-m''} dx ds/s \\
\le\; & Ct^{-n-m''} \int_0^{2t} s^{b+m''} (1+s)^{-b-m} ds/s \\
& + Ct^{b''} \int_0^{+\infty} s^{b+m''} (t+s)^{-n-b-b''-m-m''} ds/s \\
\le\; & Ct^{-n-m}.
\end{aligned}
$$

If $\max\{1, t\} \le |y|$, then

$$u(y,t) \le Ct^{b''} \iint_{Q(B(0,|y|/2))} s^{b+m''} \left(1 + |x| + s\right)^{-n-b-m} |y|^{-n-b''-m''} dx ds/s$$

$$+ Ct^{b''} \iint_{Q(B(0,2|y|))\setminus Q(B(0,|y|/2))} s^{b+m''} |y|^{-n-b-m} \left(|x-y| + t + s\right)^{-n-b''-m''} dx ds/s$$

$$+ Ct^{b''} \iint_{Q(B(0,2|y|))^c} s^{b+m''} \left(|x| + s\right)^{-2n-b-b''-m-m''} dx ds/s$$

$$\leq Ct^{b''}|y|^{-n-b''-m''}\int_0^{|y|/2}s^{b+m''}(1+s)^{-b-m}ds/s$$

$$+Ct^{b''}|y|^{-n-b-m}\int_0^{2|y|}s^{b+m''}(t+s)^{-b''-m''}ds/s$$

$$+Ct^{b''}\int_0^{+\infty}s^{b+m''}(s+|y|)^{-n-b-b''-m-m''}ds/s$$

$$\leq C\,(t/|y|)^b\,|y|^{-n-m}.$$

$\square$

Lemma 7.2. *Let*

$$0 < b < b',\ \ 0 < m < m'. \tag{7.8}$$

Let $k(y,t)$ be a measurable function defined on $\mathbf{R}_+^{n+1}$ such that

$$|k(y,t)| \leq h_{b,m}(y,t). \tag{7.9}$$

Let $\phi \in \Lambda_{b'}$,

$$\|\phi\|_{\mathcal{B},b',m'} \leq 1 \tag{7.10}$$

and

$$f(x) = \iint k(y,t)(\phi)_t(x-y)dydt/t.$$

Then,

$$\|f\|_{\mathcal{B},b,m} \leq C_{7.3}(b,b',m,m',n). \tag{7.11}$$

Proof. By (7.10) ϕ can be written in the form

$$\phi(x) = \sum_{j \in \mathbf{N}_0} 2^{-jm'}(\phi_j)_{2^j}$$

with

$$\phi_j \in \mathcal{B}_{b'}.$$

First we assume

$$|k(y,t)| \leq t^b \chi_{Q(B(0,1))}(y,t). \tag{7.12}$$

Then,

$$
\begin{aligned}
f(x) &= \sum_{j \in \mathbf{N}_0} 2^{-jm'} \left\{ \iint_{B(0,1) \times (0,2^{-j}]} k(y,t)(\phi_j)_{2^j t}(x-y)dydt/t \right. \\
&\quad \left. + \sum_{i=1}^{j} \iint_{B(0,1) \times (2^{-j+i-1}, 2^{-j+i}]} \cdots\cdots\cdots\cdots dydt/t \right\} \\
&= \sum_{j \in \mathbf{N}_0} 2^{-jm'} \left\{ (7.14)_{j,0} + \sum_{i=1}^{j} (7.14)_{j,i} \right\}, \quad \text{say} \\
&= \sum_{i \in \mathbf{N}_0} \sum_{j=i}^{\infty} 2^{-jm'} (7.14)_{j,i} \\
&= \sum_{i \in \mathbf{N}_0} (7.15)_i, \quad \text{say.}
\end{aligned}
\tag{7.13}
$$

It is easy to see that

$$
\operatorname{supp} (7.14)_{j,i} \subset B(0, 2^i + 1), \qquad (i = 0, 1, \cdots, j),
$$

$$
\begin{aligned}
\|(7.14)_{j,i}\|_{\Lambda_b} &\leq \iint_{B(0,1) \times (2^{-j+i-1}, 2^{-j+i}]} t^b \, \|(\phi_j)_{2^j t}\|_{\Lambda_b} \, dydt/t \\
&\leq C(b, b', n) 2^{-in-jb} \qquad (i = 1, \cdots, j),
\end{aligned}
$$

and that

$$
\begin{aligned}
\|(7.14)_{j,0}\|_{\Lambda_b} &= \left\| \iint_{Q(B(0,1))} k(y, 2^{-j}t)(\phi_j)_t(\cdot - y)dydt/t \right\|_{\Lambda_b} \\
&\leq C(b, b', n) 2^{-jb} \quad \text{by Lemma 1.11.}
\end{aligned}
$$

So, we have

$$
\operatorname{supp} (7.15)_i \subset B(0, 2^i + 1), \tag{7.16}
$$

$$
\|(7.15)_i\|_{\Lambda_b} \leq C(b, b', m', n) 2^{-i(m'+n+b)}. \tag{7.17}
$$

Next, we remove the restriction (7.12). Assume (7.9). Then, $k(y,t)$ can be written in the form

$$
k(y,t) = \sum_{s \in \mathbf{N}_0} 2^{-s(n+m)} k_s(2^{-s}y, 2^{-s}t)
$$

with

$$
|k_s(y,t)| \leq C(b, m, n) t^b \chi_{Q(B(0,1))}(y,t).
$$

So,

$$f(x) = \sum_{s \in \mathbf{N}_0} 2^{-s(n+m)} \iint k_s(2^{-s}y, 2^{-s}t)(\phi)_t(x-y)dydt/t$$

$$= \sum_{s \in \mathbf{N}_0} 2^{-s(n+m)} \iint k_s(y,t)(\phi)_t(2^{-s}x - y)dydt/t$$

$$= \sum_{s \in \mathbf{N}_0} 2^{-s(n+m)} g_s(x), \quad \text{say}. \tag{7.18}$$

By (7.13)–(7.17), $g_s(x)$ can be written in the form

$$g_s(x) = \sum_{i \in \mathbf{N}_0} (7.19)_{s,i}$$

with

$$\text{supp}\,(7.19)_{s,i} \subset B\left(0, (2^i + 1)2^s\right), \tag{7.20}$$

$$\|(7.19)_{s,i}\|_{\Lambda_b} \leq C(b, b', m', n)2^{-i(m'+n+b)}2^{-sb}. \tag{7.21}$$

Substituting this into (7.18) gives

$$f = \sum_{s \in \mathbf{N}_0} 2^{-s(n+m)} \sum_{i \in \mathbf{N}_0} (7.19)_{s,i}$$

$$= \sum_{j \in \mathbf{N}_0} \sum_{i=0}^{j} 2^{-(j-i)(n+m)} (7.19)_{j-i,i}$$

$$= \sum_{j \in \mathbf{N}_0} (7.22)_j, \quad \text{say}.$$

By (7.20), (7.21) and $m' > m$ we have

$$\text{supp}\,(7.22)_j \subset B(0, 2^{j+1}),$$

$$\|(7.22)_j\|_{\Lambda_b} \leq C(m, m', b, b', n)2^{-j(m+n+b)}.$$

This implies (7.11).

Lemma 7.3. *Assume (7.8). Let $k(y,t)$ be a measurable function defined on $\mathbf{R}_+^{n+1}$ such that*

$$\sup_{(y,t) \in \mathbf{R}_+^{n+1}} |k(y,t)| / h_{b',m'}(y,t) < +\infty.$$

Let $\delta > 0$, $\phi \in \Lambda_{b'}$, $\|\phi\|_{\mathcal{B},b',m'} < \infty$ and

$$\psi(x) = \iint_{\mathbf{R}_+^{n+1}} k(y,t)(\phi)_t(x-y)dydt/t,$$

$$\psi_\delta(x) = \iint_{B(0,1/\delta) \times (\delta,1/\delta)} k(y,t)(\phi)_t(x-y)dydt/t.$$

Then,

$$\|\psi - \psi_\delta\|_{\mathcal{B},b,m} \to 0 \quad (\delta \to +0).$$

This is immediate from Lemma 7.2 and from

$$\sup\left\{\frac{h_{b',m'}(y,t)}{h_{b,m}(y,t)} : (y,t) \in \mathbf{R}_+^{n+1}\setminus (B(0,1/\delta) \times (\delta, 1/\delta))\right\} \to 0 \; (\delta \to +0).$$

Lemma 7.4. *Let*

$$0 < b \le b' < b'', \;\; 0 < m \le m' < m''. \tag{7.23}$$

Assume $\phi \in \Lambda_{b''}$, $\|\phi\|_{\mathcal{B},b'',m''} < \infty$, $\psi \in \Lambda_{b'}$, $\|\psi\|_{\mathcal{B},b',m'} < \infty$ *and* (7.4). *Then, there exists a measurable function* $\mathsf{k}(y,t)$ *defined on* $\mathbf{R}_+^{n+1}$ *such that*

$$|\mathsf{k}(y,t)| \le C_{7.4}\,(b,b',b'',m,m',m'',\|\phi\|_{\mathcal{B},b'',m''},n)$$
$$\times \min\left\{\|\psi\|_{\mathcal{B},b,m}h_{b,m}(y,t), \|\psi\|_{\mathcal{B},b',m'}h_{b',m'}(y,t)\right\}, \quad (7.24)$$

$$\psi(x) = \iint \mathsf{k}(y,t)(\phi)_t(x-y)dydt/t. \tag{7.25}$$

Proof. Let $\Psi_0(x)$ and $\Psi_1(x)$ be as in Definition 3.4. Let

$$\tilde{\phi}(x) = \Psi_0(x)(\phi)_{2^{-K}}(x), \tag{7.26}$$

where $K \in \mathbf{N}$ is sufficiently large to be determined later. By (7.4) we may assume

$$\int \tilde{\phi}(x)dx > 1/2. \tag{7.27}$$

First we assume that

$$\operatorname{supp} \psi \subset B(0,1). \tag{7.28}$$

Then, Lemma 6.2 with $(\int \tilde{\phi}dx)^{-1}\tilde{\phi}$ in place of ϕ and with $\varepsilon = 1/2$ implies that there exists $\mathsf{k}_0(y,t)$ such that

$$\mathsf{k}_0(y,t) = 0 \;\text{ on } \mathbf{R}_+^{n+1}\setminus (B(0,3/2) \times (0,1/2]), \tag{7.29}$$

$$|\mathsf{k}_0(y,t)| \le 4C_{6.2}(b,b',1/2,n)\|\phi\|_{L^1}\min\left\{\|\psi\|_{\Lambda_b}t^b, \|\psi\|_{\Lambda_{b'}}t^{b'}\right\}, \tag{7.30}$$

$$\psi(x) = \iint \mathsf{k}_0(y,t)(\tilde{\phi})_t(x-y)dydt/t. \tag{7.31}$$

By (7.31), (7.26) and (3.12)

$$\iint \mathsf{k}_0(y,2^K t)(\phi)_t(x-y)dydt/t - \psi(x)$$
$$= \sum_{j \in \mathbf{N}} \iint \mathsf{k}_0(y,t)\left(\Psi_1(2^{-j}\cdot)(\phi)_{2^{-K}}(\cdot)\right)_t(x-y)dydt/t$$
$$= \sum \iint \mathsf{k}_0(y,t)\left(\Psi_1(2^{-j-K}\cdot)\phi(\cdot)\right)_{2^{-K}t}(x-y)dydt/t$$

$$= \sum_{j \in \mathbb{N}} \left\{ \iint_{B(0,3/2) \times (0,1]} k_0(y, 2^{-j}t) \left(\Psi_1(2^{-j-K} \cdot)\phi(\cdot) \right)_{2^{-K-j}t} (x-y) dy dt/t \right.$$

$$\left. + \sum_{i=2}^{j} \iint_{B(0,3/2) \times (1/2,1]} k_0(y, 2^{-j+i-1}t)(\cdots\cdots)_{2^{-K-j+i-1}t}(x-y) dy dt/t \right\}$$

$$= \sum \left\{ (7.32)_{j,1} + \sum_{i=2}^{j} (7.32)_{j,i} \right\} \quad \text{say,}$$

$$= \sum_{i \in \mathbb{N}} \sum_{j=i}^{\infty} (7.32)_{j,i} = - \sum_{i \in \mathbb{N}} \psi_i(x), \quad \text{say.}$$

It is clear that

$$\text{supp} \, (7.32)_{j,i} \subset B(0, 2^i). \qquad (7.33)$$

Since

$$2^{(j+K)m''} \|\phi\|_{\mathcal{B}, b'', m''}^{-1} \left(\Psi_1(2^{-j-K} \cdot)\phi(\cdot) \right)_{2^{-j-K}} \in \mathcal{B}_{b''}, \qquad (7.34)$$

$$\left| k_0(y, 2^{-j}t) \right| \leq 4C_{6.2} \|\phi\|_{L^1} \|\psi\|_{\Lambda_b} 2^{-jb} t^b \quad \text{by (7.30)},$$

Lemma 1.11 implies

$$\|(7.32)_{j,1}\|_{\Lambda_b} \leq C_{1.3}(b'', b, n) 2^{-(j+K)m''} \|\phi\|_{\mathcal{B}, b'', m''}$$
$$\times C(b,n) C_{6.2} \|\phi\|_{L^1} \|\psi\|_{\Lambda_b} 2^{-jb}. \qquad (7.35)$$

For $i = 2, 3, \cdots, j$ we have

$$\|(7.32)_{j,i}\|_{\Lambda_b}$$

$$\leq \iint_{B(0,3/2) \times (1/2,1]} \left| k_0(y, 2^{-j+i-1}t) \right|$$
$$\times \left\| \left(\Psi_1(2^{-j-K} \cdot)\phi(\cdot) \right)_{2^{-j-K} 2^{i-1}t} \right\|_{\Lambda_b} dy dt/t$$

$$\leq \iint_{B(0,3/2) \times (1/2,1]} 4C_{6.2} \|\phi\|_{L^1} \|\psi\|_{\Lambda_b} (2^{-j+i})^b$$
$$\times C(b,b'',n) 2^{-(j+K)m''} \|\phi\|_{\mathcal{B}, b'', m''} 2^{-i(n+b)} dy dt/t$$
$$\text{by (7.30) and (7.34)}$$

$$\leq C(b, b', b'', n) \|\phi\|_{L^1} \|\psi\|_{\Lambda_b} \|\phi\|_{\mathcal{B}, b'', m''} 2^{-(j+K)m''-jb-in}. \qquad (7.36)$$

By the same reason as (7.35) and (7.36), we have

$$\|(7.32)_{j,i}\|_{\Lambda_{b'}} \leq C(b, b', b'', n) \|\phi\|_{L^1} \|\psi\|_{\Lambda_b} \|\phi\|_{\mathcal{B}, b'', m''} 2^{-(j+K)m''-jb'-in}$$
$$\text{for } i = 1, 2, \cdots, j.$$

Thus,

$$\text{supp }\psi_i \subset B(0, 2^i), \tag{7.37}$$

$$\|\psi_i\|_{\Lambda_b} \leq A\|\psi\|_{\Lambda_b} 2^{-i(n+m''+b)}, \tag{7.38}$$

$$\|\psi_i\|_{\Lambda_{b'}} \leq A\|\psi\|_{\Lambda_{b'}} 2^{-i(n+m''+b')}, \tag{7.39}$$

where

$$A = 2^{-Km''} C(b, b', b'', n)\|\phi\|_{L^1}\|\phi\|_{\mathcal{B}, b'', m''}. \tag{7.40}$$

So far we have shown

$$\psi(x) = \iint k_0(y, 2^K t)(\phi)_t(x - y) dy dt / t + \sum_{i_1 \in \mathbf{N}} \psi_{i_1}(x) \tag{7.41}$$

where k_0 and ψ_{i_1} satisfy (7.29)–(7.30) and (7.37)–(7.39).

Next we apply the same argument with dilation to each ψ_{i_1}. Then, we get k_{i_1} and $\{\psi_{i_1,i_2}\}_{i_2 \in \mathbf{N}}$ such that

$$\psi_{i_1}(x) = \iint k_{i_1}(y, 2^K t)(\phi)_t(x - y) dy dt / t + \sum_{i_2 \subset \mathbf{N}} \psi_{i_1,i_2}(x), \tag{7.41}_{i_1}$$

$$k_{i_1}(y, t) = 0 \text{ on } \mathbf{R}_+^{n+1} \setminus \left(B(0, 2^{i_1-1} \cdot 3) \times (0, 2^{i_1-1}] \right), \tag{7.29}_{i_1}$$

$$|k_{i_1}(y, t)| \leq 4C_{6.2}(b, b', 1/2, n)\|\phi\|_{L^1} A 2^{-i_1(n+m'')}$$

$$\times \min \left\{ \|\psi\|_{\Lambda_b} 2^{-i_1 b} t^b, \|\psi\|_{\Lambda_{b'}} 2^{-i_1 b'} t^{b'} \right\}, \tag{7.30}_{i_1}$$

$$\text{supp }\psi_{i_1,i_2} \subset B(0, 2^{i_1+i_2}), \tag{7.37}_{i_1,i_2}$$

$$\|\psi_{i_1,i_2}\|_{\Lambda_b} \leq A^2\|\psi\|_{\Lambda_b} 2^{-(i_1+i_2)(n+m''+b)}, \tag{7.38}_{i_1,i_2}$$

$$\|\psi_{i_1,i_2}\|_{\Lambda_{b'}} \leq A^2\|\psi\|_{\Lambda_{b'}} 2^{-(i_1+i_2)(n+m''+b')}. \tag{7.39}_{i_1,i_2}$$

Substituting $(7.41)_{i_1}$ into (7.41) implies

$$\psi(x) = \iint \left\{ k_0(y, 2^K t) + \sum_{i_1 \in \mathbf{N}} k_{i_1}(y, 2^K t) \right\} (\phi)_t(x - y) dy dt / t$$

$$+ \sum_{i_1 \in \mathbf{N}} \sum_{i_2 \in \mathbf{N}} \psi_{i_1,i_2}(x). \tag{7.41}_{(1)}$$

Repeating this process, we get $\{k_{i_1,\cdots,i_j}\}_{j \in \mathbf{N}, i_1,\cdots,i_j \in \mathbf{N}}$ and $\{\psi_{i_1,\cdots,i_j}\}_{j \in \mathbf{N}, i_1,\cdots,i_j \in \mathbf{N}}$ such that

$$\psi(x) = \iint \left\{ k_0(y, 2^K t) + \sum_{u=1}^{j-1} \sum_{i_1,\cdots,i_u \in \mathbf{N}} k_{i_1,\cdots,i_u}(y, 2^K t) \right\} (\phi)_t(x - y) dy dt / t$$

$$+ \sum_{i_1,\cdots,i_j \in \mathbf{N}} \psi_{i_1,\cdots,i_j}(x), \tag{7.41}_{(j)}$$

$$k_{i_1,\cdots,i_u}(y,t) = 0 \text{ on } \mathbf{R}^{n+1}_+ \backslash \left(B(0, 2^{i_1+\cdots+i_u-1} \cdot 3) \times (0, 2^{i_1+\cdots+i_u-1}] \right),$$
$$(7.29)_{i_1,\cdots,i_u}$$

$$|k_{i_1,\cdots,i_u}(y,t)| \le 4C_{6.2}(b,b',1/2,n)\|\phi\|_{L^1} \cdot A^u \cdot 2^{-(i_1+\cdots+i_u)(n+m'')}$$
$$\times \min \left\{ \|\psi\|_{\Lambda_b} 2^{-(i_1+\cdots+i_u)b} t^b, \|\psi\|_{\Lambda_{b'}} 2^{-(i_1+\cdots+i_u)b'} t^{b'} \right\},$$
$$(7.30)_{i_1,\cdots,i_u}$$

$$\text{supp } \psi_{i_1,\cdots,i_j} \subset B(0, 2^{i_1+\cdots+i_j}), \qquad (7.37)_{i_1,\cdots,i_j}$$

$$\|\psi_{i_1,\cdots,i_j}\|_{\Lambda_b} \le A^j \|\psi\|_{\Lambda_b} 2^{-(i_1+\cdots+i_j)(n+m''+b)}, \qquad (7.38)_{i_1,\cdots,i_j}$$

$$\|\psi_{i_1,\cdots,i_j}\|_{\Lambda_{b'}} \le A^j \|\psi\|_{\Lambda_{b'}} 2^{-(i_1+\cdots+i_j)(n+m''+b')}. \qquad (7.39)_{i_1,\cdots,i_j}$$

Then, since $\|\psi_{i_1,\cdots,i_j}\|_{L^\infty} \le C2^{(i_1+\cdots+i_j)b}\|\psi_{i_1,\cdots,i_j}\|_{\Lambda_b}$, by taking K so large that $A < 1$ (recall (7.40)), we get that

$$\psi(x) = \lim_{j\to\infty} \iint \left\{ k_0(y, 2^K t) + \sum_{u=1}^{j-1} \sum_{i_1,\cdots,i_u \in \mathbf{N}} k_{i_1+\cdots+i_u}(y, 2^K t) \right\}(\phi)_t(x-y)dydt/t$$

converges uniformly with respect to $x \in \mathbf{R}^n$. Put

$$k(y,t) = k_0(y, 2^K t) + \sum_{u\in\mathbf{N}} \sum_{i_1,\cdots,i_u \in \mathbf{N}} k_{i_1,\cdots,i_u}(y, 2^K t).$$

Then,

$$\psi(x) = \iint k(y,t)(\phi)_t(x-y)dydt/t, \qquad (7.42)$$

$$|k(y, 2^{-K}t)| \le |k_0(y,t)| + \sum_{s\in\mathbf{N}:2^{s+1}\ge|y|+t} \sum_{u=1}^{s} \sum_{\substack{i_1+\cdots+i_u=s, \\ i_1,\cdots,i_u\in\mathbf{N}}} 4C_{6.2}(b,b',1/2,n)\|\phi\|_{L^1}$$
$$\times A^u 2^{-s(n+m'')} \min \left\{ \|\psi\|_{\Lambda_b} 2^{-sb} t^b, \|\psi\|_{\Lambda_{b'}} 2^{-sb'} t^{b'} \right\}$$
$$\text{by } (7.29)_{i_1,\cdots,i_u} \text{ and } (7.30)_{i_1,\cdots,i_u}$$

$$= |\cdots| + \sum_{s\in\mathbf{N}:\cdots} \sum_{u=1}^{s} \binom{s-1}{u-1} 4C_{6.2}\|\phi\|_{L^1} A^u 2^{-s(n+m'')} \min\{\cdots,\cdots\}$$

$$\le 4C_{6.2}\|\phi\|_{L^1} \min \left\{ \|\psi\|_{\Lambda_b} t^b, \|\psi\|_{\Lambda_{b'}} t^{b'} \right\} \chi_{B(0,3/2)\times(0,1/2]}(y,t)$$
$$+ \sum_{s\in\mathbf{N}:2^{s+1}\ge|y|+t} 4C_{6.2}\|\phi\|_{L^1} A(1+A)^{s-1} 2^{-s(n+m'')}$$
$$\times \min \left\{ \|\psi\|_{\Lambda_b} 2^{-sb} t^b, \|\psi\|_{\Lambda_{b'}} 2^{-sb'} t^{b'} \right\}$$

$$\le C(b,b',m',m'',n)\|\phi\|_{L^1} \min \left\{ \|\psi\|_{\Lambda_b} h_{b,m''-\varepsilon_1}(y,t), \|\psi\|_{\Lambda_{b'}} h_{b',m''-\varepsilon_1}(y,t) \right\}$$
$$(7.43)$$

if K is sufficiently large depending on b, b', b'', m', m'', $\|\phi\|_{B,b'',m''}$ and n so that

$$A < \min\{1, 2^{\varepsilon_1} - 1\} \quad (\text{recall } (7.40)), \quad \text{where } \varepsilon_1 = \frac{m'' - m'}{2}.$$

Next we remove the restriction (7.28). Then, ψ can be written in the form

$$\psi(x) = \sum_{j \in \mathbf{N}_0} (\psi_{(j)})_{2^j}(x)$$

with

$$\operatorname{supp} \psi_{(j)} \subset B(0,1),$$
$$\|\psi_{(j)}\|_{\Lambda_b} \leq 2^{-jm}\|\psi\|_{B,b,m},$$
$$\|\psi_{(j)}\|_{\Lambda_{b'}} \leq 2^{-jm'}\|\psi\|_{B,b',m'}.$$

Applying the above results ((7.42) and (7.43)) with dilation to each $(\psi_{(j)})_{2^j}$ gives $k_{(j)}(y,t)$ such that

$$(\psi_{(j)})_{2^j}(x) = \iint k_{(j)}(y,t)(\phi)_t(x-y)dydt/t \qquad (7.42)_j$$

and

$$\left|k_{(j)}(y, 2^{-K}t)\right| \leq C(b, b', m', m'', n)\|\phi\|_{L^1}$$

$$\times 2^{-jn} \min\left\{2^{-jm}\|\psi\|_{B,b,m}h_{b,m''-\varepsilon_1}(2^{-j}y, 2^{-j}t),\right.$$

$$\left. 2^{-jm'}\|\psi\|_{B,b',m'}h_{b',m''-\varepsilon_1}(2^{-j}y, 2^{-j}t)\right\}.$$
$$(7.43)_j$$

Letting

$$k(y,t) = \sum_{j \in \mathbf{N}_0} k_{(j)}(y,t)$$

gives (7.24)–(7.25) because

$$\sum_{j \in \mathbf{N}_0} 2^{-j(n+m)}h_{b,m''-\varepsilon_1}(2^{-j}y, 2^{-j}t) \leq C(b, m', m'', n)h_{b,m}(y,t).$$

$\square$

Applying lemma 7.4 to $\psi = \phi$ gives

Lemma 7.5. *Let*

$$0 < b'' < b''', \quad 0 < m'' < m'''. \tag{7.44}$$

Let $\phi \in \Lambda_{b'''}$ and $\|\phi\|_{\mathcal{B},b''',m'''} < \infty$. Assume (7.4). Then there exists a measurable function $\tilde{k}(y,t)$ defined on $\mathbf{R}_+^{n+1}$ such that

$$\left|\tilde{k}(y,t)\right| \le C_{7.5}\,(b'',b''',m'',m''',\|\phi\|_{\mathcal{B},b''',m'''},n)\,h_{b'',m''}(y,t), \tag{7.45}$$

$$\phi(x) = \iint \tilde{k}(y,t)(\phi)_t(x-y)dydt/t, \tag{7.46}$$

where

$$C_{7.5} = C_{7.4}\,(b'',b'',b''',m'',m'',m''',\|\phi\|_{\mathcal{B},b''',m'''},n)\,\|\phi\|_{\mathcal{B},b'',m''}.$$

Lemma 7.6. *Let*

$$0 < b \le b' < b''', \quad 0 < m \le m' < m'''. $$

Let $\phi \in \Lambda_{b'''}$, $\|\phi\|_{\mathcal{B},b''',m'''} < \infty$, $\psi \in \Lambda_{b'}$ and $\|\psi\|_{\mathcal{B},b',m'} < \infty$. Assume (7.4). Let $E \subset \mathbf{R}_+^{n+1}$ be a measurable set such that

$$\operatorname{dens}(E) \le C_{7.6}\,(b,b',b''',m,m',m''',\|\phi\|_{\mathcal{B},b''',m'''},n)\,. \tag{7.47}$$

Then, there exists a measurable function $k(y,t)$ defined on $\mathbf{R}_+^{n+1}$ such that

$$k(y,t) = 0 \quad on \ E, \tag{7.48}$$

$$|k(y,t)| \le C_{7.7}\,(b,b',b''',m,m',m''',\|\phi\|_{\mathcal{B},b''',m'''},n)$$
$$\times \min\left\{\|\psi\|_{\mathcal{B},b,m}h_{b,m}(y,t), \|\psi\|_{\mathcal{B},b',m'}h_{b',m'}(y,t)\right\}, \tag{7.49}$$

$$\psi(x) = \iint k(y,t)(\phi)_t(x-y)dydt/t. \tag{7.50}$$

Proof. Put

$$\varepsilon = \min\left\{(b''' - b')/2, (m''' - m')/2\right\},$$
$$b'' = b''' - \varepsilon, \quad m'' = m''' - \varepsilon.$$

Applying Lemma 7.4 gives $k_1(y,t)$ such that

$$|k_1(y,t)| \le C_{7.4}\,(b,b',b''',m,m',m''',\|\phi\|_{\mathcal{B},b''',m'''},n)$$
$$\times \min\left\{\|\psi\|_{\mathcal{B},b,m}h_{b,m}(y,t), \|\psi\|_{\mathcal{B},b',m'}h_{b',m'}(y,t)\right\}, \tag{7.24}_1$$

$$\psi(x) = \iint k_1(y,t)(\phi)_t(x-y)dydt/t. \tag{7.25}_1$$

Applying Lemma 7.5 gives $\tilde{k}(y,t)$ with (7.45)–(7.46). Applying dilation and translation to (7.46) gives

$$(\phi)_t(x-y) = \iint t^{-n}\tilde{k}\left((z-y)/t, s/t\right)(\phi)_s(x-z)dzds/s. \qquad (7.46)^*$$

Let

$$I_1 = \iint k_1(y,t)\chi_{E^c}(y,t)(\phi)_t(x-y)dydt/t, \qquad (7.51)$$

$$J_1 = \iint k_1(y,t)\chi_E(y,t)(\phi)_t(x-y)dydt/t. \qquad (7.52)$$

Let

$$k_2(z,s) = \iint k_1(y,t)\chi_E(y,t)t^{-n}\tilde{k}\left((z-y)/t, s/t\right)dydt/t.$$

Then, (7.45), $(7.24)_1$ and Lemma 7.1 imply that

$$
\begin{aligned}
|k_2(z,s)| \;\leq\;& C_{7.4}\left(b,b',b''',m,m',m''',\|\phi\|_{\mathcal{B},b''',m'''},n\right) \\
&\times C_{7.5}\left(b'',b''',m'',m''',\|\phi\|_{\mathcal{B},b''',m'''},n\right)\mathrm{dens}(E) \\
&\times \min\left\{ C_{7.2}(b,b'',m,m'',n)\|\psi\|_{\mathcal{B},b,m}h_{b,m}(z,s), \right. \\
&\qquad\qquad \left. C_{7.2}(b',b'',m',m'',n)\|\psi\|_{\mathcal{B},b',m'}h_{b',m'}(z,s)\right\}.
\end{aligned}
$$

So, by taking $C_{7.6}$ so that

$$C_{7.6} \leq 2^{-1}\left\{C_{7.5}\cdot\left(C_{7.2}(b,b'',m,m'',n)+C_{7.2}(b',b'',m',m'',n)\right)\right\}^{-1}$$

we get

$$|k_2(z,s)| \leq 2^{-1}C_{7.4}\min\left\{\|\psi\|_{\mathcal{B},b,m}h_{b,m}(z,s), \|\psi\|_{\mathcal{B},b',m'}h_{b',m'}(z,s)\right\}$$

and

$$\psi(x) = I_1 + J_1 = I_1 + \iint k_2(z,s)(\phi)_s(x-z)dzds/s,$$

where the last equality follows from substituting $(7.46)^*$ into (7.52).

Repeating the same argument as in the proof of Lemma 6.4, we get $\{k_j(y,t)\}_{j\in\mathbf{N}}$ such that

$$|k_j(y,t)| \leq 2^{-j+1}C_{7.4}\left(b,b',b''',m,m',m''',\|\phi\|_{\mathcal{B},b''',m'''},n\right)$$

$$\times \min\left\{\|\psi\|_{\mathcal{B},b,m}h_{b,m}(y,t), \|\psi\|_{\mathcal{B},b',m'}h_{b',m'}(y,t)\right\}, \qquad (7.24)_j$$

$$\psi(x) = I_1 + I_2 + \cdots + I_{j-1} + \iint k_j(y,t)(\phi)_t(x-y)dydt/t, \qquad (7.25)_j$$

where

$$I_i = \iint k_i(y,t)\chi_{E^c}(y,t)(\phi)_t(x-y)dydt/t.$$

Put

$$k(y,t) = \sum_{j\in\mathbf{N}} k_j(y,t)\chi_{E^c}(y,t).$$

Then, (7.48) is clear. (7.49) follows from $(7.24)_j$. (7.50) follows from $(7.25)_j$ and $(7.24)_j$. $\qquad\square$

Lemma 7.7. *Let*

$$0 < b, \ 0 < b_0, \ 0 < m, \ 0 < m_0,$$
$$b''' > \max\{b, b_0\} \ and \ m''' > \max\{m, m_0\}. \tag{7.53}$$

Let $\phi \in \mathcal{S}$ *and* $\psi \in \mathcal{S}$. *Assume* (7.3)–(7.4). *Let* $f \in \mathcal{S}'$ *and*

$$\sup\{|\langle \eta, f\rangle_{\mathcal{S}}| \,/\, \|\eta\|_{\mathcal{B}, b_0, m_0} : \eta \in \mathcal{S}\} < +\infty. \tag{7.54}$$

Then,

$$|\langle \psi, f\rangle_{\mathcal{S}}|^q \ \leq \ C_{7.8}\,(b, b_0, b''', m, m_0, m''', q, \|\phi\|_{\mathcal{B}, b''', m'''}, n)\, \|\psi\|_{\mathcal{B}, b, m}^q$$
$$\times \iint |f * (\phi)_t(x)|^q \, h_{b,m}(x, t)^q t^{n(q-1)}\, dx dt/t. \tag{7.55}$$

Proof. Let

$$b' = (\max\{b, b_0\} + b''')/2, \ m' = (\max\{m, m_0\} + m''')/2. \tag{7.56}$$

Let $\{T_j\}$ be as in the proof of Lemma 6.5. Let

$$E = \bigcup_j \{(y, t) \in T_j : C_{6.7}(n)C_{7.6}\,|f * (\phi)_t(y)|^q$$
$$\geq \iint_{T_j} |f * (\phi)_s(z)|^q \, dz ds/|T_j|\}. \tag{7.57}$$

Note that E satisfies the condition (7.47) if $C_{6.7}(n)$ (> 0) is small enough. Applying Lemma 7.6 with the above E and with $\phi(x)$ replaced by $\phi(-x)$ gives $k(y, t)$ that satisfies (7.48), (7.49) and

$$\psi(x) = \iint k(y, t)(\phi)_t(y - x)dydt/t. \tag{7.50}'$$

By (7.49) we have

$$\sup_{(y,t)\in\mathbf{R}_+^{n+1}} |k(y, t)| \,/h_{b',m'}(y, t) < +\infty. \tag{7.58}$$

Let $\delta > 0$ and

$$\psi_\delta(x) = \iint_{B(0,1/\delta)\times(\delta,1/\delta)} k(y, t)(\phi)_t(y - x)dydt/t.$$

Then, by Lemma 7.3, (7.58), $b' > b_0$ and $m' > m_0$, we get

$$\|\psi - \psi_\delta\|_{\mathcal{B}, b_0, m_0} \to 0 \ (\delta \to +0).$$

Therefore, (7.54) implies

$$\langle \psi, f \rangle_{\mathcal{S}} = \lim_{\delta \to +0} \langle \psi_\delta, f \rangle_{\mathcal{S}}$$

$$= \lim_{\delta \to +0} \iint_{B(0,1/\delta) \times (\delta,1/\delta)} k(y,t) f * (\phi)_t(y) dy dt / t.$$

So, by the same argument as (6.41) we get

$$|\langle \psi, f \rangle_{\mathcal{S}}|^q \leq \left\{ \iint |k(y,t)| |f * (\phi)_t(y)| dy dt / t \right\}^q$$

$$\leq \sum_j \left\{ \iint_{T_j} \cdots\cdots\cdots\cdots dy dt / t \right\}^q \qquad \text{by (7.3)}$$

$$\leq \sum_j \left\{ \iint_{T_j} |k(y,t)| \left((C_{6.7} C_{7.6})^{-1} \iint_{T_j} |f * (\phi)_s(z)|^q \frac{dz ds}{|T_j|} \right)^{1/q} \frac{dy dt}{t} \right\}^q$$

$$\text{by (7.48) and (7.57)}$$

$$\leq \sum_j (C_{6.7} C_{7.6})^{-1} \iint_{T_j} |f * (\phi)_s(z)|^q dz ds |T_j|^{-1}$$

$$\times \left\{ \iint_{T_j} C_{7.7} \|\psi\|_{\mathcal{B},b,m} h_{b,m}(y,t) dy dt / t \right\}^q \qquad \text{by (7.49)}$$

$$\leq \sum_j (C_{6.7} C_{7.6})^{-1} \iint_{T_j} |f * (\phi)_s(z)|^q dz ds t_j^{-n-1}$$

$$\times (C_{7.7} \|\psi\|_{\mathcal{B},b,m})^q C h_{b,m}(y_j,t_j)^q t_j^{nq}$$

$$\text{where } (y_j,t_j) \text{ is the center of } T_j$$

$$\leq \sum_j (C_{6.7} C_{7.6})^{-1} C_{7.7}^q \|\psi\|_{\mathcal{B},b,m}^q C(b,m,q,n)$$

$$\times \iint_{T_j} |f * (\phi)_s(z)|^q h_{b,m}(z,s)^q s^{nq-n} dz ds / s,$$

which implies the desired result. $\qquad\qquad\qquad\qquad\qquad\qquad\qquad\square$

Proof of Theorem 7.1. Since $f \in \mathcal{S}'$, there exists b_0 and m_0 such that (7.54) holds. Take b''' and m''' so that (7.53) holds. Then, Lemma 7.7 implies

$$\sup \left\{ |\langle \psi, f \rangle|_{\mathcal{S}} / \|\psi\|_{\mathcal{B},b,m} : \psi \in \mathcal{S} \right\}$$

$$\leq C_{7.8} \left(b, b_0, b''', m, m_0, m''', q, \|\phi\|_{\mathcal{B},b''',m'''}, n \right)^{1/q} I, \quad (7.59)$$

where

$$I^q = \iint |f * (\phi)_t(x)|^q h_{b,m}(x,t)^q t^{n(q-1)} dx dt / t.$$

If $I < +\infty$, then (7.59) implies

$$\sup \left\{ |\langle \psi, f \rangle|_{\mathcal{S}} / \|\psi\|_{\mathcal{B},b,m} : \psi \in \mathcal{S} \right\} < +\infty. \quad (7.60)$$

Thus, applying Lemma 7.7 with

$$b_0 = b, \quad b''' = a, \quad m_0 = m, \quad m''' = m + \varepsilon$$

and with (7.54) replaced by (7.60) implies that

$$\sup \{|\langle \psi, f \rangle|_{\mathcal{S}} / \|\psi\|_{\mathcal{B},b,m} : \psi \in \mathcal{S}\}$$
$$\leq C_{7.8} \left(b, b, a, m, m, m + \varepsilon, q, \|\phi\|_{\mathcal{B},a,m+\varepsilon}, n\right)^{1/q} I$$

which implies (7.5).

If $I = +\infty$, then (7.5) is clear. $\qquad\qquad\qquad\qquad\qquad\square$

If we impose certain conditions on $f(\in \mathcal{S}')$, then ϕ and ψ need not be the elements of $\mathcal{S}$. We have the following.

Lemma 7.8. *Let* $0 < b_0 < b' < b''$, $0 < b \leq b'$, $0 < m_0 < m' < m'''$ *and* $0 < m \leq m'$. *Let* $\phi \in \Lambda_{b'''}$, $\|\phi\|_{\mathcal{B},b''',m'''} < \infty$, $\psi \in \Lambda_{b'}$ *and* $\|\psi\|_{\mathcal{B},b',m'} < \infty$. *Assume* (7.3)–(7.4). *Let* f *be a linear functional on*

$$\{\eta \in \Lambda_{b_0} : \|\eta\|_{\mathcal{B},b_0,m_0} < \infty\} \tag{7.61}$$

such taht

$$\sup \{|\langle \eta, f \rangle| / \|\eta\|_{\mathcal{B},b_0,m_0} : \eta \in (7.61)\} < \infty. \tag{7.62}$$

Then

$$|\langle \psi, f \rangle|^q \leq C_{7.9} \left(b, b_0, b', b''', m, m_0, m', m''', q, \|\phi\|_{\mathcal{B},b''',m'''}, n\right) \|\psi\|^q_{\mathcal{B},b,m}$$
$$\times \iint |f * (\phi)_t(x)|^q h_{b,m}(x,t)^q t^{n(q-1)} dx\, dt/t.$$

The proof of Lemma 7.8 is completely the same with Lemma 7.7. (Note that the mapping $(y,t) \in \mathbf{R}^{n+1}_+ \mapsto (\phi)_t(y - \cdot) \in (7.61)$ is continuous by $\|\phi\|_{\mathcal{B},b''',m'''} < \infty$, $b''' > b_0$ and $m''' > m_0$.)

Next, note that if $|\alpha| < a$ and $m > 0$, then

$$\|D^\alpha_x \phi\|_{\mathcal{B},a-|\alpha|,m+|\alpha|} \leq C(a,m,n)\|\phi\|_{\mathcal{B},a,m} \tag{7.63}$$

(This is easy.)

Theorem 7.2. *Assume* (7.2)–(7.3). *Let* $\phi \in \Lambda_a$, $\|\phi\|_{\mathcal{B},a,m+\varepsilon} < \infty$, $\int \phi\, dx = 1$, $\psi \in \Lambda_a$, $\|\psi\|_{\mathcal{B},a,m} < \infty$ *and*

$$f \in \bigcup_{p \in [1,\infty]} L^p. \tag{7.64}$$

Let α *be a multi-index such that*

$$|\alpha| < a. \tag{7.65}$$

Then

$$\left| \int f(x) D_x^\alpha \psi(x) dx \right|^q \leq C(a, b, m, \varepsilon, q, \|\phi\|_{\mathcal{B},a,m+\varepsilon}, n) \|\psi\|_{\mathcal{B},b,m}^q$$

$$\times \int\int |f * D_x^\alpha(\phi)_t(x)|^q \, h_{b,m}(x,t)^q t^{n(q-1)} \frac{dxdt}{t}. \quad (7.66)$$

(*Of course, α can be $(0, \cdots, 0)$.*)

Proof. Let

$$b_0 = \left(\lim_{\delta \to +0} [a - \delta] + a \right) / 2 \text{ and } m_0 = m/2. \quad (7.67)$$

Consider the linear functional;

$$g : \eta \in \{\eta \in \Lambda_{b_0} : \|\eta\|_{\mathcal{B},b_0,m_0} < \infty\} = (7.68) \mapsto \int f(x) D_x^\alpha \eta(x) dx. \quad (7.69)$$

Then (7.63)–(7.65) imply

$$\sup \{|\langle \eta, g \rangle| / \|\eta\|_{\mathcal{B},b_0,m_0} : \eta \in (7.68)\} < \infty.$$

Thus,

$$\text{the left-hand side of } (7.66) = |\langle \psi, g \rangle|^q$$
$$\leq C_{7.9} \left(b, b_0 \text{ in } (7.67), (\max\{b, b_0\} + a)/2, a, m, m_0 \text{ in } (7.67), \right.$$
$$\left. m, m + \varepsilon, q, \|\phi\|_{\mathcal{B},a,m+\varepsilon}, n \right) \|\psi\|_{\mathcal{B},b,m}^q$$
$$\times \int\int |g * (\phi)_t(x)|^q \, h_{b,m}(x,t)^q t^{n(q-1)} dxdt/t \text{ by Lemma 7.8}$$
$$= \text{the right-hand side of } (7.66).$$

$\square$

Theorem 7.3. *Let*

$$b > 0, \ m > 0, \ q \in (0, 1] \text{ and } \psi \in \mathcal{S}.$$

Let $f \in \mathcal{S}'$, $\mathcal{F}f \in C(\mathbf{R}^n, \mathbf{C})$ and

$$\frac{|\mathcal{F}f(\xi)|}{1 + |\xi|^{n(1/q-1)}} \in L^\infty. \quad (7.70)$$

For $(x, t) \in \mathbf{R}_+^{n+1}$ let

$$u(x, t) = \mathcal{F}^{-1} \left(\mathcal{F}f(\xi) e^{-2\pi t|\xi|} \right)(x). \quad (7.71)$$

Then

$$|\langle \psi, f \rangle_\mathcal{S}|^q \leq C(b, m, q, n) \|\psi\|_{\mathcal{B},b,m}^q$$
$$\times \int\int |u(x,t)|^q \, h_{b,m}(x,t)^q t^{n(q-1)} dxdt/t. \quad (7.72)$$

Proof. Let

$$b_0 = n(1/q - 1) + n + 1. \tag{7.73}$$

Since

$$|\mathcal{F}\eta(\xi)| \leq C(b_0, n) (1 + |\xi|)^{-b_0} \|\eta\|_{\mathcal{B}, b_0, 1},$$

the condition (7.70) implies

$$\sup \left\{ \left| \int \mathcal{F}\eta(\xi) \mathcal{F}f(-\xi) d\xi \right| / \|\eta\|_{\mathcal{B}, b_0, 1} : \eta \in \Lambda_{b_0}, \ \|\eta\|_{\mathcal{B}, b_0, 1} < \infty \right\} < \infty.$$

Let

$$b''' = \max\{b_0, b\} + 2, \quad m''' = \max\{1, [m]\} + 2. \tag{7.74}$$

Choose $\{c_1, \cdots, c_{m'''}\} \subset \mathbf{R}$ so that

$$c_1 + \cdots + c_{m'''} = 1,$$

$$\sum_{j=1}^{m'''} c_j P(x, j)/(1 + |x|)^{n+m'''} \in L^\infty,$$

where $P(x, t)$ is the Poisson kernel of $\mathbf{R}_+^{n+1}$. Then

$$\left\| \sum_{j=1}^{m'''} c_j P(\cdot, j) \right\|_{\mathcal{B}, b''', m'''} < \infty. \tag{7.75}$$

Then, Lemma 7.8 with

$$\phi(x) = \sum_{j=1}^{m'''} c_j P(x, j)$$

implies

$$
\begin{aligned}
|\langle \psi, f \rangle_S|^q \ \leq \ & C_{7.9} \left(b, b_0 \text{ in } (7.73), \max\{b, b_0\} + 1, b''' \text{ in } (7.74), m, \right. \\
& \left. 1, \max\{1, [m]\} + 1, m''' \text{ in } (7.74), q, (7.75), n \right) \|\psi\|_{\mathcal{B}, b, m}^q \\
& \times \iint \left| \sum_{j=1}^{m'''} c_j u(x, jt) \right|^q h_{b,m}(x, t)^q t^{n(q-1)} dx dt/t \\
\leq \ & C(b, m, q, n) \|\psi\|_{\mathcal{B}, b, m}^q \iint |u(x, t)|^q h_{b,m}(x, t)^q t^{n(q-1)} dx dt/t.
\end{aligned}
$$

$\square$

Remark 7.1. Let $p \in (0, 1]$, $u(x, t)$ be harmonic on $\mathbf{R}_+^{n+1}$,

$$\sup_{t > 0} \|u(\cdot, t)\|_{L^p} < \infty$$

and $q \in (0, p]$. Then, Corollary 29.1 in Appendix implies that u has the form (7.71) with $f \in \mathcal{S}'$ that satisfies (7.70). So, in this case (7.72) holds.

VIII. Hardy-Littlewood-Fefferman-Stein type inequalities, 3

(You can skip this section if you are not interested in Section 10.)

Theorem 8.1. *Let* $f \in \bigcup_{p \in [1, +\infty]} L^p$,

$$0 < b, \ 0 < \varepsilon < m, \tag{8.1}$$

$$q \in (0, 1], \tag{8.2}$$

$$\{\phi_1, \cdots, \phi_J\} \subset \Lambda_{b+\varepsilon}, \ \sum_{i=1}^{J} \|\phi_i\|_{\mathcal{B}, b+\varepsilon, m} < \infty, \tag{8.3}$$

$$\mathcal{F}\phi_i \in C^{[n+b+m+2]}\left(\mathbf{R}^n \backslash \{0\}\right), \ (i = 1, \cdots, J), \tag{8.4}$$

$$\sup\left\{\sum_{i=1}^{J} |\mathcal{F}\phi_i(t\xi)| : t > 0\right\} > 1 \ for \ any \ \xi \in \mathbf{R}^n \backslash \{0\}, \tag{8.5}$$

$$\psi \in \Lambda_b, \ \|\psi\|_{\mathcal{B}, b, m} < +\infty, \tag{8.6}$$

$$\int \psi(x) x^\alpha dx = 0 \ if \ |\alpha| < m. \tag{8.7}$$

Then,

$$\left| \int f(x)\psi(x)dx \right|^q \le C_{8.1}\left(b, m, \varepsilon, q, \{\phi_i\}_{i=1}^J\right) \|\psi\|_{\mathcal{B}, b, m}^q$$

$$\times \sum_{i=1}^{J} \iint_{\mathbf{R}_+^{n+1}} |f * (\phi_i)_t(x)|^q \, h_{b, m-\varepsilon}(x, t)^q t^{n(q-1)} \frac{dxdt}{t}. \tag{8.8}$$

Lemma 8.1. *Assume* (8.1), (8.3) *and* (8.5). *Let* $k(< m)$ *be a nonnegative integer. Let* $r_0, r_1, \cdots, r_k > 0$ *and*

$$\eta_i(x) = \sum_{j_0=0}^{1} \cdots \sum_{j_k=0}^{1} \left(\prod_{s=0}^{k} (-2^{sr_s})^{j_s}\right) (\phi_i)_{2^{-r_0 j_0 - \cdots - r_k j_k}}(x). \tag{8.9}$$

(i) *Then,*

$$\int \eta_i(x) x^\alpha dx = 0 \ if \ |\alpha| \le k.$$

(ii) *If* $\min\{r_0, \cdots, r_k\}$ *is sufficiently large depending on* $\{\phi_1, \cdots, \phi_J\}$ *and* m, *then*

$$\sup\left\{\sum_{i=1}^{J} |\mathcal{F}\eta_i(t\xi)| : t > 0\right\} > 1/2 \text{ for any } \xi \in \mathbf{R}^n \backslash \{0\}.$$

Proof. (i) can be proved by the same way as Lemma 6.1. (This time, the case $\alpha = 0$ is included because of $\sum_{j_0=0}^{1} \cdot$.)

By (8.5) there exists $\varepsilon_0 > 0$ such that

$$\sup\left\{\sum_{i=1}^{J} |\mathcal{F}\phi_i(t\xi)| : t \in (\varepsilon_0, 1/\varepsilon_0)\right\} > 1 \text{ for any } \xi \in S^{n-1}. \tag{8.10}$$

Note that

$$\eta_i(x) = (-2^{0r_0})(-2^{1r_1}) \cdots (-2^{kr_k})(\phi_i)_{2^{-r_0 - \cdots - r_k}}(x) + \{\text{other terms in (8.9)}\}$$
$$= (-1)^{k+1} 2^{r_1 + \cdots + kr_k} \{(\phi_i)_{2^{-r_0 - \cdots - r_k}} + (8.11)\}, \quad \text{say.}$$

Since

$$\mathcal{F}\phi_i(\xi) \to 0 \quad (|\xi| \to \infty)$$

by (8.3), it follows that the Fourier transform of (8.11) is very small on

$$\mathbf{R}^n \backslash B(0, 2^{r_0 + \cdots + r_k} \varepsilon_0)$$

if $\min\{r_0, \cdots, r_k\}$ is very large. Therefore, if $\min\{r_0, \cdots, r_k\}$ is very large, then

$$\sup\left\{\sum_{i=1}^{J} |\mathcal{F}\eta_i(t\xi)| : t \in (2^{r_0 + \cdots + r_k} \varepsilon_0, 2^{r_0 + \cdots + r_k}/\varepsilon_0)\right\}$$

$$\geq 2^{r_1 + 2r_2 + \cdots + kr_k} \sup\left\{\sum |\mathcal{F}\phi_i(2^{-r_0 - \cdots - r_k} t\xi)| - 1/2 :\right.$$
$$\left. t \in (2^{r_0 + \cdots + r_k} \varepsilon_0, 2^{r_0 + \cdots + r_k}/\varepsilon_0)\right\}$$

$$\geq 2^{r_1 + 2r_2 + \cdots + kr_k}/2 \text{ by (8.10)}$$

for any $\xi \in S^{n-1}$, which implies (ii). $\qquad\qquad\square$

It is easy to see that it is enough to show (8.8) with $\{\phi_i\}_{i=1}^{J}$ replaced by $\{\eta_i\}_{i=1}^{J}$ in (8.9). Therefore, we can impose the following condition (8.12) on $\{\phi_i\}_{i=1}^{J}$:

$$\int \phi_i(x) x^\alpha dx = 0 \text{ if } |\alpha| < m. \tag{8.12}$$

Next, applying Lemma 5.7 with $\{\psi_1, \cdots, \psi_J\} = \{\mathcal{F}\phi_1, \cdots, \mathcal{F}\phi_J\}$ (recall (8.10)), we get $\{\Psi_1, \cdots, \Psi_J\}$ in (5.13). Put

$$\{\Phi_1, \cdots, \Phi_J\} = \{\mathcal{F}^{-1}\Psi_1, \cdots, \mathcal{F}^{-1}\Psi_J\}.$$

Then, by (5.14), (5.15) and (8.4), we have

$$\int_0^{+\infty} \sum_{i=1}^J \mathcal{F}\phi_i(t\xi)\mathcal{F}\Phi_i(t\xi)dt/t = 1 \text{ for any } \xi \in \mathbf{R}^n\backslash\{0\}, \qquad (8.13)$$

$$\text{supp } \mathcal{F}\Phi_i \subset B\left(0, C_{8.2}\left(\{\phi_i\}\right)\right)\backslash B(0, C_{8.2}^{-1}), \qquad (8.14)$$

$$|D_x^\alpha \Phi_i(x)| \le C_{8.3}\left(\alpha, \{\phi_i\}\right)(1+|x|)^{-[n+b+m+2]} \text{ for any } \alpha. \qquad (8.15)$$

Lemma 8.2. *Assume* (8.1), (8.6) *and* (8.7). *Let* $\Phi \in C^{[m-\varepsilon+1]}(\mathbf{R}^n)$,

$$\text{supp } \mathcal{F}\Phi \not\ni 0, \qquad (8.16)$$

$$|D_x^\alpha \Phi(x)| \le (1+|x|)^{-[m-\varepsilon+1]-n-b-\varepsilon} \text{ if } |\alpha| \le [m-\varepsilon+1]. \qquad (8.17)$$

Then,

$$|\psi * (\Phi)_t(x)| \le C_{8.4}(b, m, \varepsilon, n)\|\psi\|_{\mathcal{B},b,m}h_{b,m-\varepsilon}(x,t). \qquad (8.18)$$

Proof. Since the condition (8.17) implies $\|\Phi\|_{\mathcal{B},m-\varepsilon,b+\varepsilon} \le C\|\Phi\|_{\mathcal{B},[m-\varepsilon+1],b+\varepsilon} \le C(b, m, \varepsilon, n)$, applying Lemma 3.5 to Φ gives

$$\{v_j\}_{j\in\mathbf{N}} \subset \mathcal{B}_{m-\varepsilon}^b \qquad (8.19)$$

such that

$$\Phi(x) = C \sum_{j\in\mathbf{N}_0} 2^{-j(b+\varepsilon)}(v_j)_{2^j}(x).$$

First, assume

$$\psi \in \mathcal{B}_b^{m-\varepsilon}. \qquad (8.20)$$

Then,

$$
\begin{aligned}
|\psi * (\Phi)_t(x)| &\le C \sum_{j\in\mathbf{N}_0} 2^{-j(b+\varepsilon)}\left|\int \psi(x-y)(v_j)_{2^j t}(y)dy\right| \\
&\le C\sum_{j} 2^{-j(b+\varepsilon)}\|\psi\|_{\Lambda_b}(2^j t)^b \text{ by (8.19) and Lemma 1.1} \\
&\le Ct^b \text{ by (8.20).} \qquad (8.21)
\end{aligned}
$$

If $|x| + t > 2$, then

$$
\begin{aligned}
|\psi * (\Phi)_t(x)| &\le C \sum_{j\in\mathbf{N}_0 : 2^j t \ge |x|-1} 2^{-j(b+\varepsilon)}\left|\int \psi(x-y)(v_j)_{2^j t}(y)dy\right| \\
&\le C\sum_{j\in\mathbf{N}_0 : \cdots} 2^{-j(b+\varepsilon)}\|(v_j)_{2^j t}\|_{\Lambda_{m-\varepsilon}} \text{ by (8.20)} \\
&\le C\sum_{j\in\mathbf{N}_0 : \cdots} 2^{-j(b+\varepsilon)}(2^j t)^{-n-m+\varepsilon} \text{ by (8.19)} \\
&\le Ct^{-n-m+\varepsilon}(1+|x|/t)^{-n-m-b} \le Ch_{b+\varepsilon,m-\varepsilon}(x,t). \qquad (8.22)
\end{aligned}
$$

Thus, combining (8.21)–(8.22) gives

$$|\psi * (\Phi)_t(x)| \leq C h_{b,m-\varepsilon}(x,t) \tag{8.23}$$

for all $(x,t) \in \mathbf{R}_+^{n+1}$.

Next, we remove the restriction (8.20). Assume (8.6)–(8.7). Then, applying Lemma 3.5 to ψ gives

$$\{u_j\}_{j \in \mathbf{N}_0} \subset \mathcal{B}_b^{m-\varepsilon}$$

such that

$$\psi(x) = C\|\psi\|_{\mathcal{B},b,m} \sum_{j \in \mathbf{N}_0} 2^{-jm}(u_j)_{2^j}(x).$$

Applying (8.23) with dilation to each u_j gives

$$\left|2^{-jm}(u_j)_{2^j} * (\Phi)_t(x)\right| \leq 2^{-j(n+m)} C h_{b,m-\varepsilon}(2^{-j}x, 2^{-j}t).$$

Summing up the above with respect to j gives (8.18). $\square$

Lemma 8.3. *Assume* (8.1)–(8.7) *and* (8.12). *Then, there exist measurable functions* $\{k_1^i(y,t)\}_{i=1,\cdots,J}$ *and* $\{\tilde{k}^{i,j}(y,t)\}_{i,j=1,\cdots,J}$ *defined on* $\mathbf{R}_+^{n+1}$ *such that*

$$\sum_{i=1}^{J} |k_1^i(y,t)| \leq C_{8.5}\,(b,m,\varepsilon,\{\phi_i\})\,\|\psi\|_{\mathcal{B},b,m} h_{b,m-\varepsilon}(y,t), \tag{8.24}$$

$$\psi(x) = \sum_{i=1}^{J} \iint k_1^i(y,t)(\phi_i)_t(x-y)\,dy\,dt/t, \tag{8.25}$$

$$\sum_{i=1}^{J}\sum_{j=1}^{J} \left|\tilde{k}^{i,j}(y,t)\right| \leq C_{8.6}\,(b,m,\varepsilon,\{\phi_i\})\,h_{b+\varepsilon/4,m-\varepsilon/2}(y,t), \tag{8.26}$$

$$\phi_i(x) = \sum_{j=1}^{J} \iint \tilde{k}^{i,j}(y,t)(\phi_j)_t(x-y)\,dy\,dt/t. \tag{8.27}$$

Proof. Let $\{\Phi_i\}$ be as in (8.13)–(8.15). Put

$$k_1^i(y,t) = \psi * (\Phi_i)_t(y),$$
$$\tilde{k}^{i,j}(y,t) = \phi_i * (\Phi_j)_t(y).$$

Then, (8.13) implies (8.25) and (8.27). Lemma 8.2 implies (8.24). Lemma 8.2 with ψ, b, m and ε replaced by ϕ_i, $b+\varepsilon/4$, $m-\varepsilon/4$ and $\varepsilon/4$ implies

$$\begin{aligned}
\left|\tilde{k}^{i,j}(y,t)\right| &\leq C(b,m,\varepsilon,\Phi_j)\|\phi_i\|_{\mathcal{B},b+\varepsilon/4,m-\varepsilon/4} h_{b+\varepsilon/4,m-\varepsilon/2}(y,t) \\
&\leq C\left(b,m,\varepsilon,\{\phi_i\}_{i=1}^{J}\right) h_{b+\varepsilon/4,m-\varepsilon/2}(y,t),
\end{aligned}$$

because $\|\phi_i\|_{\mathcal{B},b+\varepsilon/4,m-\varepsilon/2} \leq C\|\phi_i\|_{\mathcal{B},b+\varepsilon,m} < \infty$. $\square$

Lemma 8.4. *Assume (8.1)–(8.7) and (8.12). Let $E \subset \mathbf{R}_+^{n+1}$ be a measurable set such that*

$$\operatorname{dens}(E) \le C_{8.7}\left(b, m, \varepsilon, \{\phi_i\}\right). \tag{8.28}$$

Then, there exist measurable functions $\{k^i(y,t)\}_{i=1,\cdots,J}$ defined on $\mathbf{R}_+^{n+1}$ such taht

$$k^i(y,t) = 0 \ \ on \ \ E, \tag{8.29}$$

$$\sum_{i=1}^J \left|k^i(y,t)\right| \le C_{8.8}\left(b, m, \varepsilon, \{\phi_i\}\right) \|\psi\|_{\mathcal{B},b,m} h_{b,m-\varepsilon}(y,t), \tag{8.30}$$

$$\psi(x) = \sum_{i=1}^J \iint k^i(y,t)(\phi_i)_t(x-y)dydt/t. \tag{8.31}$$

Proof. Applying Lemma 8.3 gives $\{\tilde{k}^{i,j}(y,t)\}_{i,j=1,\cdots,J}$ such that (8.26)–(8.27) hold. Applying dilation and translation gives

$$(\phi_i)_t(x-y) = t^{-n} \sum_{j=1}^J \iint \tilde{k}^{i,j}\left((z-y)/t, s/t\right)(\phi_j)_s(x-z)dzds/s. \tag{8.27*}$$

Applying Lemma 8.3 gives $\{k_1^i(y,t)\}_{i=1,\cdots,J}$ that satisfy (8.24)–(8.25).
 Put

$$I_1 = \sum_{i=1}^J \iint k_1^i(y,t)\chi_{E^c}(y,t)(\phi_i)_t(x-y)dydt/t,$$

$$J_1 = \sum_{i=1}^J \iint k_1^i(y,t)\chi_E(y,t)(\phi_i)_t(x-y)dydt/t.$$

Substituting (8.27)* into J_1 gives

$$J_1 = \sum_{j=1}^J \iint \left\{ \sum_{i=1}^J \iint k_1^i(y,t)\chi_E(y,t)t^{-n}\tilde{k}^{i,j}\left((z-y)/t, s/t\right)dydt/t \right\}$$

$$\times (\phi_j)_s(x-z)dzds/s$$

$$= \sum_{j=1}^J \iint k_2^j(z,s)(\phi_j)_s(x-z)dzds/s, \ \ \text{say.}$$

Then, by (8.24), (8.26) and Lemma 7.1 we get

$$\sum_j \left|k_2^j(y,t)\right| \le C_{7.2}(b, b+\varepsilon/4, m-\varepsilon, m-\varepsilon/2, n)\operatorname{dens}(E)$$

$$\times C_{8.5}\|\psi\|_{\mathcal{B},b,m} C_{8.6} h_{b,m-\varepsilon}(y,t).$$

So, by taking $C_{8.7}$ (> 0) so small that $C_{7.2}C_{8.7}C_{8.6} < 1/2$, we get

$$\sum_{i=1}^{J} \left| \mathsf{k}_2^i(y,t) \right| \le 2^{-1}C_{8.5}\|\psi\|_{B,b,m}h_{b,m-\varepsilon}(y,t) \tag{8.24}_2$$

and

$$\psi(x) = I_1 + \sum_{i=1}^{J} \iint \mathsf{k}_2^i(y,t)(\phi_i)_t(x-y)dydt/t. \tag{8.25}_2$$

Repeating this argument gives $\{\mathsf{k}_j^i(y,t)\}_{i=1,\cdots,J; j \in \mathbf{N}}$ such that

$$\sum_{i=1}^{J} \left| \mathsf{k}_j^i(y,t) \right| \le 2^{-j+1}C_{8.5}\|\psi\|_{B,b,m}h_{b,m-\varepsilon}(y,t), \tag{8.24}_j$$

$$\psi(x) = I_1 + \cdots + I_{j-1} + \sum_{i=1}^{J} \iint \mathsf{k}_j^i(y,t)(\phi_i)_t(x-y)dydt/t, \tag{8.25}_j$$

where

$$I_s = \sum_{i=1}^{J} \iint \mathsf{k}_s^i(y,t)\chi_{E^c}(y,t)(\phi_i)_t(x-y)dydt/t.$$

Put

$$\mathsf{k}^i(y,t) = \sum_{j \in \mathbf{N}} \mathsf{k}_j^i(y,t)\chi_{E^c}(y,t).$$

Then, (8.29) is clear. $(8.24)_j$ implies (8.30) with $C_{8.8} = 2C_{8.5}$.
Letting $j \to \infty$ in $(8.25)_j$ implies (8.31).   □

Proof of Theorem 8.1. Let $\{T_j\}_{j \in \mathbf{N}}$ be as in the proof of Lemma 6.5. Let

$$E = \bigcup_{j} \left\{ (y,t) \in T_j : C_{6.7}(n)C_{8.7} \left(\sum_{i=1}^{J} |f * (\phi_i)_t(y)| \right)^q \right.$$

$$\left. \ge \iint_{T_j} \left(\sum_{i=1}^{J} |f * (\phi_i)_s(z)| \right)^q dzds/|T_j| \right\}. \tag{8.32}$$

Note that E satisfies the condition (8.28) if $C_{6.7}$ is small enough.
Applying Lemma 8.4 with the above E and with $\phi(x)$ replaced by $\phi(-x)$ we
get $\{\mathsf{k}^i(y,t)\}_{i=1,\cdots,J}$ that satisfies (8.29), (8.30) and

$$\psi(x) = \sum_{i=1}^{J} \iint \mathsf{k}^i(y,t)(\phi_i)_t(y-x)dydt/t. \tag{8.31}'$$

Then,

$$\left| \int f(x)\psi(x)dx \right|^q$$

$$= \left| \sum_{i=1}^{J} \iint k^i(y,t) f*(\phi_i)_t(y)dydt/t \right|^q$$

by (8.31)', $f \in \bigcup_{p\in[1,+\infty]} L^p$ and (8.30)

$$\leq \sum_{j} \left\{ \iint_{T_j} \left(\sum_{i=1}^{J} |k^i(y,t)| \right) \left(\sum_{i=1}^{J} |f*(\phi_i)_t(y)| \right) dydt/t \right\}^q \quad \text{by (8.2)}$$

$$\leq \sum_{j} (C_{6.7}C_{8.7})^{-1} \iint_{T_j} \left(\sum_{i=1}^{J} |f*(\phi_i)_s(z)| \right)^q dzds/|T_j|$$

$$\times \left\{ \iint_{T_j} C_{8.8}\|\psi\|_{\mathcal{B},b,m} h_{b,m-\varepsilon}(y,t)dydt/t \right\}^q$$

by (8.29), (8.32) and (8.30)

$$\leq \sum (C_{6.7}C_{8.7})^{-1} C_{8.8}^q \|\psi\|_{\mathcal{B},b,m}^q C(b,m-\varepsilon,q,n)$$

$$\times \iint_{T_j} \left(\sum_{i=1}^{J} |f*(\phi_i)_s(z)| \right)^q h_{b,m-\varepsilon}(y,s)^q s^{nq-n}dzds/s,$$

which implies (8.8). $\qquad\qquad\qquad\qquad\qquad\qquad\qquad\qquad\qquad\quad \square$

IX. Grand maximal functions from radial maximal functions

Theorem 9.1. *Let* $f \in \mathcal{D}'$, $0 < b < a$, $q > n/(n+b)$, $\phi \in \mathcal{D}$, $\text{supp}\,\phi \in B(0,1)$ *and* $\int \phi \, dx = 1$. *Then,*

$$G_b f(x) \leq C(a,b,q,\|\phi\|_{\Lambda_a}, n) M_q(N_{\phi,0} f)(x), \qquad (9.1)$$

$$\|G_b f\|_{L^q} \leq C(a,b,q,\|\phi\|_{\Lambda_a}, n)\|N_{\phi,0} f\|_{L^q}. \qquad (9.2)$$

Theorem 9.2. *Let* $f \in \mathcal{S}'$, $0 < b < a$, $0 < m$, $q > n/(n+b)$, $\phi \in \mathcal{S}$ *and* $\int \phi \, dx = 1$. *Then,*

$$G_b f(x) \leq C(a,b,m,q,\|\phi\|_{\mathcal{B},a,m}, n) M_q(N_{\phi,0} f)(x), \qquad (9.3)$$

$$\|G_b f\|_{L^q} \leq C(a,b,m,q,\|\phi\|_{\mathcal{B},a,m}, n)\|N_{\phi,0} f\|_{L^q}. \qquad (9.4)$$

Theorem 9.3. *Let* $0 < b < a$, $0 < m$, $q > n/(n+b)$, $f \in \bigcup\limits_{\substack{|\alpha|<a \\ p\in[1,\infty]}} D_x^\alpha L^p$,

$\phi \in \Lambda_a$, $\|\phi\|_{\mathcal{B},a,m} < \infty$ *and* $\int \phi \, dx = 1$. *Then,* (9.3)–(9.4) *hold.*

Remark 9.1. The derivative D_x^α in Theorem 9.3 is taken in the sense of distributions. Though ϕ does not belong to $\mathcal{S}$, (7.63) implies that if $|\alpha| < a$, then $D_x^\alpha \phi \in \bigcap\limits_{p\in[1,\infty]} L^p$. So, if $g \in \bigcup\limits_{p\in[1,\infty]} L^p$, then $D_x^\alpha g * (\phi)_t(x)$ $(= g * D_x^\alpha(\phi)_t(x))$ makes sense. Thus, $N_{\phi,0}(D_x^\alpha g)(x)$ makes sense.

We prove Theorem 9.2.

Proof of (9.3). We may assume $q \in (n/(n+b), 1]$. Let $\psi \in \mathcal{B}_b \cap \mathcal{D}$. Theorem 7.1 implies

$$|\langle \psi, f \rangle_{\mathcal{S}}|^q \leq C_{7.1}(a,b,m/2,m/2,q,\|\phi\|_{\mathcal{B},a,m}, n)\|\psi\|_{\mathcal{B},b,m/2}^q$$

$$\times \iint |f * (\phi)_t(y)|^q h_{b,m/2}(y,t)^q t^{n(q-1)} dy \, dt/t$$

$$\leq C_{7.1}(\cdots)C \iint N_{\phi,0} f(y)^q h_{b,m/2}(y,t)^q t^{n(q-1)} dy \, dt/t$$

$$\leq C_{7.1}(\cdots)C$$

$$\times \int_0^\infty M((N_{\phi,0} f)^q)(0) t^{bq+n(q-1)} (1+t)^{-(n+b+m/2)q+n} dt/t$$

$$\leq C_{7.1}(\cdots)C(b,m,q,n) M((N_{\phi,0} f)^q)(0)$$

$$\text{by } (n+b)q > n \text{ and } m > 0. \qquad (9.5)$$

Let $x \in \mathbf{R}^n$ and $t > 0$. Define $\widetilde{f} \in \mathcal{S}'$ by $\langle \psi, \widetilde{f} \rangle_{\mathcal{S}} = \langle (\psi)_t(\cdot - x), f \rangle_{\mathcal{S}}$. Then, by (9.5)

$$
\begin{aligned}
|\langle (\psi)_t(\cdot - x), f \rangle_{\mathcal{S}}|^q &= |\langle \psi, \widetilde{f} \rangle_{\mathcal{S}}|^q \leq C_{7.1} C M((N_{\phi,0}\widetilde{f})^q)(0) \\
&= C_{7.1} C M((N_{\phi,0}f)^q)(x)
\end{aligned}
$$

which implies (9.3). □

Proof of (9.4). Take $q' \in (n/(n+b), q)$. Then (9.3) implies

$$
G_b f(x) \leq C M_{q'}(N_{\phi,0}f)(x).
$$

So,

$$
\|G_b f\|_{L^q} \leq C \|M_{q'}(N_{\phi,0}f)\|_{L^q} \leq C \|N_{\phi,0}f\|_{L^q},
$$

by $q > q'$ and by Corollary 0.4. □

By repeating the same argument, Theorems 9.1 and 9.3 follow from Theorem 6.1 and Theorem 7.2, respectively.

Remark 9.2. Combining Theorems 9.1 (or 9.2), 4.1 and 3.2, we get that if $f \in \mathcal{D}'$, $\psi \in \mathcal{D}$, $\int \phi\,dx = 1$ (or $f \in \mathcal{S}', \phi \in \mathcal{S}, \int \phi\,dx = 1$) and if $q \in (0,1]$, then

$$
c(q, \phi)\|f\|_{H^q} \leq \|N_{\phi,0}f\|_{L^q} \leq C(q, \phi)\|f\|_{H^q}. \tag{9.6}
$$

Combining Theorems 9.3, 4.1 and 3.2, we get that if $0 < a$, $0 < m$, $f \in \bigcup_{\substack{|\alpha| < a \\ P \in [1,\infty]}} D_x^\alpha L^p$, $\phi \in \Lambda_a$, $\|\phi\|_{\mathcal{B},a,m} < \infty$, $\int \phi\,dx = 1$ and if $q \in (n/(n+a), 1]$, then (9.6) holds.

Notes. This kind of argument for the case ϕ is the Poisson kernel is in C. Fefferman–E. M. Stein [72] p. 170. The case of general ϕ is in A. Uchiyama [80b, 85b, 86]. (See also A. Miyachi [87].)

X. S-functions from g-functions

(The statements of the results in this section are complicated. But, these results will not be used in later sections. You can skip this section.)

In this section we will show that L^q-norms of $S_{\psi,\delta}f$ are essentially independent of the choice of ψ and $\delta \geq 0$ if ψ satisfies certain conditons. The point is the fact that this includes the case $\delta = 0$. (The "g-function" in the title means $S_{\psi,0}f$.) We write

$$\chi(x) = \chi_{B(0,1)}(x). \tag{10.0}$$

For the definition of $S_{\psi,\delta,m}$ recall Definition 3.6. In the notation $M(f * (\phi)_t)(x)$, the convolution $*$ and the maximal operator M are taken with respect to the variable $x \in \mathbf{R}^n$, with $t \in (0, +\infty)$ fixed.

Theorem 10.1. *Let* $0 < b < a$, $\delta, \delta' \geq 0$, $f \in \mathcal{D}'$, $\phi \in \mathcal{D}$, $\operatorname{supp}\phi \subset B(0,1)$, $\int \phi\, dx = 1$, $\psi \in \mathcal{D}$ *and* $\operatorname{supp}\psi \subset B(0,1)$. *Let* $m' < b$ *and* $q \in (n/(n+b-m'), \infty)$. *Then*

$$S_{\psi,\delta,m'}f(x) \leq C(a,b,m',q,\|\phi\|_{\Lambda_a},n)(1+\delta)^{n+b-m'}\|\psi\|_{\Lambda_b}$$
$$\times \left\{ \int_0^{+\infty} M(|f * (\phi)_t|^q * (\chi)_t)(x)^{2/q} t^{2m'} \frac{dt}{t} \right\}^{1/2}, \tag{10.1}$$

$$\|S_{\psi,\delta,m'}f\|_{L^q} \leq C(a,b,\delta,\delta',m',q,\|\phi\|_{\Lambda_a},n)\|\psi\|_{\Lambda_b}\|S_{\phi,\delta',m'}f\|_{L^q}. \tag{10.2}$$

Corollary 10.1. *Let* a, b, δ, δ', f, ϕ *and* ψ *be as in Theorem 10.1. Let* α *be a multi-index with* $|\alpha| < b$. *Let* $q \in (n/(n+b-|\alpha|), \infty)$. *Then,*

$$\|S_{D_x^\alpha\psi,\delta}f\|_{L^q} \leq C(a,b,\delta,\delta',\alpha,q,\|\phi\|_{\Lambda_a},n)\|\psi\|_{\Lambda_b}\|S_{D_x^\alpha\phi,\delta'}f\|_{L^q}. \tag{10.3}$$

Theorem 10.2. *Let* $0 < b < a$, $0 < m$, $0 < \varepsilon$, $\delta, \delta' \geq 0$, $f \in \mathcal{S}'$, $\phi \in \mathcal{S}$, $\int \phi\, dx = 1$ *and* $\psi \in \mathcal{S}$. *Let* $m' \in (-m, b)$ *and* $q \in (n/(n+b-m'), \infty)$. *Then*

$$S_{\psi,\delta,m'}f(x) \leq C(a,b,m,\varepsilon,m',q,\|\phi\|_{\mathcal{B},a,m+\varepsilon},n)(1+\delta)^{n+b-m'}\|\psi\|_{\mathcal{B},b,m}$$
$$\times \left\{ \int_0^{+\infty} M(|f * (\phi)_t|^q * (\chi)_t)(x)^{2/q} t^{2m'} \frac{dt}{t} \right\}^{1/2}, \tag{10.4}$$

$$\|S_{\psi,\delta,m'}f\|_{L^q} \leq C(a,b,m,\varepsilon,\delta,\delta',m',q,\|\phi\|_{\mathcal{B},a,m+\varepsilon},n)$$
$$\times \|\psi\|_{\mathcal{B},b,m}\|S_{\phi,\delta',m'}f\|_{L^q}. \tag{10.5}$$

Corollary 10.2. *Let* $a, b, m, \varepsilon, \delta, \delta', f, \phi$ *and* ψ *be as in Theorem 10.2. Let* α *be a multi-index with* $|\alpha| < b$. *Let* $q \in (n/(n + b - |\alpha|), \infty)$. *Then,*

$$\left\| S_{D_x^\alpha \psi, \delta} f \right\|_{L^q} \leq C\left(a, b, m, \varepsilon, \delta, \delta', \alpha, q, \|\phi\|_{\mathcal{B}, a, m+\varepsilon}, n\right) \|\psi\|_{\mathcal{B}, b, m} \left\| S_{D_x^\alpha \phi, \delta'} f \right\|_{L^q}.$$

(10.6)

Theorem 10.3. *Let* $0 < b < a$, $0 < m$, $0 < \varepsilon$ *and* $\delta, \delta' \geq 0$. *Let*

$$f \ \in \ \bigcup_{p \in [1, +\infty]} L^p,$$

$$\phi \ \in \ \Lambda_a, \ \|\phi\|_{\mathcal{B}, a, m+\varepsilon} < +\infty, \ \int \phi \, dx = 1,$$

$$\psi \ \in \ \Lambda_a, \ \|\psi\|_{\mathcal{B}, a, m} < +\infty.$$

Let α *be a multi-index such that* $|\alpha| < a$. *Let* $m' \in (-m, b)$ *and* $q \in (n/(n + b - m'), \infty)$. *Then*

$$S_{\psi, \delta, m'} D_x^\alpha f(x) \leq C(a, b, m, \varepsilon, m', \alpha, q, \|\phi\|_{\mathcal{B}, a, m+\varepsilon}, n) (1 + \delta)^{n+b-m'} \|\psi\|_{\mathcal{B}, b, m}$$

$$\times \left\{ \int_0^{+\infty} M(|D_x^\alpha f * (\phi)_t|^q * (\chi)_t)(x)^{2/q} t^{2m'} \, dt/t \right\}^{1/2},$$

(10.4)′

$$\left\| S_{\psi, \delta, m'} D_x^\alpha f \right\|_{L^q} \leq C(a, b, m, \varepsilon, \delta, \delta', m', \alpha, q, \|\phi\|_{\mathcal{B}, a, m+\varepsilon}, n) \|\psi\|_{\mathcal{B}, b, m}$$

$$\times \left\| S_{\phi, \delta', m'} D_x^\alpha f \right\|_{L^q}.$$

(10.5)′

(Of course α *can be* $(0, \cdots, 0)$. *The derivative* $D_x^\alpha f$ *is taken in the sense of distributions. Though* ϕ *and* ψ *do not belong to* S, *(7.63) implies* $D_x^\alpha \phi, D_x^\alpha \psi \in \bigcap_{1 \leq p \leq \infty} L^p$. *So* $D_x^\alpha f * (\phi)_t(x) \ (= f * D_x^\alpha(\phi)_t(x))$ *and* $D_x^\alpha f * (\psi)_t(x)$ *make sense.)*

Corollary 10.3. *Let* $a, b, m, \varepsilon, \delta, \delta', f, \phi$ *and* ψ *be as in Theorem 10.3. Let* α *be a multi-index such that* $|\alpha| < b$. *Let* $q \in (n/(n + b - |\alpha|), \infty)$. *Then,*

$$\left\| S_{D_x^\alpha \psi, \delta} f \right\|_{L^q} \leq C(a, b, m, \varepsilon, \delta, \delta', \alpha, q, \|\phi\|_{\mathcal{B}, a, m+\varepsilon}, n) \|\psi\|_{\mathcal{B}, b, m} \left\| S_{D_x^\alpha \phi, \delta'} f \right\|_{L^q}.$$

(10.6)′

Theorem 10.4. *Let* $0 < b$, $0 < \varepsilon < m$, $\delta, \delta' \geq 0$ *and* $f \in \bigcup_{p \in [1, +\infty]} L^p$. *Assume (8.3)–(8.7). Let* $m' \in (-m + \varepsilon, b)$ *and* $q \in (n/(n + b - m'), \infty)$. *Then,*

$$S_{\psi, \delta, m'} f(x) \ \leq \ C\left(b, m, m', q, \{\phi_i\}_{i=1}^J\right) (1 + \delta)^{n+b-m'} \|\psi\|_{\mathcal{B}, b, m}$$

$$\times \left\{ \sum_{i=1}^J \int_0^{+\infty} M(|f * (\phi_i)_t|^q * (\chi)_t)(x)^{2/q} t^{2m'} \, dt/t \right\}^{1/2},$$

(10.7)

$$\left\| S_{\psi, \delta, m'} f \right\|_{L^q} \ \leq \ C\left(b, m, \delta, \delta', m', q, \{\phi_i\}_{i=1}^J\right)$$

$$\times \|\psi\|_{\mathcal{B}, b, m} \sum_{i=1}^J \left\| S_{\phi_i, \delta', m'} f \right\|_{L^q}.$$

(10.8)

First we prepare a couple of lemmas.

Lemma 10.A . Let $\{f_j\}_{j\in\mathbf{N}} \subset L^1_{\mathrm{loc}}$, $r \in (1, +\infty)$ and $p \in (1, +\infty)$. Then,

$$\int \left(\sum_{j\in\mathbf{N}} M f_j(x)^r \right)^{p/r} dx \le C(p,r,n) \int \left(\sum_{j\in\mathbf{N}} |f_j(x)|^r \right)^{p/r} dx.$$

For the proof of Lemma 10.A see C. Fefferman–E. M. Stein [71] or J. García-Cuerva–Rubio de Francia [85] p. 497.

Lemma 10.B . Let $u(x,t)$ be a continuous function defined on $\mathbf{R}_+^{n+1}$. Let $d\mu$ be a positive measure on $(0, +\infty)$. Let $r \in (1, +\infty)$ and $p \in (1, +\infty)$. Let

$$Mu(\cdot,t)(x) = \sup_{s>0} \int_{B(x,s)} |u(y,t)| dy / |B(x,s)|.$$

Then,

$$\int \left\{ \int_0^{+\infty} Mu(\cdot,t)(x)^r d\mu(t) \right\}^{p/r} dx \le C(p,r,n) \int \left\{ \int_0^{+\infty} |u(x,t)|^r d\mu(t) \right\}^{p/r} dx.$$

Lemma 10.B is a continuous version of Lemma 10.A and is an easy consequence of Lemma 10.A.

Lemma 10.1. Let $u(x,t) \in C\left(\mathbf{R}_+^{n+1}\right)$. Let $0 < p < \infty$, $0 < q < \min\{2,p\}$ and $\delta > 0$. Then

$$\int \left\{ \int_0^{+\infty} M(|u(\cdot,t)|^q * (\chi)_t)(x)(x)^{2/q} dt/t \right\}^{p/2} dx \le C(p,q,n) \|S_0 u\|_{L^p}^p, \quad (10.9)$$

$$\int \left\{ \int_0^{+\infty} M(|u(\cdot,t)|^q * (\chi)_t)(x)(x)^{2/q} dt/t \right\}^{p/2} dx$$

$$\le C(p,q,n)(1+\delta)^{np/q} \|S_\delta u\|_{L^p}^p. \quad (10.10)$$

Proof. Since

$$M(|u(\cdot,t)|^q * (\chi)_t)(x) \le C(n) M(|u(\cdot,t)|^q)(x),$$

we have

the left-hand side of (10.9)

$$\le C \int \left\{ \int_0^{+\infty} M(|u(\cdot,t)|^q)(x)^{2/q} dt/t \right\}^{(p/q)/(2/q)} dx$$

$$\le C \int \left\{ \int_0^{+\infty} |u(x,t)|^{q(2/q)} dt/t \right\}^{(p/q)/(2/q)} dx \qquad \text{by Lemma 10.B}$$

$$= \text{the right-hand side of (10.9).}$$

For the proof of (10.10) note that if $\delta \in (0, 1]$, then

$$M(|u(\cdot, t)|^q * (\chi)_t)(x) \le C(n) M(|u(\cdot, t)|^q * (\chi)_{\delta t})(x)$$

and that if $\delta \in (1, +\infty)$, then

$$|u(\cdot, t)|^q * (\chi)_t(x) \le C(n) \delta^n |u(\cdot, t)|^q * (\chi)_{\delta t}(x).$$

So,

the left-hand side of (10.10)

$$\le C(n)(1 + \delta)^{np/q} \int \left\{ \int_0^{+\infty} M(|u(\cdot, t)|^q * (\chi)_{\delta t})(x)^{2/q} \frac{dt}{t} \right\}^{(p/q)/(2/q)} dx$$

$$\le C(p, q, n)(1 + \delta)^{np/q} \int \left\{ \int_0^{+\infty} (|u(\cdot, t)|^q * (\chi)_{\delta t})(x)^{2/q} \frac{dt}{t} \right\}^{(p/q)/(2/q)} dx$$

by Lemma 10.B

$$\le C(p, q, n)(1 + \delta)^{np/q} \int \left\{ \int_0^{+\infty} |u(\cdot, t)|^2 * (\chi)_{\delta t}(x) \frac{dt}{t} \right\}^{p/2} dx$$

$$= \text{the right-hand side of (10.10)}.$$

$\square$

Lemma 10.2. *Let* $b > 0$, $m' \in \mathbf{R}$, $\delta > 0$, $f \in \mathcal{D}'$, $\psi \in \mathcal{D}$ *and* $\operatorname{supp} \psi \subset B(0, 1)$. *Then,*

$$\begin{aligned} S_{\psi, \delta, m'} f(x) &\le C(n)(1 + \delta)^{n+b-m'} \|\psi\|_{\Lambda_b} \\ &\times \sup \{ S_{\eta, 0, m'} f(x) : \eta \in \mathcal{D} \cap B_b \}. \end{aligned} \tag{10.11}$$

Proof. By translation we may assume $x = 0$. For $y \in B(0, \delta)$ let

$$\tilde{\psi}_y(z) = (1 + \delta)^n \psi((1 + \delta)z + y). \tag{10.12}$$

Then,

$$\operatorname{supp} \tilde{\psi}_y \subset B(0, 1), \tag{10.13}$$

$$\left\| \tilde{\psi}_y \right\|_{\Lambda_b} \le (1 + \delta)^{n+b} \|\psi\|_{\Lambda_b}, \tag{10.14}$$

$$f * (\psi)_t(ty) = f * (\tilde{\psi}_y)_{(1+\delta)t}(0). \tag{10.15}$$

Thus,

$$\iint_{\Gamma(0,\delta)} |f*(\psi)_t(y)|^2 \, t^{2m'} (\delta t)^{-n} dy dt/t$$

$$= \delta^{-n} \int_{B(0,\delta)} dy \int_0^{+\infty} |f*(\psi)_t(ty)|^2 \, t^{2m'} dt/t$$

$$= \delta^{-n} \int_{B(0,\delta)} dy \int_0^{+\infty} \left|f*(\tilde\psi_y)_{(1+\delta)_t}(0)\right|^2 t^{2m'} dt/t \qquad \text{by (10.15)}$$

$$= \delta^{-n}(1+\delta)^{-2m'} \int_{B(0,\delta)} dy \int_0^{+\infty} \left|f*(\tilde\psi_y)_t(0)\right|^2 t^{2m'} dt/t$$

$$\leq \{\text{the right-hand side of (10.11) with } x=0\}^2 \qquad \text{by (10.13)--(10.14).}$$

$\square$

Proof of (10.1). We may assume $q \in (n/(n+b-m'),1]$, $x=0$ and $\|\psi\|_{\Lambda_b}=1$. Applying Theorem 6.1 with $\varepsilon=1$ gives

$$|f*\psi(0)|^q \leq C_{6.1} \iint_{B(0,2)\times(0,1)} |f*(\phi)_t(y)|^q \, t^{q(n+b)-n} dy dt/dt$$

$$\leq C \int_0^1 M(|f*(\phi)_t|^q * (\chi)_t)(0) t^{q(n+b)-n} dt/t.$$

Applying dilation gives

$$|f*(\psi)_s(0)|^q \leq C \int_0^1 M(|f*(\phi)_{st}|^q * (\chi)_{st})(0) t^{q(n+b)-n} dt/t.$$

So,

$$\int_0^{+\infty} |f*(\psi)_s(0)|^2 \, s^{2m'} \frac{ds}{s}$$

$$\leq C \int_0^{+\infty} \left\{ \int_0^1 M(|f*(\phi)_{st}|^q * (\chi)_{st})(0) t^{q(n+b)-n} \frac{dt}{t} \right\}^{2/q} s^{2m'} \frac{ds}{s}$$

$$\leq C \left\{ \int_0^1 t^{q(n+b)-n} \left(\int_0^{+\infty} M(|f*(\phi)_{st}|^q * (\chi)_{st})(0)^{2/q} s^{2m'} \frac{ds}{s} \right)^{q/2} \frac{dt}{t} \right\}^{2/q}$$

$$\text{by Minkowski's inequality}$$

$$= C \left\{ \int_0^1 t^{q(n+b-m')-n} \frac{dt}{t} \right\}^{2/q} \left\{ \int_0^{+\infty} M(|f*(\phi)_s|^q * (\chi)_s)(0)^{2/q} s^{2m'} \frac{ds}{s} \right\}$$

$$= C \int_0^{+\infty} M(\cdots)(0)^{2/q} s^{2m'} \frac{ds}{s} \qquad \text{by } q(n+b-m') > n.$$

Thus, we get (10.1) for the case $\delta=0$. The case $\delta>0$ follows from the case $\delta=0$ and from Lemma 10.2. $\square$

Proof of (10.2). Take $q' \in (n/(n+b-m'), \min\{q,2\})$. Then, (10.1) implies

$$S_{\psi,\delta,m'}f(x) \leq C\left(a,b,\delta,m',q',\|\phi\|_{\Lambda_a},n\right)\|\psi\|_{\Lambda_b}$$

$$\times \left\{\int_0^{+\infty} M(|f*(\phi)_t|^{q'}*(\chi)_t)(x)^{2/q'}t^{2m'}\,dt/t\right\}^{1/2}.$$

So,

$$\|S_{\psi,\delta,m'}f\|_{L^q}^q \leq (C\|\psi\|_{\Lambda_b})^q \int \left\{\int_0^{+\infty} M(|f*(\phi)_t|^{q'}*(\chi)_t)(x)^{2/q'}t^{2m'}\frac{dt}{t}\right\}^{\frac{q}{2}}\,dx,$$

which combined with "Lemma 10.1 with $u(x,t) = t^{m'}|f*(\phi)_t(x)|$ and with q,p and δ replaced by q',q and δ', respectively" implies (10.2).   □
(10.3) is immediate from (10.2),

$$S_{D_x^\alpha \psi,\delta}f = S_{\psi,\delta,|\alpha|}D_x^\alpha f \tag{10.16}$$

and from taking $m' = |\alpha|$.

Lemma 10.3. *Let $b > 0$, $m > 0$, $m' \in \mathbf{R}$, $\delta > 0$, $f \in \mathcal{S}'$ and $\psi \in \mathcal{S}$. Then,*

$$S_{\psi,\delta,m'}f(x) \leq C(b,m,n)(1+\delta)^{n+b-m'}\|\psi\|_{\mathcal{B},b,m}$$
$$\times \sup\{S_{\eta,0,m'}f(x) : \eta \in \mathcal{S}, \|\eta\|_{\mathcal{B},b,m} \leq 1\}.$$

Proof. *This follows from the same argument as Lemma 10.2 and from the fact that if $y \in B(0,\delta)$ and if $\tilde{\psi}_y$ is defind by the same formula as (10.12) with $\psi \in \mathcal{S}$, then*

$$\left\|\tilde{\psi}_y\right\|_{\mathcal{B},b,m} \leq C(b,m,n)(1+\delta)^{n+b}\|\psi\|_{\mathcal{B},b,m}.$$

□

Proof of (10.4). We may assume $q \in (n/(n+b-m'),1]$, $x = 0$ and $\|\psi\|_{\mathcal{B},b,m} = 1$. Applying Theorem 7.1 gives

$$|f*\psi(0)|^q \leq C_{7.1}\iint |f*(\phi)_t(y)|^q h_{b,m}(y,t)^q t^{n(q-1)}\,dy\,dt/t$$

$$= C\int_0^{+\infty} t^{bq+n(q-1)}(1+t)^{-(n+b+m)q+n}\,dt/t$$

$$\times \int_{\mathbf{R}^n} |f*(\phi)_t(y)|^q (1+t)^{-n}(1+|y|/(1+t))^{-(n+b+m)q}\,dy$$

$$\leq C\int_0^{+\infty} M(|f*(\phi)_t|^q*(\chi)_t)(0)t^{q(n+b)-n}(1+t)^{-(n+b+m)q+n}\,dt/t.$$

Applying dilation gives

$$|f*(\psi)_s(0)|^q \leq C\int_0^{+\infty} M(|f*(\phi)_{st}|^q*(\chi)_{st})(0)$$
$$\times t^{q(n+b)-n}(1+t)^{-(n+b+m)q+n}\,dt/t.$$

So,

$$\int_0^{+\infty} |f * (\psi)_s(0)|^2 \, s^{2m'} \, ds/s$$

$$\leq C \int_0^{+\infty} \left\{ \int_0^{+\infty} M(|f * (\phi)_{st}|^q * (\chi)_{st})(0) \right.$$

$$\left. \times t^{q(n+b)-n}(1+t)^{-(n+b+m)q+n} dt/t \right\}^{2/q} s^{2m'} \, ds/s$$

$$\leq C \left\{ \int_0^{+\infty} t^{q(n+b)-n}(1+t)^{-(n+b+m)q+n} \right.$$

$$\left. \times \left(\int_0^{+\infty} M(|f * (\phi)_{st}|^q * (\chi)_{st})(0)^{2/q} s^{2m'} \, ds/s \right)^{q/2} dt/t \right\}^{2/q}$$

$$= C \left\{ \int_0^{+\infty} t^{q(n+b-m')-n}(1+t)^{-(n+b+m)q+n} dt/t \right\}^{2/q}$$

$$\times \left\{ \int_0^{+\infty} M(|f * (\phi)_s|^q * (\chi)_s)(0)^{2/q} s^{2m'} \, ds/s \right\}$$

$$= C \int_0^{\infty} M(\cdots)(0)^{2/q} s^{2m'} \, ds/s$$

by $q(n+b-m') > n$ and $m+m' > 0$.

Thus, we get (10.4) for the case $\delta = 0$. The case $\delta > 0$ follows from the case $\delta = 0$ and from Lemma 10.3. □

(10.5) follows from (10.4) by the same way as (10.2) followed from (10.1). (10.6) is immediate from (10.5) and (10.16).

Theorem 10.3 follows from the same argument as Theorem 10.2 with f replaced by $D_x^\alpha f$ and with Theorem 7.1 replaced by Theorem 7.2. Corollary 10.3 is immediate from (10.5)' and from (10.16) by taking $m' = |\alpha|$.

(10.7) for the case $\delta = 0$ follows from the same argument as (10.4) with Theorem 7.1 replacced by Theorem 8.1. (10.7) for the case $\delta > 0$ follows from the case $\delta = 0$ and from Lemma 10.4 below, which can be proved by the same way as Lemma 10.3. (We omit its proof.) (10.8) follows from (10.7) by the same way as (10.2) followed from (10.1).

Lemma 10.4. *Let* $b > 0$, $m' \in \mathbf{R}$, $\delta > 0$ *and* $f \in \bigcup_{p \in [1,+\infty]} L^p$. *Assume* (8.6)–(8.7). *Then,*

$$S_{\psi,\delta,m'} f(x) \leq C(b,m,n)(1+\delta)^{n+b-m'} \|\psi\|_{\mathcal{B},b,m}$$

$$\times \sup \left\{ S_{\eta,0,m'} f(x) : \eta \in \Lambda_b, \|\eta\|_{\mathcal{B},b,m} \leq 1, \right.$$

$$\left. \int \eta(x) x^\alpha dx = 0 \ \text{if} \ |\alpha| < m \right\}.$$

Remark 10.1. Let

$$\psi \in \mathcal{S} \text{ and } \int \psi \, dx = 1. \qquad (10.17)$$

If

$$\delta \geq 0, \ m' \in \mathbf{N}, \ 0 < q \leq 1, \ f \in \mathcal{S}', \ \phi \in \mathcal{S} \text{ and } \int \phi \, dx = 1,$$

then

$$C(\delta, m', q, \phi, \psi) \sum_{\alpha: |\alpha| = m'} \left\| S_{D_x^\alpha \phi, \delta} f \right\|_{L^q} \geq \sum_{\alpha: |\alpha| = m'} \left\| S_{D_x^\alpha \psi, 1} f \right\|_{L^q}$$

by (10.6) (with its $a, b, m, \varepsilon, \delta$ and δ' replaced by
$n/q - n + m' + 2, \ n/q - n + m' + 1, 1, 1, 1$ and δ, respectively)

$$\geq c(m', q, \psi) \left\| f - P_f \right\|_{H^q} \qquad \text{for some polynomial } P_f \text{ of degree } < m'$$

by Remarks 5.3–5.4.

(In particular, if $m' = 1$, then P_f is a constant c_f.) (Theorem 3.3 implies the converse:

$$C(\delta, m', q, \phi) \left\| f - P_f \right\|_{H^q} \geq \sum_{\alpha: |\alpha| = m'} \left\| S_{D_x^\alpha \phi, \delta} f \right\|_{L^q} .)$$

Similarly if we assume (10.17) and if

$$a > 1, \ m > 0, \ \delta \geq 0, \ m' \in \mathbf{N}, \ m' < a, \ q \in (n/(n + a - m'), 1],$$

$$f \in \bigcup_{p \in [1, \infty]} L^p, \ \phi \in \Lambda_a, \ \|\phi\|_{B, a, m} < \infty \text{ and } \int \phi \, dx = 1,$$

then

$$C(\delta, m', q, \phi, \psi) \sum_{\alpha: |\alpha| = m'} \left\| S_{D_x^\alpha \phi, \delta} f \right\|_{L^q} \geq \sum_{\alpha: |\alpha| = m'} \left\| S_{D_x^\alpha \psi, 1} f \right\|_{L^q}$$

by (10.6)′ (with its $a, b, m, \varepsilon, \delta$ and δ' replaced by
$a, (a + (n/q - n + m'))/2, m/2, m/2, 1$ and δ, respectively)

$$\geq \begin{cases} c(q, \psi) \|f\|_{H^q} & \text{if } f \in \bigcup_{p \in [1, \infty]} L^p, \\ c(q, \psi) \|f - c_f\|_{H^q} & \text{if } f \in L^\infty \end{cases} \qquad (10.18)$$

by Remark 5.1.
    Similarly, if we assume (10.17) and if

$$a > 0, \ m > 0, \ \delta \le 0, \ q \in (n/(n+a), 1],$$

$$f \in \bigcup_{p \in [1,\infty]} L^p, \ \{\phi_1, \cdots, \phi_J\} \subset \Lambda_a,$$

$$\sum_{i=1}^{J} \|\phi_i\|_{\mathcal{B},a,m} < \infty, \ \{\mathcal{F}\phi_i\} \subset C^{[n+a+m+2]}\left(\mathbf{R}^n \backslash \{0\}\right),$$

$$\sup\left\{\sum_{i=1}^{J} |\mathcal{F}\phi_i(t\xi)| : t > 0\right\} > 1 \ \text{for any } \xi \in \mathbf{R}^n \backslash \{0\},$$

then

$$C(m, \delta, q, \{\phi_i\}, \psi) \sum_{i=1}^{J} \|S_{\phi_i, \delta} f\|_{L^q} \ge \sum_{\alpha: |\alpha| = [m]+1} \|S_{D_x^\alpha \psi, 1} f\|_{L^q}$$

by (10.8) (with its $b, \varepsilon, m, \delta, \delta', m'$ and ψ replaced by $a - \min\{a - (n/q - n), m\}/2, \min\{a - (n/q - n), m\}/2, m, 1, \delta, 0$ and $D_x^\alpha(\psi$ in (10.17)), respectively)

$$\ge \begin{cases} c(m, q, \psi)\|f\|_{H^q} & \text{if } f \in \bigcup_{p \in [1,+\infty)} L^p, \\ c(m, q, \psi)\|f - c_f\|_{H^q} & \text{if } f \in L^\infty \end{cases} \tag{10.19}$$

by Remark 5.1.

Remark 10.2. Let

$$\delta \ge 0, \ q \in (0,1], \ f \in \mathcal{D}', \phi \in \mathcal{D}, \ \int \phi \, dx = 1$$

and

$$\sum_{j=1}^{n} \left\|S_{D_{x_j}\phi, \delta} f\right\|_{L^q} < \infty.$$

Then, Corollary 10.1 (with $\psi = \phi$) implies

$$\sum_{j=1}^{n} \|S_{\phi,1,1} D_{x_j} f\|_{L^q} = \sum_{j=1}^{n} \left\|S_{D_{x_j}\phi, 1} f\right\|_{L^q} \le C(q, \delta, \phi) \sum_{j=1}^{n} \left\|S_{D_{x_j}\phi, \delta} f\right\|_{L^q} < \infty.$$

So, Remark 6.2 implies $\{D_{x_1} f, \cdots, D_{x_n} f\} \subset \mathcal{S}'$. This implies $f \in \mathcal{S}'$. Therefore, the first statement of Remark 10.1 implies the existence of a constant c_f such that

$$\|f - c_f\|_{H^q} \le C(q, \delta, \phi) \sum_{j=1}^{n} \left\|S_{D_{x_j}\phi, \delta} f\right\|_{L^q}. \tag{10.20}$$

Notes. The equivalence between $\|S_{\phi,0}f\|_{L^q}$ and $\|S_{\phi,\delta}f\|_{L^p}$ with $\phi = (\nabla p)(\cdot, 1)$ is in C. Fefferman–E. M. Stein [72], where $P(x,t)$ is the Poisson kernel and where $\nabla = (D_t, D_{x_1}, \cdots, D_{x_n})$. The case of general ϕ is in A. Uchiyama [85a, 85b]. The idea to use Lemma 10.A in the estimate of $S_{\phi,\delta}f$ is in R. Fefferman–E. M. Stein [82].

XI. Good λ inequalities for nontangential maximal functions and S-functions of harmonic functions

For a harmonic function $u(x,t)$ on $\mathbf{R}_+^{n+1}$, let $N_\delta u$ be as in (0.2) and let

$$S_\delta(t|\nabla u|)(x)$$
$$= \left\{ \Gamma\left(\frac{n+2}{2}\right) \pi^{-n/2} \iint_{\Gamma(x,\delta)} t^2 \, |\nabla u(y,t)|^2 \, (\delta t)^{-n} dy dt/t \right\}^{1/2},$$

where $\nabla = (D_t, D_{x_1}, \cdots, D_{x_n})$. We will show the following precise relations between $N_\delta u$ and $S_{\delta'}(t|\nabla u|)$.

Theorem 11.1. *Let $u(x,t)$ be a harmonic function defined on $\mathbf{R}_+^{n+1}$. Let $0 < \delta < \delta'$, $\lambda > 0$ and $\gamma > 1$. Then,*

$$|\{x \in \mathbf{R}^n : S_\delta(t|\nabla u|)(x) > \gamma\lambda, \ N_{\delta'}u(x) \le \lambda\}|$$
$$\le C(\delta, \delta', n)e^{-\gamma^2/C(\delta,\delta',n)} |\{x \in \mathbf{R}^n : S_\delta(t|\nabla u|)(x) > \lambda\}|, \quad (11.1)$$
$$|\{x \in \mathbf{R}^n : N_\delta u(x) > \gamma\lambda, \ S_{\delta'}(t|\nabla u|)(x) \le \lambda\}|$$
$$\le C(\delta, \delta', n)e^{-\gamma/C(\delta,\delta',n)} |\{x \in \mathbf{R}^n : N_\delta u(x) > \lambda\}|. \quad (11.2)$$

Lemma 11.1. *Let $\|f\|_{\mathrm{BMO}} \le 1$, $\lambda > 0$ and $\mu \in \mathbf{R}$. Then*

$$|\{x \in \mathbf{R}^n : f(x) > \mu + \lambda\}| \le C(n)e^{-\lambda/C(n)} |\{x \in \mathbf{R}^n : f(x) > \mu\}|. \quad (11.3)$$

Proof. We may assume

$$\mu = 0. \quad (11.4)$$

Put

$$E = \{x \in \mathbf{R}^n : f(x) > 0\}. \quad (11.5)$$

We may assume $|E| < \infty$. Let $\{I_j\}$ be the family of the maximal dyadic cubes in $\mathbf{R}^n$ such that

$$|I_j \cap E| \, / \, |I_j| > 1/2. \quad (11.6)$$

Then, the condition $\|f\|_{\mathrm{BMO}} \le 1$ and the maximality of I_j imply $\mathrm{av}(f, I_j) \le C(n)$. So, Lemma 1.9 implies

$$|\{x \in I_j : f(x) > \lambda\}| \le Ce^{-(\lambda-C')/C} |I_j|. \quad (11.7)$$

Thus,

the left-hand side of (11.3) with (11.4)

$$= \sum_j |\{x \in I_j : f(x) > \lambda\}| \qquad \text{by Corollary 0.3}$$

$$\leq \sum_j Ce^{-(\lambda-C')/C} |I_j| \qquad \text{by (11.7)}$$

$$\leq Ce^{-(\lambda-C')/C} \times 2|E| \qquad \text{by (11.6)}$$

$$= \text{the right-hand side of (11.3) with (11.4).}$$

$\square$

In the following part of this section we assume all the conditons of Therem 11.1. Let

$$\Omega = \{x \in \mathbf{R}^n : N'_\delta u(x) > 1\}, \quad \Omega^c = \mathbf{R}^n \backslash \Omega, \tag{11.8}$$

$$W = \bigcup_{x \in \Omega^c} \Gamma(x, \delta). \tag{11.9}$$

Claim 1.
$$|\nabla u(y,t)| \chi_W(y,t) \leq C(\delta, \delta', n)/t. \tag{11.10}$$

Proof. Let $(y,t) \in W$. Then there exists $x \in \Omega^c$ such that $(y,t) \in \Gamma(x,\delta)$. So,

$$\Gamma(x, \delta') \supset \Big\{ (z,s) \in \mathbf{R}^{n+1}_+ : \left(|z-y|^2 + (s-t)^2 \right)^{1/2}$$

$$\leq 2^{-1/2} \min \left\{ (\delta' - \delta)t, \ (1 - \delta/\delta')t \right\} \Big\} = B_0, \ \text{say.}$$

Thus, the harmonicity of u implies

$$|\nabla u(y,t)| \leq C \{\text{radius of } B_0\}^{-1} \sup \{|u(y,t)| : (y,t) \in B_0\}$$
$$\leq Ct^{-1} N_{\delta'} u(x) \leq Ct^{-1} \qquad \text{by } x \in \Omega^c.$$

$\square$

Claim 2.
$$\left\| t|\nabla u|^2 \chi_W \, dy dt \right\|_c \leq C(\delta, \delta', n). \tag{11.11}$$

Proof. Take any ball $B = B(x_B, r_B)$. Let $\varepsilon > 0$. Let

$$\theta(x) = \max \{ \text{dist}(x, \Omega^c)/\delta, \ \text{dist}(x, B)/\delta, \ \varepsilon \}. \tag{11.12}$$

Then, θ satisfies

$$|\theta(x) - \theta(y)| \leq |x - y|/\delta. \tag{11.13}$$

Let

$$\mathcal{R} = \{(x,t) \in \mathbf{R}^{n+1}_+ : x \in \mathbf{R}^n, \ \theta(x) < t < r_B\}. \tag{11.14}$$

(See Fig. 11.1.) Then

$$\{B \times (\varepsilon, r_B)\} \cap W \subset \mathcal{R} \subset \{((1+\delta)B) \times (\varepsilon, r_B)\} \cap W.$$

Let

$$\partial^+ = \{(x, r_B) \in \partial\mathcal{R} : x \in \mathbf{R}^n\},$$
$$\partial^- = \{(x, t) \in \partial\mathcal{R} : x \in \mathbf{R}^n, \ t < r_B\}.$$

Then, noting (11.13) and applying Green's theorem to each connected component of $\mathcal{R}$, we get

$$\iint_{\mathcal{R}} |\nabla u|^2 t \, dy \, dt = \iint_{\mathcal{R}} 2^{-1} \Delta(u^2) t \, dy \, dt$$
$$\leq \int_{\partial^+} \{|u|^2 + |u||\nabla u|t\} \, dx + \int_{\partial^-} \{|u|^2 + |u||\nabla u|t\} \, d\sigma \leq C|B|$$

by $|u| + t|\nabla u| \leq C$ on $\partial\mathcal{R}$ (recall (11.10)) and by (11.13), where

$$d\sigma = \left\{1 + \sum_{i=1}^{n} (\partial\theta/\partial x_i)^2\right\}^{1/2} dx.$$

Thus, letting $\varepsilon \to +0$ gives

$$\iint_{Q(B) \cap W} |\nabla u|^2 t \, dy \, dt \leq C|B|.$$

$\square$

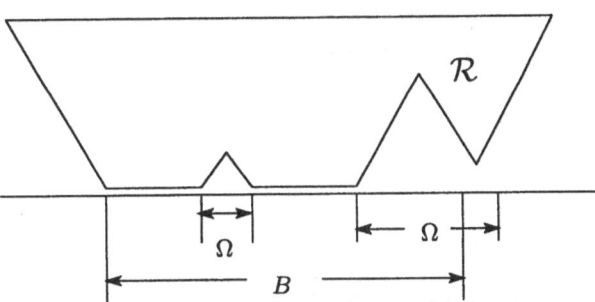

Fig. 11.1: The region $\mathcal{R} = \{(x, t) \in \mathbf{R}_+^{n+1} : x \in \mathbf{R}^n, \ \theta(x) < t < r_B\}$.

Proof of (11.1). We may assume $\lambda = 1$. We may assume $|\{S_\delta(t|\nabla u|)(x) > 1\}| < \infty$. Since $S_\delta(t|\nabla u|\chi_W)(x) \leq S_\delta(t|\nabla u|)(x)$, we may assume $S_\delta(t|\nabla u|\chi_W)(x) \not\equiv \infty$. Then, (11.10)–(11.11) and Lemma 1.12 imply

$$\left\|S_\delta\left(t|\nabla u|\chi_W\right)^2\right\|_{\mathrm{BMO}} \leq C. \tag{11.15}$$

Note that
$$S_\delta(t|\nabla u|)(x) = S_\delta(t|\nabla u|\chi_W)(x) \text{ on } \Omega^c.$$

Therefore,

$$
\begin{aligned}
&|\{S_\delta(t|\nabla u|)(x) > \gamma, \ N_{\delta'}u(x) \le 1\}| \\
&= |\{S_\delta(t|\nabla u|\chi_W)(x) > \gamma, \ N_{\delta'}u(x) \le 1\}| \\
&\le |\{S_\delta(t|\nabla u|\chi_W)(x)^2 > \gamma^2\}| \\
&\le C e^{-(\gamma^2-1)/C} |\{S_\delta(t|\nabla u|\chi_W)(x)^2 > 1\}| \ \text{ by (11.15) and Lemma 11.1} \\
&\le C e^{-(\gamma^2-1)/C} |\{S_\delta(t|\nabla u|)(x) > 1\}|.
\end{aligned}
$$

$\square$

Let
$$\omega = \{x \in \mathbf{R}^n : S_{\delta'}(t|\nabla u|)(x) > 1\}, \ \omega^c = \mathbf{R}^n \backslash \omega, \qquad (11.16)$$
$$w = \bigcup_{x \in \omega^c} \Gamma(x, \delta). \qquad (11.17)$$

Claim 3.

$$|\nabla u(y,t)| \, \chi_w(y,t) \le C(\delta, \delta', n)/t, \qquad (11.18)$$
$$\left\||t|\nabla u|^2 \chi_w \, dy dt\right\|_C \le C(\delta, \delta', n). \qquad (11.19)$$

Proof. Since (11.19) follows from Lemma 5.4, we explain only (11.18). Let $(y, t) \in w$. Then, there exists an $x \in \omega^c$ such that $(y, t) \in \Gamma(x, \delta)$. So,

$$
\begin{aligned}
B_0 \ = \ &\Big\{(z, s) \in \mathbf{R}^{n+1} : \left(|z - y|^2 + (s - t)^2\right)^{1/2} \\
&\le 2^{-1/2} \min\left\{(\delta' - \delta)t, \ (1 - \delta/\delta')t\right\}\Big\} \subset \Gamma(x, \delta').
\end{aligned}
$$

Thus,

$$
\begin{aligned}
|\nabla u(y,t)|^2 \ &\le C \iint_{B_0} |\nabla u(z,s)|^2 \, dzds/t^{n+1} \qquad \text{by the harmonicity of } \nabla u \\
&\le C S_{\delta'}(t|\nabla u|)(x)^2/t^2 \le C/t^2,
\end{aligned}
$$

which implies (11.18). $\square$

Claim 4. Let $N_\delta(u\chi_w)(x) \not\equiv \infty$. Then
$$\|N_\delta(u\chi_w)\|_{\text{BMO}} \le C(\delta, \delta', n). \qquad (11.20)$$

Proof. Take any ball $B = B(x_B, r_B)$. In the following proof, we will fix this B and will show
$$\inf_{c \in \mathbf{R}} \inf_B \int_B |N_\delta(u\chi_w)(x) - c| \, dx/|B| \le C. \qquad (11.21)$$

For $t > 0$ let

$$n'_t(x) = N_\delta\left(u\chi_w\chi_{\mathbf{R}^n \times (t, +\infty)}\right)(x),$$

$$n''_t(x) = N_\delta\left(u\chi_w\chi_{\mathbf{R}^n \times (0, t]}\right)(x).$$

Then, $N_\delta(u\chi_w)(x) = \max\{n'_t(x), n''_t(x)\}$. The condition $N_\delta(u\chi_w) \not\equiv \infty$, (11.18) and the geometric property of w imply $n'_t(x) < \infty$ for all x and t and

$$|n'_t(x) - n'_t(y)| \leq C|x - y|/t. \tag{11.22}$$

So, if $N_\delta(u\chi_w)(x) \equiv n'_{r_B}(x)$ on B, then (11.21) is clear. Thus, we may assume $n''_{r_B}(x) \not\equiv 0$ on B, in particular, we may assume

$$(x_B, r_B(1 + 2\delta)/\delta) \in w. \tag{11.23}$$

Put

$$A = u(x_B, r_B(1 + 2\delta)/\delta).$$

Since (11.23) implies

$$n''_{r_B(1+2\delta)/\delta}(x) \geq |A|$$

and since (11.22) implies

$$\left|n'_{r_B(1+2\delta)/\delta}(x) - n'_{r_B(1+2\delta)/\delta}(y)\right| \leq C \text{ for } x, y \in B,$$

for the proof of (11.21) it is enough to show

$$\int_B \left(n''_{r_B(1+2\delta)/\delta}(x) - |A|\right) dx \leq C|B|. \tag{11.24}$$

We will write

$$n''(x) \text{ for } n''_{r_B(1+2\delta)/\delta}(x)$$

and will let $\mu\ (> 1)$ be large enough.

Since (11.18), (11.23) and the geometrical property of w imply

$$\sup\left\{|u(y, t) - A| : (y, t) \in w \cap \bigcup_{x \in B} \Gamma(x, \delta), \ t \in [r_B/3, r_B(1 + 2\delta)/\delta)\right\}$$

$$\leq C, \tag{11.25}$$

we have that if $x \in B$ and if

$$n''(x) - |A| > \mu,$$

then there exists $(y_x, t_x) \in w \cap \Gamma(x, \delta, r_B/3)$ such that

$$|u(y_x, t_x)| - |A| > \mu.$$

So, the geometrical property of w and (11.18) imply

$$B(x, \delta t_x) \times \{3t_x\} \subset w,$$
$$|u(z, 3t_x)| - |A| > \mu - C \qquad \text{for any } z \in B(x, \delta t_x).$$

Thus, we can take a finite number of disjoint balls $\{B(x_i, 2\delta t_i)\}_{i=1}^m$ such that

$$t_i < r_B/3, \ B(x_i, \delta t_i) \times \{3t_i\} \subset w, \tag{11.26}$$
$$|u(z, 3t_i)| - |A| > \mu - C \text{ for any } z \in B(x_i, \delta t_i), \tag{11.27}$$
$$|\{x \in B : n''(x) - |A| \geq \mu\}| \leq C \left| B \cap \bigcup_{i=1}^m B(x_i, \delta t_i) \right|. \tag{11.28}$$

For $i = 1, 2, \cdots, m$ let

$$\theta_i(x) = \begin{cases} 3t_i & \text{on } B(x_i, \delta t_i), \\ 6t_i - 3|x - x_i|/\delta & \text{on } B(x_i, 2\delta t_i) \backslash B(x_i, \delta t_i), \\ 0 & \text{otherwise.} \end{cases}$$

Let $\varepsilon = 3 \min \{t_1, \cdots, t_m\}$ and

$$\theta(x) = \max \{\theta_1(x), \ \theta_2(x), \ \cdots, \ \theta_m(x), \ \text{dist}(x, \omega^c)/\delta, \ \text{dist}(x, B)/\delta, \ \varepsilon\}.$$

Then, θ satisfies

$$|\theta(x) - \theta(y)| \leq 3|x - y|/\delta, \tag{11.29}$$
$$|\{x \in B : |u(x, \theta(x))| - |A| \geq \mu - C\}| \geq \left| B \cap \bigcup_{i=1}^m B(x_i, t_i) \right|. \tag{11.30}$$

Let

$$\mathcal{R} = \left\{ (x, t) \in \mathbf{R}_+^{n+1} : x \in \mathbf{R}^n, \ \theta(x) < t < r_B \right\}.$$

(See Fig. 11.2.) Then,

$$\mathcal{R} \subset \{((1 + \delta)B) \times (0, r_B)\} \cap w. \tag{11.31}$$

Let

$$\partial^+ = \{(x, r_B) \in \partial \mathcal{R} : x \in \mathbf{R}^n\},$$
$$\partial^- = \{(x, t) \in \partial \mathcal{R} : x \in \mathbf{R}^n, \ t < r_B\}.$$

By (11.25)

$$|u - A| \leq C \text{ on } \partial^+. \tag{11.32}$$

Therefore, noticing (11.29) and applying Green's theorem to each connected component of $\mathcal{R}$, we get

$$\frac{1}{2}\int_{\partial-}|u-A|^2dx \le \iint_{\mathcal{R}}|\nabla u|^2tdydt$$

$$+\int_{\partial+}\left\{|u-A|^2+|u-A||\nabla u|t\right\}dx + \int_{\partial-}|u-A||\nabla u|td\sigma$$

$$\le \iint_{\{((1+\delta)B)\times(0,r_B)\}\cap w}|\nabla u|^2tdydt + C\int_{\partial+}dx + C\int_{\partial-}|u-A|d\sigma$$

by (11.31), (11.32) and (11.18)

$$\le C|B| + C|B|^{1/2}\left\{\int_{\partial-}|u-A|^2dx\right\}^{1/2}$$

by (11.19) and (11.29), where

$$d\sigma = \left\{1+\sum_{i=1}^{n}(\partial\theta/\partial x_i)^2\right\}^{1/2}dx.$$

Thus,

$$\int_{\partial-}|u-A|^2dx \le C|B|. \tag{11.33}$$

Thus,

$$\mu^2\left|\{x \in B : n''(x) - |A| > \mu\}\right|$$
$$\le \mu^2 C\left|\{x \in B : |u(x,\theta(x))| - |A| \ge \mu - C\}\right| \quad \text{by (11.28) and (11.30)}$$
$$\le C|B| \qquad \text{by (11.33)}$$

which implies (11.24). □

Proof of (11.2). We may assume $\lambda = 1$. We may assume $|\{N_\delta u(x) > 1\}| < \infty$. Since $N_\delta(u\chi_w)(x) \le N_\delta u(x)$, we may assume $N_\delta(u\chi_w) \not\equiv +\infty$. Then, we have (11.20). Note that

$$N_\delta(u\chi_w)(x) = N_\delta u(x) \quad \text{on } w^c.$$

So,

$$|\{N_\delta u(x) > \gamma, \ S_{\delta'}u(x) \le 1\}|$$
$$= |\{N_\delta(u\chi_w)(x) > \gamma, \ S_{\delta'}u(x) \le 1\}| \le |\{N_\delta(u\chi_w)(x) > \gamma\}|$$
$$\le Ce^{-(\gamma-1)/C}|\{N_\delta(u\chi_w)(x) > 1\}| \quad \text{by (11.20) and Lemma 11.1.}$$
$$Ce^{-(\gamma-1)/C}|\{N_\delta u(x) > 1\}|.$$

□

Notes. Distribution function inequalities between the S-functions and the nontangential maximal functions of harmonic functions defined on $\mathbf{R}_+^{n+1}$ were first given by D. L. Burkholder–R. F. Gundy [72]. (See also R. F. Gundy–R. Wheeden [74] and C. Fefferman–E. M. Stein [72] p. 161.) R. Fefferman–R. F.

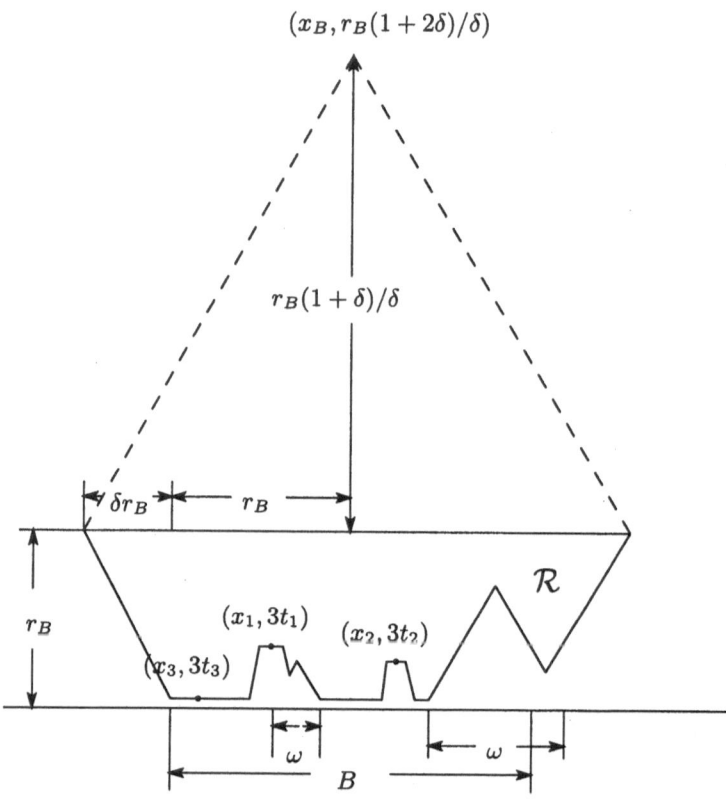

Fig. 11.2: The region $\mathcal{R} = \big\{ (x,t) \in \mathbf{R}_+^{n+1} : x \in \mathbf{R}^n, \ \theta(x) < t < r_B \big\}$.

Gundy–M. Silverstein–E. M. Stein [82] refined these estimates. The argument in this section is a refinement of the above papers and is in T. Murai–A. Uchiyama [86].

Good λ inequalities have been obtain for many operators. For example, see R. Hunt [72], R. Coifman [72], B. Muckenhoupt–R. Wheeden [74] R. Coifman–C. Fefferman [74] and T. Murai [88].

XII. A direct proof of

$$\left| \int_{\mathbf{R}^n} f(x)g(x)dx \right| \le C(n) \, \|S_1(tD_t u)\|_{L^1} \, \|g\|_{\mathrm{BMO}},$$

$$\text{(12.1)}$$

where

$$f, \; g \in L^2, \tag{12.2}$$

$$u(x,t) = \int_{\mathbf{R}^n} P(x-y,t)f(y)dy. \tag{12.3}$$

(In (12.3), $P(x,t)$ denotes the Poisson kernel.) Since $\|S_1(tD_t u)\|_{L^1}$ dominates $\|f\|_{H^1}$ by (10.19), (12.1) follows from Lemma 2.2 (with $p=1$) and Theorem 2.2. In this section, we explain C. Fefferman's direct proof of (12.1), which is one of the oldest proofs of his H^1-BMO duality theorem. (Another one of the oldest proofs will be explained in Section 19.)

We must emphasize that C. Fefferman's discovery of (12.1) was much earlier than the introduction of the atomic H^p which we defined in Section 2. In fact, his discoveries of (12.1) (and (19.1)) caused the introduction of the atomic H^p.

The argument in this section works even if we replace the Poisson kernel in (12.3) by very general kernels.

Lemma 12.1. *Let $g \in$ BMO. Let $\phi \in \mathcal{B}_1^0$ and*

$$v(x,t) = g * (\phi)_t(x). \tag{12.4}$$

Then

$$\left\| v(x,t)^2 dx dt/t \right\|_{\mathcal{C}} \le C_{12.1}(n)\|g\|_{\mathrm{BMO}}^2.$$

Proof. Let $B = B(x_B, r_B)$. Then

$$\iint_{Q(B)} g * (\phi)_t(x)^2 \frac{dx dt}{t} = \iint_{Q(B)} ((g - \mathrm{av}(g, 2B))\chi_{2B}) * (\phi)_t(x)^2 \frac{dx dt}{t}$$

$$\le \iint_{\mathbf{R}_+^{n+1}} \cdots \frac{dx dt}{t} \le C \int_{2B} (g(x) - \mathrm{av}(g, 2B))^2 \, dx \quad \text{by Lemma 3.3}$$

$$\le C\|g\|_{\mathrm{BMO}}^2 |B| \qquad \text{by Lemma 1.9.}$$

$\square$

Lemma 12.2. *Let μ be a positive measure on $\mathbf{R}_+^{n+1}$ such that $\|\mu\|_C \le 1$. Let*

$$C_{12.2}(n) = \Gamma\left((n+2)/2\right)\pi^{-n/2}.$$

Then, there exists a measurable function $h : \mathbf{R}^n \to (0, +\infty]$ such that

$$C_{12.2}\int t^{-n}\chi_{\Gamma(x,1,h(x))}(y,t)dx \ge 1/2 \text{ for any } (y,t) \in \mathbf{R}_+^{n+1}, \qquad (12.5)$$

$$C_{12.2}\iint_{\Gamma(x,1,h(x))} t^{-n}d\mu(y,t) \le 2^{n+1} \text{ for any } x \in \mathbf{R}^n. \qquad (12.6)$$

Proof. Let

$$h(x) = \sup\left\{h > 0 : C_{12.2}\iint_{\Gamma(x,1,h)} t^{-n}d\mu(y,t) \le 2^{n+1}\right\}.$$

Then, (12.6) is clear. Let $B = B(x_B, r_B)$ be any ball. Then, the condition $\|\mu\|_C \le 1$ implies

$$C_{12.2}\int_B dx \iint_{\Gamma(x,1,r_B)} t^{-n}d\mu(y,t) \le \iint_{Q(2B)} d\mu \le 2^n|B|.$$

Thus,

$$\left|\left\{x \in B : C_{12.2}\iint_{\Gamma(x,1,r_B)} t^{-n}d\mu(y,t) > 2^{n+1}\right\}\right| \le |B|/2.$$

So,

$$|\{x \in B : h(x) \ge r_B\}|/|B| \ge 1/2. \qquad (12.7)$$

So,

$$C_{12.2}\int r_B^{-n}\chi_{\Gamma(x,1,h(x))}(x_B, r_B)dx$$

$$= |\{x \in B : h(x) > r_B\}|/|B|$$

$$\ge \frac{|\{x \in (1+\varepsilon)B : h(x) \ge (1+\varepsilon)r_B\}| - |(1+\varepsilon)B\setminus B|}{|(1+\varepsilon)B|} \quad \text{if } \varepsilon > 0$$

$$\ge \frac{1}{2} - \frac{|(1+\varepsilon)B\setminus B|}{|(1+\varepsilon)B|} \quad \text{by (12.7)}$$

$$\to \frac{1}{2} \quad (\varepsilon \to +0). \qquad (12.8)$$

Since the point $(x_B, r_B) \in \mathbf{R}_+^{n+1}$ is arbitrary, (12.8) implies (12.5). $\qquad \square$

Now, we begin the proof of (12.1) under the condition (12.2). Take $\phi \in L^1$ such that $c_{12.3}(\phi)\phi \in \mathcal{B}_1^0$ for some $c_{12.3} > 0$ and

$$\int_0^\infty \mathcal{F}((D_t P)(\cdot, 1))(t\xi)\mathcal{F}\phi(t\xi)dt/t = 1 \text{ for any } \xi \in \mathbf{R}^n \backslash \{0\}.$$

We may assume $g \in \mathrm{BMO} \cap L^2$. Let $u(x,t)$ and $v(x,t)$ be as in (12.3)–(12.4). Applying Lemma 12.2 to $d\mu = v(x,t)^2 dxdt/t$ gives $h(x) > 0$ that satisfies (12.5) and

$$C_{12.2} \iint_{\Gamma(x,1,h(x))} t^{-n}v(y,t)^2 dydt/t \leq 2^{n+1} C_{12.1} c_{12.3}^{-2} \|g\|_{\mathrm{BMO}}^2. \qquad (12.6)'$$

Then

$$\left| \int f(x)g(x)dx \right| = \left| \iint D_t u(y,t) v(y,t) dydt \right| \text{ by Plancherel's theorem}$$

$$\leq 2C_{12.2} \int dx \iint_{\Gamma(x,1,h(x))} t^{-n} |D_t u(y,t)| \, |v(y,t)| \, dydt \text{ by (12.5)}$$

$$\leq 2 \int dx \left\{ C_{12.2} \iint_{\Gamma(x,1)} t^{1-n} |D_t u(y,t)|^2 \, dydt \right\}^{1/2}$$

$$\times \left\{ C_{12.2} \iint_{\Gamma(x,1,h(x))} t^{-n} v(y,t)^2 dydt/t \right\}^{1/2}$$

$$\leq 2 \|S_1(tD_t u)\|_{L^1} (2^{n+1} C_{12.1})^{1/2} c_{12.3}^{-1} \|g\|_{\mathrm{BMO}} \text{ by (12.6)'}.$$

$\square$

Remark 12.1. Since we have already obtained (12.1) in previous sections, in this section we have imposed a strong condition (12.2).

Remark 12.2. Once (12.1) is proved under the condition (12.2), then it is easy to show (12.1) under the condition like

$$f \in \mathcal{S}_0 \text{ and } g \in \mathrm{BMO}. \qquad (12.2)'$$

We will show it. Let $g \in \mathrm{BMO}$. Let $r > 1$. Then

$$\int_{|x|>r} (1 + |x|)^{-n-1} |g(x) - \mathrm{av}(g, B(0,r))| \, dx$$

$$\leq \int_{|x|>r} (1 + |x|)^{-n-1} |g(x)| \, dx + |\mathrm{av}(g, B(0,r))| \int_{|x|>r} (1 + |x|)^{-n-1} dx$$

$$\leq \cdots + (|\mathrm{av}(g, B(0,1))| + C\|g\|_{\mathrm{BMO}} \log(2+r)) \, r^{-1} \text{ by Remark1.3}$$

$$\to 0 \, (r \to +\infty) \text{ by Lemma 1.10.}$$

Let $\phi \in \mathcal{D}$ and $\phi(x) \equiv 1$ near $x = 0$. Let $f \in \mathcal{S}_0$. Then, $fg \in L^1$ by Lemma 1.10 and

$$\left|\int_{\mathbf{R}^n} f(x)g(x)dx\right| = \left|\int f(x)(g(x) - \mathrm{av}(g, B(0,r)))dx\right|$$

$$\left|\lim_{r \to +\infty}\int f(x)\left\{\phi(x/r)(g(x) - \mathrm{av}(g, B(0,r)))\right\}dx\right| \quad \text{by the above estimate}$$

$$\le \mathrm{limsup}_{r \to +\infty}C\,\|S_1(tD_tu)\|_{L^1}\,\|\{\phi(x/r)(g(x) - \mathrm{av}(g, B(0,r)))\}\|_{\mathrm{BMO}}$$
$$\text{by "(12.1) under (12.2)"}$$

$$\le C\,\|S_1(tD_tu)\|_{L^1}\,\|g\|_{\mathrm{BMO}} \qquad \text{by (1.9).}$$

Remark 12.3. If $g \in \mathrm{BMO}$ and if

$$v(x,t) = \int P(x - y, t)g(y)dy, \quad (x,t) \in \mathbf{R}_+^{n+1}, \tag{12.9}$$

then

$$\left\||t\,|\nabla v(x,t)|^2\,dxdt\right\|_\mathcal{C} \le C\|g\|_{\mathrm{BMO}}^2, \tag{12.10}$$

where $\nabla = \nabla_{t,x} = (D_t, D_{x_1}, \cdots, D_{x_n})$.

Proof. If $\phi \in \mathcal{B}_1^0$ and $j \in \mathbf{N}$, then for any ball B

$$\iint_{Q(B)} |(\phi)_{2^j t} * g(x)|^2\,dxdt/t$$

$$= \iint_{Q(B)} |(\phi)_t * g(x)|^2\,dxdt/t + \iint_{B \times (r_B,\ 2^j r_B)} |(\phi)_t * g(x)|^2\,dxdt/t$$

$$\le C\|g\|_{\mathrm{BMO}}^2|B| + Cj\|g\|_{\mathrm{BMO}}^2|B|$$

by Lemma 12.1 and by $|(\phi)_t * g(x)| \le C\|g\|_{\mathrm{BMO}}$. So

$$\left\||(\phi)_{2^j t} * g(x)|^2\,dxdt/t\right\|_\mathcal{C} \le Cj\|g\|_{\mathrm{BMO}}^2. \tag{12.11}$$

Since

$$\|(D_t P)(\cdot, 1)\|_{\mathcal{B},1,1} < \infty, \tag{12.12}$$

and since

$$\int (D_t P)(x, 1)dx = 0,$$

Lemma 3.5 implies that $(D_t P)(x, 1)$ can be written in the form

$$(D_t P)(x, 1) = C\sum_{j=0}^{\infty} 2^{-j}(\phi_j)_{2^j}(x)$$

with

$$\phi_j \in \mathcal{B}_1^0.$$

Thus, for any ball B we have

$$\left\{ |B|^{-1} \iint_{Q(B)} |tD_t P(\cdot,t) * g(x)|^2 \, dx \, dt / t \right\}^{1/2}$$

$$\leq C \sum_{j=0}^{\infty} 2^{-j} \left\{ |B|^{-1} \iint_{Q(B)} |(\phi_j)_{2^j t} * g(x)|^2 \, dx \, dt / t \right\}^{1/2}$$

$$\leq C \sum 2^{-j} (1+j)^{1/2} \|g\|_{\mathrm{BMO}} \qquad \text{by (12.11)}$$

$$\leq C \|g\|_{\mathrm{BMO}}.$$

Similarly we can get

$$\left\{ |B|^{-1} \iint_{Q(B)} |tD_{x_j} P(\cdot,t) * g(x)|^2 \, dx \, dt / t \right\}^{1/2} \leq C \|g\|_{\mathrm{BMO}}.$$

Thus, we get (12.10). $\qquad\qquad\qquad\qquad\qquad\qquad\qquad\qquad\qquad\qquad\square$

Notes. The argument in this section is in C. Fefferman–E. M. Stein [72] p. 148. In their paper, Remark 12.3 is stated in a more complete form that if $\int |g(x)| (1+|x|)^{-n-1} \, dx < \infty$ and if v is defined by (12.9), then

$$\|g\|_{\mathrm{BMO}}^2 \approx \left\| t \, |\nabla v(x,t)|^2 \, dx \, dt \right\|_C . \tag{12.13}$$

More generally, E. B. Fabes–R. L. Johnson–U. Neri [76] showed that if a harmonic function $v(x,t)$ defined on $\mathbf{R}_+^{n+1}$ satisfies

$$\left\| t \, |\nabla v(x,t)|^2 \, dx \, dt \right\|_C < \infty,$$

where $\nabla = \nabla_{t,x} = (D_t, D_{x_1}, \cdots, D_{x_n})$, then v can be written in the form of (12.9) and (12.13) holds.

A. Chang–J. M. Wilson–T. Wolff [85] obtained a very remarkable result that if $\int |g(x)| (1+|x|)^{-n-1} \, dx < \infty$, if $v(x,t)$ is defined by (12.9) and if

$$S_\delta(t \, |\nabla_x v|) \in L^\infty \tag{12.14}$$

for some $\delta > 0$, where $\nabla_x v = (D_{x_1} v, \cdots, D_{x_n} v)$, then

$$|B|^{-1} \int_B \exp\big(c_1 \, |g(x) - \mathrm{av}(g,B)|^2 \, / \, \|S_\delta(t \, |\nabla_x v|)\|_{L^\infty}^2\big) dx \leq c_2 \tag{12.15}$$

for any ball B, where $c_1 > 0$ and $c_2 < \infty$ depend on δ and n. The condition (12.14) is slightly stronger than the condition $\left\| t \, |\nabla_x v(x,t)|^2 \, dx \, dt \right\|_C < \infty$ and the conclusion (12.15) is slightly stronger than Lemma 1.9.

XIII. A direct proof of

$$\left| \int_{\mathbf{R}^n} f(x) g(x) dx \right| \leq C(n) \| N_1 u \|_{L^1} \| g \|_{\mathrm{BMO}}, \qquad (13.1)$$

where u is defined by (12.3) and

$$f \in L^1 \cap L^\infty, \ g \in \mathrm{BMO}, \ \text{and} \ \mathrm{supp}\, g \ \text{is compact}$$
$$(13.2)$$

Since $\| N_1 u \|_{L^1}$ dominates $\| f \|_{H^1}$ by Theorems 4.1 and 9.3, we have already obtained (13.1). In this section, we give a direct proof of (13.1) by modifying the argument of L. Carleson [76] and by using the ideas in N. Th. Varopoulos [77], P. W. Jones [78] and J. B. Garnett–P. W. Jones [82].

The argument in this section works even if we replace the Poisson kernel $P(x,t)$ in (12.3) by very general kernels.

Remark 13.1. Carleson's result is finer than (13.1). He showed (13.1) with $\| N_0 u \|_{L^1}$ in place of $\| N_1 u \|_{L^1}$.

Definition 13.1. For $t_0 > 0$ let $\delta_{t=t_0}$ be the measure induced by the n-dimensional Lebesgue measure on the hyperplane $t = t_0$ in $\mathbf{R}_+^{n+1}$.

Lemma 13.1. *Let $\| g \|_{\mathrm{BMO}} \leq 1$ and $\mathrm{supp}\, g \subset [0,1]^n$. Then, there exists a family of dyadic cubes $\{ I_i \}_{i \in \mathbf{N}}$ in $\mathbf{R}^n$ and constants $\{ a_i \}_{i \in \mathbf{N}}$ such that*

$$I_i \subset [0,1]^n, \qquad (13.3)$$

$$|a_i| \leq C(n), \qquad (13.4)$$

$$\left\| \sum_{i \in \mathbf{N}} |a_i|\, \chi_{I_i} \delta_{t=\ell(I_i)} \right\|_c \leq C(n), \qquad (13.5)$$

$$\left\| g - \sum_{i \in \mathbf{N}} a_i \chi_{I_i} \right\|_{L^\infty} \leq C(n). \qquad (13.6)$$

Proof. Remark 1.9 implies the existence of the constant $C_{13.1}(n)$ such that

$$|\{ x \in I : |g(x) - \mathrm{av}(g, I)| > C_{13.1} \}| < 2^{-n-1}|I| \qquad (13.7)$$

for any cube I. Then,

$$|\mathrm{av}(g, J) - \mathrm{av}(g, I)| \leq 2C_{13.1} \ \text{if} \ J \subset I \ \text{and if} \ \ell(J) = \ell(I)/2.$$

Let

$$I_1 = [0,1]^n.$$

Let $\{ I_{1,i_2} \}_{i_2=1,2,\cdots}$ be the maximal dyadic subcubes of I_1 such that

$$|\mathrm{av}(g, I_{1,i_2}) - \mathrm{av}(g, I_1)| > 2C_{13.1}.$$

Next, we define $\left\{ I_{1,i_2,\cdots,i_{k-1},i_k} \right\}_{i_k=1,2,\cdots}$ inductively to be the collection of the maximal dyadic cubes satisfying

$$I_{1,i_2,\cdots,i_{k-1},i_k} \subset I_{1,i_2,\cdots,i_{k-1}},$$
$$\left\| \mathrm{av}(g, I_{1,i_2,\cdots,i_{k-1},i_k}) - \mathrm{av}(g, I_{1,i_2,\cdots,i_{k-1}}) \right\| > 2C_{13.1}.$$

Then, (13.7) implies

$$\sum_{i_k} |I_{1,i_2,\cdots,i_{k-1},i_k}| \leq \sum_{i_k} (1 - 2^{-n-1})^{-1} \left| \left\{ x \in I_{1,i_2,\cdots,i_{k-1},i_k} : \right. \right.$$
$$\left. \left. \left| g(x) - \mathrm{av}(g, I_{1,i_2,\cdots,i_{k-1},i_k}) \right| \leq C_{13.1} \right\} \right|$$
$$\leq (1 - 2^{-n-1})^{-1} \left| \left\{ x \in I_{1,i_2,\cdots,i_{k-1}} : \right. \right.$$
$$\left. \left. \left| g(x) - \mathrm{av}(g, I_{1,i_2,\cdots,i_{k-1}}) \right| > C_{13.1} \right\} \right|$$
$$\leq (1 - 2^{-n-1})^{-1} 2^{-n-1} \left| I_{1,i_2,\cdots,i_{k-1}} \right| \leq 2^{-1} \left| I_{1,i_2,\cdots,i_{k-1}} \right| \qquad (13.8)$$

and the maximality of $\left\{ I_{1,i_2,\cdots,i_{k-1},i_k} \right\}_{i_k}$ implies

$$\left| \mathrm{av}(g, I_{1,i_2,\cdots,i_{k-1},i_k}) - \mathrm{av}(g, I_{1,i_2,\cdots,i_{k-1}}) \right| \leq 4C_{13.1}. \qquad (13.9)$$

Put

$$a_1 = \mathrm{av}(g, I_1),$$
$$a_{1,i_2,\cdots,i_{k-1},i_k} = \mathrm{av}(g, I_{1,i_2,\cdots,i_{k-1},i_k}) - \mathrm{av}(g, I_{1,i_2,\cdots,i_{k-1}}).$$

Renumbering these $\{I_{1,i_2,\cdots,i_k}\}_{k \in \mathbf{N}, i_2,\cdots,i_k \in \mathbf{N}}$ and $\{a_{1,i_2,\cdots,i_{k-1}}\}$, we get $\{I_i\}_{i \in \mathbf{N}}$ and $\{a_i\}$. Then, (13.3) is clear. (13.4) follows from (13.9) and $|\mathrm{av}(g, [0,1]^n)| \leq C$. We will show (13.5)–(13.6).

Take any ball $B = B(x_0, t_0)$. Since the repeated application of (13.8) implies

$$\sum_{j=1}^{\infty} \sum_{i_{k+1},\cdots,i_{k+j} \in \mathbf{N}} |I_{1,i_2,\cdots,i_k,i_{k+1},\cdots,i_{k+j}}|$$
$$\leq |I_{1,i_2,\cdots,i_k}|, \qquad (13.10)$$

for any $k \in \mathbf{N}$ and any $I_{1,i_2,\cdots,i_k}$, we have

$$\iint_{Q(B)} \sum_{\substack{k \in \{2,3,\cdots\} \\ I_2,\cdots,i_k \in \mathbf{N}}} |a_{1,i_2,\cdots,i_k}| \delta_{t=l}(I_{1,i_2,\cdots,i_k}) \chi_{I_1,i_2,\cdots,i_k}$$

$$\leq \sum_{\substack{I_{1,i_2,\cdots,i_k} \text{ s.t.} \\ \ell(I_{1,i_2,\cdots,i_k}) \leq t_0, \\ I_{1,i_2,\cdots,i_k} \cap B \neq \emptyset}} C |I_{1,i_2,\cdots,i_k}| \text{ by (13.4)}$$

$$\leq \sum_{\substack{\text{maximal } I_{1,i_2,\cdots,i_k} \text{ s.t.} \\ \ell(I_{1,i_2,\cdots,i_k}) \leq t_0, \\ I_{1,i_2,\cdots,i_k} \cap B \neq \emptyset}} C \cdot 2 |I_{1,i_2,\cdots,i_k}| \text{ by (13.10)}$$

$$\leq C|B|,$$

which implies (13.5). If x is a Lebesgue point of f, then

$$\mathrm{av}(f,I) \to f(x) \ (\ell(I) \to +0, I \text{ is a cube containing } x)$$

So, from the above procedure, it follows that

$$I_{1,i_2,\cdots,i_k} \not\ni x \text{ except finitely many } I_{1,i_2,\cdots,i_k}$$

and that if $J_x = I_{1,i_2,\cdots,i_k}$ is the smallest one containing x, then

$$\left| f(x) - \sum_{k \in \mathbf{N}, i_2,\cdots,i_k \in \mathbf{N}} a_{1,i_2,\cdots,i_k} \chi_{I_{1,i_2,\cdots,i_k}}(x) \right| = |f(x) - \mathrm{av}(f,J_x)| \leq 2C_{13.2},$$

which implies (13.6) with $C(n) = 2C_{13.2}$. $\qquad\square$

Lemma 13.2. *Let $\|g\|_{\mathrm{BMO}} \leq 1$ and $\mathrm{supp}\, g \subset [0,1]^n$. Then, there exists a sequence of functions $\{h_j\}_{j=-\infty}^0 \subset L^\infty([0,1]^n)$ such that*

$$\left\| \sum_{j=-\infty}^0 h_j \delta_{t=2^j} \right\|_C \leq C(n), \tag{13.11}$$

$$\left\| g - \sum_{j=-\infty}^0 h_j \right\|_{L^\infty} \leq C(n), \tag{13.12}$$

and such that the derivatives $\nabla_x h_j$, which are taken in the sense of distributions on $\mathbf{R}^n$, are measures on $\mathbf{R}^n$ satisfying

$$\left\| \sum_{j=-\infty}^0 2^j |\nabla_x h_j| \delta_{t=2^j} \right\|_C \leq C(n). \tag{13.13}$$

$$|\nabla_x h_j| \leq C(n) \times \left\{ \begin{array}{l} \text{the singular measure induced by the} \\[4pt] (n-1)\text{-dimentional Lebesgue measure on} \\[4pt] \displaystyle\bigcup_{k\in\{1,\cdots,n\},\ i\in\mathbf{Z}} \left(\mathbf{R}^{k-1} \times \{i2^j\} \times \mathbf{R}^{n-k} \right) \end{array} \right\} (13.14)$$

where

$$|\nabla_x h| = |D_{x_1} h| + \cdots + |D_{x_n} h|.$$

Proof. Let $\{I_i\}_{i\in\mathbf{N}}$ and $\{a_i\}_{i\in\mathbf{N}}$ be as in Lemma 13.1. For $j = 0, -1, -2, \cdots$ put

$$h_j(x) = \sum_{i:\ell(I_i)=2^j} a_i \chi_{I_i}(x).$$

Then, (13.11) and (13.12) are easy from (13.5) and (13.6), respectively. (13.14) is clear. (13.13) follows from the fact that

$$\iint_{Q(B(x_B, r_B))} \ell(I) |\nabla_x \chi_I| \delta_{t=\ell(I)}$$
$$\leq \begin{cases} C(n)|I| & \text{if } I \subset B(x_B, (1+\sqrt{n})r_B), \\ 0 & \text{otherwise,} \end{cases}$$

and from (13.5).   □

Next, we make these h_j's smooth.

Lemma 13.3. Let $\|g\|_{\text{BMO}} \leq 1$ and $\operatorname{supp} g \subset [1/4, 3/4]^n$. Then, there exists a sequence of functions $\{h_j\}_{j=-\infty}^{0} \subset L^\infty(\mathbf{R}^n)$ such that the distributional derivatives $\nabla_x h_j$ belong to $L^\infty(\mathbf{R}^n)$ and such that

$$\left\| \sum_{j=-\infty}^{0} h_j \delta_{t=2^j} \right\|_{\mathcal{C}} \leq C(n), \qquad (13.11)'$$

$$\left\| g - \sum_{j=-\infty}^{0} h_j \right\|_{L^\infty} \leq C(n), \qquad (13.12)'$$

$$\left\| \sum_{j=-\infty}^{0} 2^j \left| \nabla_x h_j \right| \delta_{t=2^j} \right\|_{\mathcal{C}} \leq C(n), \qquad (13.13)'$$

$$\operatorname{supp} h_j \subset [-1/4, 5/4]^n, \qquad (13.15)$$
$$2^j \|\nabla_x h_j\|_{L^\infty} \leq C(n). \qquad (13.16)$$

Proof. Let $z \in B(0, 1/4)$. Applying Lemma 13.2 to $g(\cdot + z)$ gives $\{\widetilde{h}_{j,z}\}_{j=-\infty}^{0} \subset L^\infty([0,1]^n)$ that satisfy (13.11), (13.13), (13.14) and (13.12) with $g(\cdot)$ replaced by $g(\cdot + z)$. Put

$$\widetilde{h}_{j,z}(x) = h_{j,z}(x - z).$$

Then, these $\{\widetilde{h}_{j,z}\}$ satisfy (13.11)–(13.13), (13.15) and

$$\left| \nabla_x \widetilde{h}_{j,z} \right| \leq C(n) \times \Big\{ \text{the singular measure induced by the}$$

$$(n-1)\text{-dimentional Lebesgue measure on}$$

$$\bigcup_{k\in\{1,\cdots,n\},i\in\mathbf{Z}} \left(\mathbf{R}^{k-1} \times \{i2^j + z_k\} \times \mathbf{R}^{n-k} \right) \Big\},$$

$$(13.14)'$$

where

$$z = (z_1, \cdots, z_n).$$

Put

$$h_j(x) = \int_{B(0,\ 1/4)} \widetilde{h}_{j,z}(x)dz / |B(0, 1/4)|, \quad (j = 0, -1, -2, \cdots).$$

Then, these $\{h_j\}$ satisfy $(13.11)'$–$(13.13)'$ and (13.15). (13.16) follows from $(13.14)'$.

The measurablity of $\widetilde{h}_{j,z}(x)$ as a function of (z, x) follows from its construction in Lemmas 13.1–2. $\qquad\square$

Lemma 13.4. *Let* $\|g\|_{\mathrm{BMO}} \leq 1$ *and* $\mathrm{supp}\, g$ *be compact. Then, there exists a finite signed measure* μ *on* $\mathbf{R}_+^{n+1}$ *such that*

$$\left\| g(\cdot) - \iint P(\cdot - y, t)d\mu(y, t) \right\|_{L^\infty} \leq C(n), \qquad (13.17)$$

$$\|\mu\|_{\mathcal{C}} \leq C(n). \qquad (13.18)$$

Proof. Let $\phi \in \mathcal{D}$ be such that

$$c\phi \in \mathcal{B}_2^0 \text{ for some } c > 0 \text{ and that}$$

$$\int_0^{+\infty} \mathcal{F}P(\cdot, t)(\xi)\mathcal{F}\phi(t\xi)dt/t = 1 \text{ for any } \xi \in \mathbf{R}^n\backslash\{0\}. \quad (13.19)$$

By translation and dilation, we may assume $\mathrm{supp}\, g \subset [1/4, 3/4]^n$. Let $\{h_j\}_{j=-\infty}^{0}$ be as in Lemma 13.3. Put

$$w_j(x, t) = (\phi)_t * h_j(x) - (\phi)_t * P(\cdot, 2^j) * h_j(x), \quad j = 0, -1, -2, \cdots,$$

$$(13.20)$$

$$\mu = \sum_j h_j(x)\delta_{t=2^j} + \sum_j w_j(x, t)dxdt/t. \qquad (13.21)$$

Then $\sum_j \iint |w_j| \, dx dt/t \leq \sum \int (|h_j| + 2^j |\nabla h_j|) \, dx < \infty$ will follow from (13.24), (13.26), (13.11)′, (13.13)′ and (13.15). Then (13.17) is clear from (13.12)′ and (13.19)–(13.21). Since we have (13.11)′, for the proof of (13.18) it is enough to show

$$\left\| \sum_j |w_j(x,t)| \, dx dt/t \right\|_{\mathcal{C}} \leq C. \tag{13.22}$$

First, we prepare some estimates for w_j.

Case 1 : $t < 2^j$.

Since

$$|(\phi)_t * h_j(x)| \leq C \int_{B(x,t)} \left| h_j(y) - \int_{B(x,t)} h_j(z)dz/|B(x,t)| \right| dy/|B(x,t)|$$

by $\int (\phi)_t dx = 0$, $\|(\phi)_t\|_{L^\infty} \leq Ct^{-n}$ and $\text{supp}\,(\phi)_t \subset B(x,t)$

$$\leq C \iint_{B(x,t) \times B(x,t)} |h_j(y) - h_j(z)| \, dy dz / t^{2n}$$

$$\leq Ct^{1-n} \int_{B(x,t)} |\nabla h_j(y)| \, dy, \tag{13.23}$$

we have

$$|w_j(x,t)| \leq |(\phi)_t * h_j(x)| + \left| P(\cdot, 2^j) * (\phi)_t * h_j(x) \right|$$

$$\leq Ct^{1-n} \int_{B(x,t)} |\nabla h_j(y)| \, dy + Ct \int_{\mathbf{R}^n} P(x - y, 2^j) \, |\nabla h_j(y)| \, dy. \tag{13.24}$$

In particular, (13.16) and (13.24) imply

$$|w_j(x,t)| \leq Ct2^{-j}. \tag{13.25}$$

Case 2 : $t \geq 2^j$.

If $|x| < 2t$, then

$$|(\phi)_t(x) - (\phi)_t * P(\cdot, 2^j)(x)|$$

$$\leq \int_{B(x,4t)} |(\phi)_t(x) - (\phi)_t(y)| \, P(x - y, 2^j) dy$$

$$+ |(\phi)_t(x)| \int_{B(x,4t)^c} P(x - y, 2^j) dy$$

$$\leq C \int_{B(x,4t)} t^{-n} \left(|x - y|/t\right) P(x - y, 2^j) dy + Ct^{-n}(t/2^j)^{-1}$$

$$\leq Ct^{-n-1} 2^j \log(1 + t/2^j).$$

If $|x| \geq 2t$, then

$$
\begin{aligned}
\left|(\phi)_t(x) - (\phi)_t * P(\cdot, 2^j)(x)\right| &= \left|(\phi)_t * P(\cdot, 2^j)(x)\right| \\
&\leq Ct \sup \left\{ \left|\nabla_y P(x-y, 2^j)\right| : y \in B(0, t) \right\} \\
&\leq Ct 2^{-j(n+1)} \left(1 + |x|/2^j\right)^{-n-2} \\
&\leq C(2^j/t) t^{-n} \left(1 + |x|/t\right)^{-n-1}.
\end{aligned}
$$

So,

$$
\left|(\phi)_t(x) - (\phi)_t * P(\cdot, 2^j)(x)\right| \leq C(2^j/t)^{1/2} t^{-n} \left(1 + |x|/t\right)^{-n-1}
$$

for any $x \in \mathbf{R}^n$. Then,

$$
\begin{aligned}
|w_j(x, t)| &= \left|\left((\phi)_t - (\phi)_t * P(\cdot, 2^j)\right) * h_j(x)\right| \\
&\leq C(2^j/t)^{1/2} \int_{\mathbf{R}^n} t^{-n} \left(1 + |x-y|/t\right)^{-n-1} |h_j(y)| \, dy. \quad (13.26)
\end{aligned}
$$

Now, we begin the proof of (13.22). Let $B = B(x_B, r_B)$. If $2^j \leq r_B$, then (13.24) and (13.26) imply

$$
\begin{aligned}
&\iint_{Q(B)} |w_j(x, t)| \, dx\,dt/t \\
&\leq \iint_{B \times (0, 2^j]} \frac{dx\,dt}{t} \left\{ Ct^{1-n} \int_{B(x, t)} |\nabla h_j(y)| \, dy + Ct \int_{\mathbf{R}^n} P(x-y, 2^j) |\nabla h_j(y)| \, dy \right\} \\
&\quad + \iint_{B \times (2^j, r_B]} \frac{dx\,dt}{t} C(2^j/t)^{1/2} \int_{\mathbf{R}^n} t^{-n} \left(1 + |x-y|/t\right)^{-n-1} |h_j(y)| \, dy \\
&\leq C \iint_{\mathbf{R}^n \times (0, 2^j]} \left(1 + |x_B - y|/r_B\right)^{-n-1} t |\nabla h_j(y)| \, dy\,dt/t \\
&\quad + C \iint_{\mathbf{R}^n \times (2^j, r_B]} C(2^j/t)^{1/2} \left(1 + |x_B - y|/r_B\right)^{-n-1} |h_j(y)| \, dy\,dt/t \\
&\leq C \int_{\mathbf{R}^n} \left(1 + |x_B - y|/r_B\right)^{-n-1} \left(2^j |\nabla h_j(y)| + |h_j(y)|\right) dy. \quad (13.27)
\end{aligned}
$$

Thus,

$$
\begin{aligned}
&\iint_{Q(B)} \sum_{j=-\infty}^{0} |w_j(x, t)| \, dx\,dt/t \\
&= \sum_{j: 2^j > r_B} \iint_{Q(B)} |w_j(x, t)| \, dx\,dt/t + \sum_{j: 2^j \leq r_B} \iint_{Q(B)} \cdots \\
&\leq \sum_{j: 2^j > r_B} \iint_{Q(B)} Ct 2^{-j} \, dx\,dt/t
\end{aligned}
$$

$$+ \sum_{j:2^j \le r_B} \int_{\mathbf{R}^n} C\,(1 + |x_B - y|/r_B)^{-n-1}\,(2^j\,|\nabla h_j(y)| + |h_j(y)|)\,dy$$

$$\text{by (13.25) and (13.27)}$$

$$\le C|B| + C\left\{\left\|\sum_j 2^j\,|\nabla h_j|\,\delta_{t=2^j}\right\|_C + \left\|\sum_j h_j \delta_{t=2^j}\right\|_C\right\}|B|$$

$$\le C|B| \qquad \text{by (13.11)' and (13.13)'.}$$

This concludes the proof of (13.22). $\qquad\qquad\qquad\qquad\qquad\qquad\qquad\square$

Lemma 13.5. *Let μ be a measure on $\mathbf{R}_+^{n+1}$ such that $\|\mu\|_C \le 1$. Let $u \in C(\mathbf{R}_+^{n+1})$. Then,*

$$\iint_{\mathbf{R}_+^{n+1}} |u(x,t)|\,d|\mu|(x,t) \le 2^{n+2}\,\|N_1 u\|_{L^1}.$$

Proof. Let $h : \mathbf{R}^n \to (0, +\infty]$ be as in Lemma 12.2. Then,

$$\iint |u(y,t)|\,d|\mu|(y,t) \le 2C_{12.2}\int dx \iint_{\Gamma(x,1,h(x))} t^{-n}\,|u(y,t)|\,d|\mu|(y,t)$$

$$\text{by (12.5)}$$

$$\le 2C_{12.2}\int N_1 u(x)dx \iint_{\Gamma(x,1,h(x))} t^{-n}d|\mu|(y,t)$$

$$\le 2^{n+2}\,\|N_1 u\|_{L^1} \qquad\qquad \text{by (12.6).}$$

$$\qquad\qquad\qquad\qquad\qquad\qquad\qquad\qquad\qquad\qquad\qquad\qquad\square$$

Now, we begin the proof of (13.1) under the condition (13.2) and $\|g\|_{\text{BMO}} = 1$. Let μ be as in Lemma 13.4. Then,

$$\left|\int f(x)g(x)dx\right| - C\|f\|_{L^1}$$

$$\le \left|\int f(x)dx \iint P(x-y,t)d\mu(y,t)\right| \qquad \text{by (13.17)}$$

$$= \left|\iint P(\cdot,t) * f(y)d\mu(y,t)\right| \qquad \text{by } f \in L^\infty \text{ and } \iint d|\mu| < \infty$$

$$= \left|\iint u(y,t)d\mu(y,t)\right| \qquad \text{where } u \text{ is defined by (12.3)}$$

$$\le 2^{n+2}\,\|N_1 u\|_{L^1}\,\|\mu\|_C \qquad \text{by Lemma 13.5}$$

$$\le C\,\|N_1 u\|_{L^1} \qquad \text{by (13.18).}$$

This concludes the proof of (13.1).

Remark 13.2. Refining the arguments of L. Carleson [76] and J.-O. Strömberg's unpublished work, P. W. Jones [78] showed the following :

Theorem 2.2 of P. W. Jones [78] ; Suppose $f \in \mathrm{BMO}(\mathbf{R}^n)$ and $\|f\|_{\mathrm{BMO}} \leq 1$. Then

$$f = \sum_I a_I \tag{13.28}$$

where

$$a_I \text{ is supported on } 3I, \text{ where } I \text{ is a dyadic cube,} \tag{13.29}$$

$$a_I \text{ is } C^\infty \text{ and } \|\nabla a_I\|_{L^\infty} \leq C_{13.2}(n) \|a_I\|_{L^\infty} / \ell(I), \tag{13.30}$$

$$\sum_{I \subset J} \|a_I\|_{L^\infty} |I| \leq C_{13.3}(n) |J| \text{ for every dyadic cube } J. \tag{13.31}$$

Conversely, if $f = \sum a_I$ satisfies (13.29)–(13.31), then $f \in \mathrm{BMO}(\mathbf{R}^n)$ and

$$\|f\|_{\mathrm{BMO}} \leq C_{13.4}(n) \left(1 + C_{13.2}\right) C_{13.3}.$$

Theorem 4.2 of P. W. Jones [78] ; Let $K(x)$ be a kernel satisfying

$$|K(x)| \leq A \left(1 + |x|\right)^{-n-\varepsilon} \text{ for constants } A, \varepsilon > 0 \tag{13.32}$$

and

$$\int_{\mathbf{R}^n} K(x) dx = 1. \tag{13.33}$$

Suppose $f \in \mathrm{BMO}(\mathbf{R}^n)$. Then there is a measure μ_f on $\mathbf{R}^{n+1}_+$ so that

$$f \text{ is the } K \text{ balayage of } \mu_f \tag{13.34}$$

and

$$\|\mu_f\|_c \leq C_{13.5}(A, \varepsilon, n) \|f\|_{\mathrm{BMO}}. \tag{13.35}$$

Theorem 4.3 of P. W. Jones [78] ; Let $K(x)$ be a kernel satisfying (13.32)–(13.33) and

$$K \text{ is } C^1 \text{ and } |\nabla K(x)| \leq B \left(1 + |x|\right)^{-n-\varepsilon}.$$

Suppose $f \in \mathrm{BMO}(\mathbf{R}^n)$. Then there is a measure μ_f on $\mathbf{R}^{n+1}_+$ so that (13.34) and

$$\sup |\mu|(B \times (0, +\infty))/|B| \leq C_{13.6}(A, B, \varepsilon, n) \|f\|_{\mathrm{BMO}}$$

hold, where the above supremum is taken over all balls B in $\mathbf{R}^n$.

($\sum$ in (13.28) and the balayage in (13.34) can be defined using weak* limits in BMO.) P. W. Jones [78] contains several other important theorems.

Notes. L. Carleson [76] (and A. Uchiyama [80a]) showed Lemma 13.4 with μ of the from

$$\mu(E) = \int_{\{x \in \mathbf{R}^n : (x, t(x)) \in E\}} b(x)dx \qquad \text{for any open set } E \subset \mathbf{R}^{n+1}_+,$$

$b(x)$ and $t(x)$ are measurable functions on $\mathbf{R}^n$,

$\|b\|_{L^\infty} \le C\|g\|_{\mathrm{BMO}}$, $t(x) > 0$.

The results of P. W. Jones [78] (see the last two theorems in Remark 13.2) and E. Amar–A. Bonami [79] are also better than our Lemma 13.4. Their results have

$$g(x) = \iint P(x - y, t)d\mu(y, t) \qquad (13.17)'$$

in place of (13.17). J.-O. Strömberg's unpublished work seems to have played an important role in (13.17)'. (As for Lemma 13.4 see also J. M. Wilson [88].)

Lemma 13.1 seems to be due to J. B. Garnett. Lemma 13.2 is due to N. Th. Varopoulos [77]. The idea of the proof of Lemma 13.3 is due to J. B. Garnett–P. W. Jones [82].

XIV. Subharmonicity, 1

Definition 14.1. For a function $u(x)$ $(= u(x_1, \cdots, x_n))$ and $m \in \mathbf{N}$ let

$$\nabla^m u = \left(D_{x_{i_1}} D_{x_{i_2}} \cdots D_{x_{i_m}} u \right)_{i_1, i_2, \cdots, i_m \in \{1, \cdots, n\}},$$

$$|\nabla^m u| = \left\{ \sum_{i_1=1}^{n} \sum_{i_2=1}^{n} \cdots \sum_{i_m=1}^{n} \left| D_{x_{i_1}} D_{x_{i_2}} \cdots D_{x_{i_m}} u \right|^2 \right\}^{1/2}.$$

The purpose of this section is to show the following which will be applied in Section 17.

Theorem 14.1. *Let* $m \in \mathbf{N}$ *and* $n \in \{2, 3, 4, \cdots\}$. *Let* $u(x)$ $(= u(x_1, \cdots, x_n))$ *be a real-valued harmonic function defined on an open set* $\Omega \subset \mathbf{R}^n$ *and let*

$$|\nabla^m u(x)| \neq 0 \ \text{for any} \ x \in \Omega. \tag{14.1}$$

Then, if $q \in (0, 2]$, *then*

$$\triangle \left(|\nabla^m u|^q \right) \geq \frac{q \{q(m+n-2) - (n-2)\}}{2(m+1) + n - 4} |\nabla^m u|^{q-2} \left| \nabla^{m+1} u \right|^2 \tag{14.2}$$

on Ω. *In particular, if*

$$q \in \left[\frac{n-2}{n+m-2}, +\infty \right) \cap (0, +\infty), \tag{14.3}$$

then

$$\triangle \left(|\nabla^m u|^q \right) \geq 0. \tag{14.4}$$

Almost all part of this section will be devoted to the proof of the following lemma.

Lemma 14.1. *Let* $m, n \in \{2, 3, 4, \cdots\}$. *Let*

$$\left\{ a_{i_1, i_2, \cdots, i_m} \right\}_{i_1, i_2, \cdots, i_m \in \{1, 2, \cdots, n\}} \subset \mathbf{R}$$

be such that

$$a_{i_1, \cdots, i_m} = a_{i_{\tau(1)}, \cdots, i_{\tau(m)}} \tag{14.5}$$

for any permutation τ *of* $\{1, 2, \cdots, m\}$ *and such that*

$$\sum_{i=1}^{n} a_{i,i,i_3,\cdots,i_m} = 0 \tag{14.6}$$

for any $i_3,\cdots,i_m \in \{1,2,\cdots,n\}$. *Then,*

$$\sum_{i_2=1}^{n}\cdots\sum_{i_m=1}^{n} a_{1,i_2,\cdots,i_m}^2 \le \frac{m+n-3}{2m+n-4} \sum_{i_1=1}^{n}\sum_{i_2=1}^{n}\cdots\sum_{i_m=1}^{n} a_{i_1,i_2,\cdots,i_m}^2. \tag{14.7}$$

The case $n = 2$ and the case $m = 2$ are easy.

The proof of Lemma 14.1 for the case $n = 2$. Since (14.5) and (14.6) imply

$$a_{1,i_2,i_3,\cdots,i_m} = \begin{cases} a_{2,1,i_3,\cdots,i_m} & \text{if } i_2 = 2, \\ -a_{2,2,i_3,\cdots,i_m} & \text{if } i_2 = 1, \end{cases}$$

we have

$$\sum_{i_2}\cdots\sum_{i_m} a_{1,i_2,\cdots,i_m}^2 = \sum_{i_2}\cdots\sum_{i_m} a_{2,i_2,\cdots,i_m}^2 = \frac{1}{2}\sum_{i_1}\sum_{i_2}\cdots\sum_{i_m} a_{i_1,i_2,\cdots,i_m}^2.$$

$\square$

The proof of Lemma 14.1 for the case $m = 2$.

$$\frac{\displaystyle\sum_{i=1}^{n} a_{1,i}^2}{\displaystyle\sum_{i=1}^{n}\sum_{j=1}^{n} a_{i,j}^2} = \frac{\displaystyle\sum_{i=2}^{n} a_{1,i}^2 + a_{1,1}^2}{\displaystyle\sum_{i=2}^{n}(a_{1,i}^2 + a_{i,1}^2) + \sum_{i=1}^{n} a_{i,i}^2 + \sum_{i,j\ge 2: i\ne j} a_{i,j}^2}$$

$$\le \max\left\{\frac{\displaystyle\sum_{i=2}^{n} a_{1,i}^2}{\displaystyle\sum_{i=2}^{n}(a_{1,j}^2 + a_{i,1}^2)}, \frac{a_{1,1}^2}{\displaystyle\sum_{i=1}^{n} a_{i,i}^2}\right\}$$

$$= \max\left\{\frac{1}{2}, \frac{\left(\displaystyle\sum_{i=2}^{n} a_{i,i}\right)^2}{\left(\displaystyle\sum_{i=2}^{n} a_{i,i}\right)^2 + \displaystyle\sum_{i=2}^{n} a_{i,i}^2}\right\} \qquad \text{by (14.5)–(14.6)}$$

$$\le \frac{n-1}{n} \qquad \text{by } \left(\sum_{i=2}^{n} a_{i,i}\right)^2 \le (n-1)\sum_{i=2}^{n} a_{i,i}^2.$$

$\square$

For the proof of the case "$n \ne 2$ and $m \ne 2$" of Lemma 14.1 we need a lot of preparations.

Definition 14.2. We fix the dimension $n \geq 2$. For $k \in \mathbf{N}$ let

$$\mathcal{P}_k = \left\{ \sum_{i_1=1}^{n} \sum_{i_2=1}^{n} \cdots \sum_{i_k=1}^{n} a_{i_1,i_2,\cdots,i_k} x_{i_1} x_{i_2} \cdots x_{i_k} : a_{i_1,i_2,\cdots,i_k} \in \mathbf{R} \right\},$$

$$\mathcal{P}_0 = \mathbf{R}.$$

For $k \in \mathbf{N}_0$ let

$$\mathcal{H}_k = \left\{ P \in \mathcal{P}_k : \Delta P \left(= \sum_{i=1}^{n} D_{x_i}^2 P \right) \equiv 0 \right\}.$$

Definition 14.3. Let

$$S^{n-1} = \left\{ x = (x_1, \cdots, x_n) \in \mathbf{R}^n : x_1^2 + \cdots + x_n^2 = 1 \right\},$$

let $d\sigma_{n-1}$ be the area measure of the surface S^{n-1}. Note that

$$\sigma_{n-1}\left(S^{n-1}\right) = \frac{2\pi^{n/2}}{\Gamma(n/2)}.$$

The following is clear.

Lemma 14.2. *If* $P \in \mathcal{P}_k$, *then*

$$\frac{\partial P}{\partial n} = kP \text{ on } S^{n-1}, \tag{14.8}$$

$$\int_{B(0,1)} P(x)dx = \frac{1}{k+n} \int_{S^{n-1}} P(x)d\sigma_{n-1}(x), \tag{14.9}$$

where $\dfrac{\partial}{\partial n}$ *is the outward normal derivative.*

Definition 14.4. For real-valued continuous functions P, Q defined on S^{n-1} let

$$(P,Q) = \int_{S^{n-1}} P(x)Q(x)d\sigma_{n-1}(x).$$

Lemma 14.3. *Let* $n \geq 2$, $P(x_1, \cdots, x_n) \in \mathcal{H}_{m_1}$, $Q(x_1, \cdots, x_n) \in \mathcal{H}_{m_2}$ *and* $m_1 \neq m_2$. *Then*

$$(P,Q) = 0.$$

Proof.

$$(m_1 - m_2)(P,Q) = (\partial P/\partial n, \ Q) - (P, \ \partial Q/\partial n) \qquad \text{by (14.8)}$$

$$= \int_{B(0,1)} (Q\Delta P - P\Delta Q)dx \qquad \text{by Green's theorem}$$

$$= 0 \text{ by } \Delta P \equiv \Delta Q \equiv 0.$$

$\square$

Lemma 14.4. *Let $m \in \mathbf{N}$ and $n \in \{2, 3, 4, \cdots\}$. Let*

$$\{a_{i_1, \cdots, i_m}\}_{i_1, \cdots, i_m \in \{1, \cdots, n\}}$$

satisfy (14.5)–(14.6). Let

$$P(x_1, \cdots, x_n) = \frac{1}{m!} \sum_{i_1=1}^{n} \cdots \sum_{i_m=1}^{n} a_{i_1, \cdots, i_m} x_{i_1} \cdots x_{i_m}. \tag{14.10}$$

Then,

$$P \in \mathcal{H}_m, \tag{14.11}$$

$$(P, P) = \frac{\Gamma(n/2)\sigma_{n-1}(S^{n-1})}{2^m \Gamma(m+n/2)\Gamma(m+1)} \sum_{i_1=1}^{n} \cdots \sum_{i_m=1}^{n} a_{i_1, \cdots, i_m}^2. \tag{14.12}$$

Proof. First, note that

$$P \in \mathcal{P}_m, \tag{14.13}$$

$$D_{x_{i_1}} \cdots D_{x_{i_m}} P = a_{i_1, \cdots, i_m}. \tag{14.14}$$

(14.14) and (14.6) imply

$$D_{x_{i_3}} \cdots D_{x_{i_m}} (\Delta P) \equiv 0$$

for any $i_3, \cdots, i_m \in \{1, \cdots, n\}$, which combined with (14.13) implies $\Delta P \equiv 0$, i.e. (14.11).

Next, note that

$$(P, P) = (P^2, 1) = (2m)^{-1} \left(\partial(P^2)/\partial n, 1 \right) \quad \text{(14.13) and (14.8)}$$

$$= (2m)^{-1} \int_{B(0,1)} \Delta(P^2) dx$$

$$= m^{-1} \int_{B(0,1)} \sum_{i=1}^{n} (D_{x_i} P)^2 \, dx \quad \text{by the harmonicity of } P$$

$$= m^{-1}(2m - 2 + n)^{-1} \sum_{i=1}^{n} (D_{x_i} P, D_{x_i} P) \quad \text{by (14.9).}$$

Repeating this argument once more gives

$$(P, P) = (2m\,(m + (n-2)/2))^{-1} (2(m-1)\,(m-1 + (n-2)/2))^{-1}$$

$$\times \sum_{i_1=1}^{n} \sum_{i_2=1}^{n} (D_{x_{i_1}} D_{x_{i_2}} P, \; D_{x_{i_1}} D_{x_{i_2}} P).$$

So, repeating this argument m times and using (14.14), we get (14.12). $\qquad\square$

Definition 14.5. For $\nu \in \{k/2 : k \in \mathbb{N}\}$ and for $k \in \mathbb{N}_0$ let $C_k^\nu(s)$ be the polynomial of s of degree k such that

$$(1 - 2st + t^2)^{-\nu} = \sum_{k=0}^{\infty} C_k^\nu(s)t^k \text{ if } 2\,|st| + t^2 < 1. \qquad (14.15)$$

Lemma 14.5. *Let* $\nu \in \{j/2 : j \in \mathbb{N}\}$ *and* $k \in \mathbb{N}_0$. *Then,*

$$C_k^\nu(s) = \frac{2^k \Gamma(\nu + k)}{\Gamma(k+1)\Gamma(\nu)} s^k + \text{Const.}s^{k-2} + \text{Const.}s^{k-4} + \cdots, \qquad (14.16)$$

$$C_k^\nu(1) = \frac{\Gamma(2\nu + k)}{\Gamma(k+1)\Gamma(2\nu)}, \qquad (14.17)$$

$$(1 - s^2)D_s^2 C_k^\nu(s) - (2\nu + 1)sD_s C_k^\nu(s) + k(2\nu + k)C_k^\nu(s) = 0, \qquad (14.18)$$

$$(1 - s^2)D_s C_k^\nu(s) + ksC_k^\nu(s) - (k + 2\nu - 1)C_{k-1}^\nu(s) = 0 \text{ if } k \neq 0. (14.19)$$

Proof. (14.16) and (14.17) are easy. Let

$$F(s,t) = (1 - 2st + t^2)^{-\nu}$$

Then, (14.18) and (14.19) follow from

$$(1 - s^2)D_s^2 F - (2\nu + 1)sD_s F + t^2 D_t^2 F + (2\nu + 1)tD_t F = 0$$

and

$$(1 - s^2)D_s F + (st - t^2)D_t F - 2\nu t F = 0,$$

respectively. □

Lemma 14.6.

$$C_k^\nu(s) = \frac{2^k \Gamma(\nu + k)(-1)^k \Gamma(k + 2\nu)}{\Gamma(k+1)\Gamma(\nu)\Gamma(2k + 2\nu)}(1 - s^2)^{-\nu+1/2} D_s^k \left\{(1 - s^2)^{k+\nu-1/2}\right\}. \qquad (14.20)$$

Proof. Put

$$y(s) = (1 - s^2)^{-\nu+1/2} D_s^k \left\{(1 - s^2)^{k+\nu-1/2}\right\}.$$

Then, it is not hard to show that this is a polynomial of degree k having the form

$$y(s) = (-1)^k \frac{\Gamma(2k + 2\nu)}{\Gamma(k + 2\nu)} \left\{s^k + \text{Const.}s^{k-2} + \text{Const.}s^{k-4} + \cdots\right\} \qquad (14.21)$$

and that this satisfies the differential equatoin

$$(1 - s^2)D_s^2 y(s) - (2\nu + 1)sD_s y(s) + k(2\nu + k)y(s) = 0. \qquad (14.22)$$

Put

$$p(s) = \{\text{the left-hand side of (14.20)}\}$$
$$-\{\text{the right-hand side of (14.20)}\}.$$

Then, (14.16), (14.18), (14.21) and (14.22) imply that $p(s)$ is a polynomial of the form

$$p(s) = a_{k-2}s^{k-2} + a_{k-4}s^{k-4} + a_{k-6}s^{k-6} + \cdots$$

and satisfies the differential equation (14.22). So,

$$a_i = \frac{(i+1)(i+2)}{(i-k)(i+2\nu+k)}a_{i+2} \text{ for } i = k-2,\ k-4,\ k-6,\cdots$$

with $a_k = 0$, which implies $p \equiv 0$ i.e. (14.20).   □

Lemma 14.7.

$$\int_{[-1,1]} C_k^\nu(s)C_j^\nu(s)(1-s^2)^{\nu-1/2}ds = \delta_{kj}\frac{2^{1-2\nu}\pi\Gamma(k+2\nu)}{\Gamma(k+1)(k+\nu)\Gamma(\nu)^2},$$

where δ_{kj} is Kronecker's delta.

Proof. If $i \in \{0,1,2,\cdots,k-1\}$, then

$$\int_{[-1,1]} s^i D_s^k(1-s^2)^{k+\nu-1/2}ds = 0 \tag{14.23}$$

by integrations by parts. So, the case $j \neq k$ is clear from (14.20). The case $j = k$ follows from

$$\int_{[-1,1]} C_k^\nu(s)C_k^\nu(s)(1-s^2)^{\nu-1/2}ds$$

$$= \frac{2^k\Gamma(k+\nu)}{\Gamma(k+1)\Gamma(\nu)}\int_{[-1,1]} s^k C_k^\nu(s)(1-s^2)^{\nu-1/2}ds \text{ by (14.16) and (14.23)}$$

$$= \left\{\frac{2^k\Gamma(k+\nu)}{\Gamma(k+1)\Gamma(\nu)}\right\}^2\frac{(-1)^k\Gamma(k+2\nu)}{\Gamma(2k+2\nu)}\int_{[-1,1]} s^k D_s^k(1-s^2)^{k+\nu-1/2}ds$$

$$\text{by (14.20)}$$

$$= \left\{\frac{2^k\Gamma(k+\nu)}{\Gamma(k+1)\Gamma(\nu)}\right\}^2\frac{(-1)^k\Gamma(k+2\nu)}{\Gamma(2k+2\nu)}\frac{(-1)^k k!\Gamma(k+\nu+1/2)\Gamma(1/2)}{\Gamma(k+\nu+1)}$$

and from

$$\Gamma(2k+2) = \frac{2^{2k+2\nu-1}\Gamma(k+\nu)\Gamma(k+\nu+1/2)}{\Gamma(1/2)}.$$

□

Lemma 14.8. *Let* $n \in \{3, 4, 5, \cdots\}$, $m \in \mathbf{N}_0$, $\mu \in \{0, 1, \cdots, m\}$ *and* $H_\mu(x_2, \cdots, x_n) \in \mathcal{H}_\mu$. *Then,*

$$|x|^{m-\mu} C_{m-\mu}^{\mu+(n-2)/2}(x_1/|x|) H_\mu(x_2, \cdots, x_n) \tag{14.24}$$

belongs to $\mathcal{H}_m$.

Proof. (14.16) implies that (14.24) $\in \mathcal{P}_m$. By computing the Laplacian directly we can show that

$$|x|^{-2\mu-(n-2)} H_\mu(x_2, \cdots, x_n)$$

is harmonic on $\mathbf{R}^n \backslash \{0\}$. Thus, if $t \in \mathbf{R}$, then

$$|(1, 0, \cdots, 0) - tx|^{-2\mu-(n-2)} H_\mu(x_2, \cdots, x_n) \tag{14.25}$$

is harmonic of $B\left(0, 1/|t|\right)$ $(\subset \mathbf{R}^n)$. Since

$$(14.25) = \left\{ 1 - 2 \left(t|x| \right) \left(x_1/|x| \right) + \left(t/|x| \right)^2 \right\}^{-\mu-(n-2)/2} H_\mu$$

$$= \sum_{k=0}^{\infty} C_k^{\mu+(n-2)/2} \left(x_1/|x| \right) \left(t|x| \right)^k H_\mu \quad \text{by (14.15)},$$

we get

$$(14.24) = \frac{1}{(m-\mu)!} D_t^{m-\mu} D_t (14.25) \mid_{t=0}.$$

Thus, (14.24) is harmonic. $\qquad\qquad\square$

Lemma 14.9. *Let* $n \in \{3, 4, 5, \cdots\}$, $m \in \mathbf{N}_0$ *and*

$$P(x_1, \cdots, x_n) \in \mathcal{H}_m. \tag{14.26}$$

Then, there exist

$$H_\mu(x_2, \cdots, x_n) \in \mathcal{H}_\mu \quad (\mu = 0, 1, \cdots, m) \tag{14.27}$$

such that

$$P(x_1, \cdots, x_n) = \sum_{\mu=0}^{m} |x|^{m-\mu} C_{m-\mu}^{\mu+(n-2)/2}(x_1/|x|) H_\mu(x_2, \cdots, x_n). \tag{14.28}$$

Proof. Let the degree of P as a polynomial of x_1 be $m - \mu_1$. We may assume $m - \mu_1 \geq 1$. Then, there exists a polynomial $Q_{\mu_1}(x_2, \cdots, x_n) \in \mathcal{P}_{\mu_1}$ such that

$$\{\text{the degree of } P - x_1^{m-\mu_1} Q_{\mu_1} \text{ as a polynomial of } x_1\}$$
$$\leq m - \mu_1 - 1. \tag{14.29}$$

The condition (14.26) implies

$$Q_{\mu_1} \in \mathcal{H}_{\mu_1}. \tag{14.30}$$

Put

$$H_{\mu_1}(x_2, \cdots, x_n) = C_{m-\mu_1}^{\mu_1+(n-2)/2}(1)^{-1} Q_{\mu_1}(x_2, \cdots, x_n).$$

Then, (14.26), (14.30) and Lemma 14.8 imply

$$P - |x|^{m-\mu_1} C_{m-\mu_1}^{\mu_1+(n-2)/2}(x_1/|x|) H_{\mu_1} \in \mathcal{H}_m. \tag{14.31}$$

Since

$$\left\{ \text{the coefficient of } x_1^{m-\mu_1} \text{ of } |x|^{m-\mu_1} C_{m-\mu_1}^{\mu_1+(n-2)/2}(x_1/|x|) \right\}$$
$$= C_{m-\mu_1}^{\mu_1+(n-2)/2}(1),$$

(14.29) implies

$$\{\text{the degree of (14.31) as a polynomial of } x_1\} \le m - \mu_1 - 1.$$

If (14.31) contains x_1, then we apply the same argument to (14.31) and get $\mu_2 \in \{\mu_1 + 1, \ \mu_1 + 2, \ \cdots, \ m - 1\}$ and $H_{\mu_2}(x_2, \cdots, x_n) \in \mathcal{H}_{\mu_2}$ such that

$$\Big[\text{the degree of}$$
$$\left\{ P - |x|^{m-\mu_1} C_{m-\mu_1}^{\mu_1+(n-2)/2}(x_1/|x|) H_{\mu_1}(x_2, \cdots, x_n) \right.$$
$$\left. - |x|^{m-\mu_2} C_{m-\mu_2}^{\mu_2+(n-2)/2}(x_1/|x|) H_{\mu_2}(x_2, \cdots, x_n) \right\}$$
$$\text{as a polynomial of } x_1 \Big] \le m - \mu_2 - 1.$$

Repeating this argument at most $m - \mu_1$ times, we get the decomposition (14.28). □

Lemma 14.10. *Let* $n \in \{3, 4, 5, \cdots\}$ *and* $m \in \mathbf{N}_0$. *For* $\mu = 0, 1, \cdots, m$ *let*

$$H_\mu(x_2, \cdots, x_n) \in \mathcal{H}_\mu, \tag{14.32}$$
$$W_\mu(x_1, \cdots, x_n) = |x|^{m-\mu} C_{m-\mu}^{\mu+(n-2)/2}(x_1/|x|) H_\mu(x_2, \cdots, x_n). \tag{14.33}$$

Let $\mu, \nu \in \{0, 1, \cdots, m\}$. *Then*

$$(W_\mu, W_\nu) = \delta_{\mu\nu} \frac{2^{3-2\mu-n} \pi \Gamma(m+\mu+n+2)}{\Gamma(m-\mu+1)(m-1+n/2)\Gamma(\mu+(n-2)/2)^2}$$
$$\times \int_{S^{n-2}} H_\mu(x_2, \cdots, x_n)^2 d\sigma_{n-2}, \tag{14.34}$$

$$(D_{x_1}W_\mu,\ D_{x_1}W_\nu) = \begin{cases} \delta_{\mu\nu}\dfrac{(m+\mu+n-3)^2 2^{3-2\mu-n}\pi\Gamma(m+\mu+n-3)}{\Gamma(m-\mu)(m-2+n/2)\Gamma(\mu+(n-2)/2)^2} \\ \qquad \times \int_{S^{n-2}} H_\mu(x_2,\cdots,x_n)^2 d\sigma_{n-2} \ \ \text{if } m>0, \\ 0 \qquad\qquad\qquad\qquad\qquad\qquad\quad \text{if } m=0, \end{cases}$$

$$(14.35)$$

$$\frac{(D_{x_1}W_\mu,\ D_{x_1}W_\mu)}{(W_\mu,W_\mu)} = \begin{cases} \dfrac{(m+\mu+n-3)(m-\mu)(2m+n-2)}{2m+n-4} & \text{if } m>0, \\ 0 & \text{if } m=0, \end{cases}$$

$$(14.36)$$

where we define $\dfrac{1}{\Gamma(0)} = 0$.

(14.36) follows from (14.34)–(14.35).

Proof of (14.34). Note that

$$
\begin{aligned}
(W_\mu, W_\nu) &= \int_{[-1,1]} C_{m-\mu}^{\mu+(n-2)/2}(x_1) C_{m-\nu}^{\nu+(n-2)/2}(x_1)\left(1-x_1^2\right)^{-1/2} dx_1 \\
&\quad \times \int_{\{(x_2,\cdots,x_n):x_2^2+\cdots+x_n^2=1-x_1^2\}} H_\mu(x_2,\cdots,x_n)H_\nu(x_2,\cdots,x_n)d\sigma_{n-2} \\
&= \int_{[-1,1]} C_{m-\mu}^{\mu+(n-2)/2}(x_1) C_{m-\nu}^{\nu+(n-2)/2}(x_1)\left(1-x_1^2\right)^{(n-3+\mu+\nu)/2} dx_1 \\
&\quad \times \int_{S^{n-2}} H_\mu H_\nu d\sigma_{n-2}.
\end{aligned}
$$

Thus, the desired result follows from Lemmas 14.3 and 14.7. $\square$

Proof of (14.35). If $\mu = m$, then this is clear. If $m - \mu > 0$, then

$$
\begin{aligned}
D_{x_1}&\left\{|x|^{m-\mu}C_{m-\mu}^{\mu+(n-2)/2}(x_1/|x|)\right\} \\
&= (m-\mu)|x|^{m-\mu-2}x_1 C_{m-\mu}^{\mu+(n-2)/2}(x_1/|x|) \\
&\quad + |x|^{m-\mu}\left(D_s C_{m-\mu}^{\mu+(n-2)/2}\right)(x_1/|x|)\left(|x|^{-1}-x_1^2|x|^{-3}\right) \\
&= |x|^{m-\mu-1}\Big\{(m-\mu)(x_1/|x|)C_{m-\mu}^{\mu+(n-2)/2}(x_1/|x|) \\
&\qquad + \left(1-(x_1/|x|)^2\right)\left(D_s C_{m-\mu}^{\mu+(n-2)/2}\right)(x_1/|x|)\Big\} \\
&= |x|^{m-\mu-1}(m-\mu+2\mu+(n-2)-1)C_{m-\mu-1}^{\mu+(n-2)/2}(x_1/|x|)
\end{aligned}
$$

by (14.19). So,

$$D_{x_1}W_\mu = (m+\mu+n-3)|x|^{m-1-\mu}C_{m-1-\mu}^{\mu+(n-2)/2}(x_1/|x|)H_\mu(x_2,\cdots,x_n).$$

Thus, applying (14.34) to the right-hand side of the above formula, we get (14.35). $\square$

Lemma 14.11. *Let* $n \in \{3, 4, 5, \cdots\}$ *and* $m \in \mathbf{N}$. *Then*

$$\sup\left\{\frac{(D_{x_1}P, D_{x_1}P)}{(P, P)} : P \in \mathcal{H}_m\right\} = \frac{(m + n - 3)m(2m + n - 2)}{2m + n - 4}.$$

Proof. Let

$$P_0(x_1, \cdots, x_n) = |x|^m C_m^{(n-2)/2}(x_1/|x|).$$

Then, (14.36) implies

$$\frac{(D_{x_1}P_0, D_{x_1}P_0)}{(P_0, P_0)} = \frac{(m + n - 3)m(2m + n - 2)}{2m + n - 4}. \tag{14.37}$$

On the other hand, Lemma 14.9 implies that any $P(x_1, \cdots, x_n) \in \mathcal{H}_m$ can be written in the form

$$P = \sum_{\mu=0}^{m} W_\mu$$

with (14.33) and (14.27). Thus,

$$\frac{(D_{x_1}P, D_{x_1}P)}{(P, P)} = \frac{\displaystyle\sum_{\mu=0}^{m-1}(D_{x_1}W_\mu, D_{x_1}W_\mu)}{\displaystyle\sum_{\mu=0}^{m}(W_\mu, W_\mu)} \quad \text{by (14.34)–(14.35)}$$

$$\leq \max_{0 \leq \mu \leq m-1}\frac{(D_{x_1}W_\mu, D_{x_1}W_\mu)}{(W_\mu, W_\mu)} = \frac{(m + n - 3)m(2m + n - 2)}{2m + n - 4} \quad \text{by (14.36)}.$$

$$\tag{14.38}$$

So, (14.37) and (14.38) imply the desired result. $\qquad\qquad\square$

Proof of Lemma 14.1 for the case $n \geq 3$. Let $P(x_1, \cdots, x_n)$ be as in (14.10). Then, by (14.5)

$$D_{x_1}P = \frac{1}{(m-1)!}\sum_{i_2=1}^{n}\cdots\sum_{i_m=1}^{n}a_{1,i_2,\cdots,i_m}x_{i_2}\cdots x_{i_m}.$$

Thus, (14.12) imples

$$\frac{(D_{x_1}P, D_{x_1}P)}{(P, P)} = \frac{\dfrac{\Gamma(n/2)\sigma_{n-1}(S^{n-1})}{2^{m-1}\Gamma(m-1+n/2)\Gamma(m)}\displaystyle\sum_{i_2}\cdots\sum_{i_m}a_{1,i_2,\cdots,i_m}^2}{\dfrac{\Gamma(n/2)\sigma_{n-1}(S^{n-1})}{2^{m}\Gamma(m+n/2)\Gamma(m+1)}\displaystyle\sum_{i_1}\sum_{i_2}\cdots\sum_{i_m}a_{i_1,i_2,\cdots,i_m}^2}$$

$$= 2(m-1+n/2)m\frac{\displaystyle\sum_{i_2}\cdots\sum_{i_m}a_{1,i_2,\cdots,i_m}^2}{\displaystyle\sum_{i_1}\sum_{i_2}\cdots\sum_{i_m}a_{i_1,i_2,\cdots,i_m}^2},$$

which combined with Lemma 14.11 implies (14.7). $\qquad\qquad\square$

Lemma 14.12. *Assume all the conditions of Lemma 14.1. Let*

$$\{b_{i_2,\cdots,i_m}\}_{i_2,\cdots,i_m \in \{1,\cdots,n\}} \subset \mathbf{R}.$$

Then,

$$\sum_{i_1=1}^{n}\left\{\sum_{i_2=1}^{n}\cdots\sum_{i_m=1}^{n} a_{i_1,i_2,\cdots,i_m} b_{i_2,\cdots,i_m}\right\}^2$$

$$\leq \frac{m+n-3}{2m+n-4}\left\{\sum_{i_1=1}^{n}\sum_{i_2=1}^{n}\cdots\sum_{i_m=1}^{n} a_{i_1,i_2,\cdots,i_m}^2\right\}\left\{\sum_{i_2=1}^{n}\cdots\sum_{i_m=1}^{n} b_{i_2,\cdots,i_m}\right\}.$$

$$(14.39)$$

Proof. Let

$$(v_{i,j})_{i,j \in \{1,\cdots,n\}}$$

be any orthogonal matrix and let

$$a'_{i_1,\cdots,i_m} = \sum_{j_1=1}^{n}\cdots\sum_{j_m=1}^{n} a_{j_1,\cdots,j_m} v_{i_1,j_1}\cdots v_{i_m,j_m}.$$

Then, this $\{a'_{i_1,\cdots,i_m}\}$ satisfies (14.5), (14.6),

$$\sum_{i_1=1}^{n}\cdots\sum_{i_m=1}^{n} a'^2_{i_1,i_2,\cdots,i_m} = \sum_{i_1=1}^{n}\cdots\sum_{i_m=1}^{n} a_{i_1,\cdots,i_m}^2, \qquad (14.40)$$

$$\sum_{i_2=1}^{n}\cdots\sum_{i_m=1}^{n} a'^2_{1,i_2,\cdots,i_m} = \sum_{i_2=1}^{n}\cdots\sum_{i_m=1}^{n}\left\{\sum_{j_1=1}^{n} a_{j_1,i_2,\cdots,i_m} v_{1,j_1}\right\}^2.$$

$$(14.41)$$

Applying Lemma 14.1 to $\{a'_{i_1,\cdots,i_m}\}$ and using (14.40)–(14.41), we get

$$\sum_{i_2}\cdots\sum_{i_m}\left\{\sum_{j_1} a_{j_1,i_2,\cdots,i_m} v_{1,j_1}\right\}^2 \leq \frac{m+n-3}{2m+n-4}\sum_{i_1}\sum_{i_2}\cdots\sum_{i_m} a_{i_1,i_2,\cdots,i_m}^2,$$

which implies

$$\left\{\sum_{i_2}\cdots\sum_{i_m}\sum_{j_1} a_{j_1,i_2,\cdots,i_m} v_{1,j_1} b_{i_2,\cdots,i_m}\right\}^2$$

$$\leq \frac{m+n-3}{2m+n-4}\left\{\sum_{i_1}\sum_{i_2}\cdots\sum_{i_m} a_{i_1,i_2,\cdots,i_m}^2\right\}\left\{\sum_{i_2}\cdots\sum_{i_m} b_{i_2,\cdots,i_m}^2\right\}.$$

Since $(v_{1,1},\cdots,v_{1,n}) \in \mathbf{R}^n$ is an arbitrary unit vector, the above inequality implies (14.39). $\qquad\qquad\square$

Proof of Theorem 14.1 Assume (14.1). Note that

$$\Delta\left\{\sum_{i_1=1}^{n}\cdots\sum_{i_m=1}^{n}\left(D_{x_{i_1}}\cdots D_{x_{i_m}}u\right)^2\right\}^{q/2}$$

$$= q\{\cdots\}^{q/2-2}\left[\{\cdots\}\left\{\sum_{i_0=1}^{n}\sum_{i_1=1}^{n}\cdots\sum_{i_m=1}^{n}\left(D_{x_{i_0}}D_{x_{i_1}}\cdots D_{x_{i_m}}u\right)^2\right\}\right.$$

$$\left.-(2-q)\sum_{i_0=1}^{n}\left\{\sum_{i_1=1}^{n}\cdots\sum_{i_m=1}^{n}\left(D_{x_{i_1}}\cdots D_{x_{i_m}}u\right)\left(D_{x_{i_0}}D_{x_{i_1}}\cdots D_{x_{i_m}}u\right)\right\}^2\right].$$

(14.42)

Put

$$a_{i_0,i_1,\cdots,i_m} = D_{x_{i_0}}D_{x_{i_1}}\cdots D_{x_{i_m}}u(x). \tag{14.43}$$

Then, by the harmonicity of u, $\{a_{i_0,i_1,\cdots,i_m}\}$ satisfies (14.5) and (14.6) with $\{1,\cdots,m\}$ replaced by $\{0,1,\cdots,m\}$. So, applying Lemma 14.12 with $\{1,\cdots,m\}$ replaced by $\{0,1,\cdots,m\}$, with (14.43) and with

$$b_{i_1,\cdots,i_m} = D_{x_{i_1}}\cdots D_{x_{i_m}}u(x),$$

and assuming $q \in (0,2]$, we get

$$(14.42) = q\left\{\sum_{i_1}\cdots\sum_{i_m}b^2_{i_1,\cdots,i_m}\right\}^{q/2-2}$$

$$\times\left[\left\{\sum_{i_1}\cdots\sum_{i_m}b^2_{i_1,\cdots,i_m}\right\}\left\{\sum_{i_0}\sum_{i_1}\cdots\sum_{i_m}a^2_{i_0,i_1,\cdots,i_m}\right\}\right.$$

$$\left.-(2-q)\sum_{i_0}\left\{\sum_{i_1}\cdots\sum_{i_m}b_{i_1,\cdots,i_m}a_{i_0,i_1,\cdots,i_m}\right\}^2\right]$$

$$\geq q\{\cdots\}^{q/2-1}\left\{\sum_{i_0}\sum_{i_1}\cdots\sum_{i_m}a^2_{i_0,i_1,\cdots,i_m}\right\}\left\{1-(2-q)\frac{(m+1)+n-3}{2(m+1)+n-4}\right\}$$

which implies (14.2).

Since $\nabla^m u$ is an $\mathbf{R}^{mn}$-valued harmonic function, $|\nabla^m u|^q$ is subharmonic if $q \geq 1$. So, (14.4) for $q \geq 1$ holds. (14.4) for $q \in \left[\dfrac{n-2}{n+m-2}, 1\right) \cap (0,1)$ follows from (14.2). $\square$

Remark 14.1. When $n = 2$, the condition (14.3) causes no problem. When $n \geq 3$, in order for (14.4) to hold we cannot remove the condition (14.3). We explain it. Let

$$n \geq 3, \quad \Omega = \mathbf{R}^n\setminus\{0\} \quad \text{and} \quad u(x) = |x|^{-n+2}.$$

Then, for any $x \in \Omega$, any $r > 0$ and any orthogonal matrix

$$V = (v_{i,j})_{i,j \in \{1,2,\cdots,n\}}$$

we have

$$|(\nabla^m u)(x)|^2 = |\nabla^m (u(Vx))|^2$$

$$= \sum_{i_1} \cdots \sum_{i_m} \left\{ \sum_{j_1} \cdots \sum_{j_m} (D_{x_{j_1}} \cdots D_{x_{j_m}} u)(Vx) v_{j_1,i_1} \cdots v_{j_m,i_m} \right\}^2$$

$$= \sum_{j_1} \cdots \sum_{j_m} \left((D_{x_{j_1}} \cdots D_{x_{j_m}} u)(Vx) \right)^2 = |(\nabla^m u)(Vx)|^2$$

and

$$(\nabla^m u)(rx) = r^{-n+2-m} \nabla^m u(x).$$

These imply

$$|(\nabla^m u)(x)| = C(m,n)|x|^{-n+2-m} \quad \text{with} \quad C(m,n) \neq 0.$$

So, in this case $\triangle(|\nabla^m u|^q) \geq 0$ on Ω only if (14.3) holds.

Finally we give a condition in order for a set of functions $\{u_{j_1,\cdots,j_m}\}_{j_1,\cdots,j_m \in \{1,\cdots,n\}}$ to be written in the form of

$$\nabla^m u = (D_{x_{j_1}} \cdots D_{x_{j_m}} u) = (u_{j_1,\cdots,j_m})$$

with some harmonic function u.

Theorem 14.2. *Let $n \in \{2,3,4,\cdots\}$. Let $\Omega = \{x = (x_1,\cdots,x_n) \in \mathbf{R}^n : x_1 > 0\}$. Let $\{u_j(x)\}_{j=1,\cdots,n} \subset C^1(\Omega)$ and let*

$$u_{j,i} = D_{x_i} u_j. \tag{14.44}$$

Assume that

$$u_{j,i} \equiv u_{i,j} \quad \text{on } \Omega \text{ for any } i,j \in \{1,\cdots,n\}. \tag{14.45}$$

Then there exists $u(x) \in C^2(\Omega)$ such that

$$\nabla u \equiv (u_j)_{j=1,\cdots,n} \quad \text{on } \Omega. \tag{14.46}$$

In particular

$$\triangle u \equiv \sum_{j=1}^{n} D_{x_j} u_j \quad \text{on } \Omega. \tag{14.47}$$

Proof. The conditon (14.45) implies that if $i < j$ and

$$(a_1, \cdots, a_n), (a_1, \cdots, a_{i-1}, b_i, a_{i+1}, \cdots, a_{j-1}, b_j, a_{j+1}, \cdots, a_n) \in \Omega,$$

then

$$\int_{a_i}^{b_i} u_i(a_1, \cdots, a_{i-1}, x_i, a_{i+1}, \cdots, a_n) dx_i$$

$$+ \int_{a_j}^{b_j} u_j(a_1, \cdots, a_{i-1}, b_i, a_{i+1}, \cdots, a_{j-1}, x_j, a_{j+1}, \cdots, a_n) dx_j$$

$$= \int_{a_j}^{b_j} u_j(a_1, \cdots, a_{j-1}, x_j, a_{j+1}, \cdots, a_n) dx_j$$

$$+ \int_{a_i}^{b_i} u_i(a_1, \cdots, a_{i-1}, x_i, a_{i+1}, \cdots, a_{j-1}, b_j, a_{j+1}, \cdots, a_n) dx_i$$

$$\tag{14.48}$$

So, if $\widetilde{C}$ is a continuous path in Ω which consists of finite number of line segments, each of which is parallel to one of the coordinate axes, then a repeated use of (14.48) implies that

" $\int_{\widetilde{C}} \sum_{j=1}^{n} u_j dx_j$ depends only on the starting point and the ending

point of $\widetilde{C}$. "

This holds for any C^1-path in Ω by approximating it by pathes of the above type.

Freeze $x_0 \in \Omega$. For each $x \in \Omega$ take a C^1-path C ($\subset \Omega$) starting from x_0 and ending at x. Put

$$u(x) = \int_C \sum_{j=1}^{n} u_j dx_j.$$

Then this is the desired one. $\qquad\qquad\qquad\qquad\qquad\qquad\qquad\qquad\qquad\square$

Theorem 14.3. *Let* $m, n \in \{2, 3, 4, \cdots\}$. *Let* $\Omega = \{x = \{x_1, \cdots, x_n\} \in \mathbf{R}^n : x_1 > 0\}$. *Let* $\{u_{j_1, \cdots, j_m}(x)\}_{j_1, \cdots, j_m \in \{1, \cdots, n\}} \subset C^1(\Omega)$ *and let*

$$u_{j_1, \cdots, j_m, j_{m+1}} = D_{x_{j_{m+1}}} u_{j_1, \cdots, j_m}. \tag{14.49}$$

Assume that

$$u_{j_1, \cdots, j_m} \equiv u_{j_{\tau(1)}, \cdots, j_{\tau(m)}} \quad \text{on } \Omega, \tag{14.50}$$

$$u_{j_1, \cdots, j_{m+1}} \equiv u_{j_{\sigma(1)}, \cdots, j_{\sigma(m+1)}} \quad \text{on } \Omega \tag{14.51}$$

for any permutation τ *of* $\{1, \cdots, m\}$ *and any permutation* σ *of* $\{1, \cdots, m+1\}$. *Then there exists* $u(x) \in C^{m+1}(\Omega)$ *such that*

$$\nabla^m u \equiv (u_{j_1,\cdots,j_m}) \quad on \ \Omega. \tag{14.52}$$

Furthermore, if

$$\sum_{j=1}^{n} u_{j,j,j_3,\cdots,j_m} \equiv 0 \ on \ \Omega \tag{14.53}$$

for any $j_3,\cdots,j_m \in \{1,\cdots,n\}$, *then we can impose the condition*

$$\Delta u \equiv 0 \ on \ \Omega. \tag{14.54}$$

Proof. We show this by the induction on m.

The case $m = 2$. Applying Theorem 14.2 to each $\{u_{i,j}\}_{j=1,\cdots,n}$ $(i = 1,\cdots,n)$ gives u_i such that

$$\nabla u_i \equiv (u_{i,j})_{j=1,\cdots,n} \quad on \ \Omega.$$

So, applying Theorem 14.2 to $\{u_i\}_{i=1,\cdots,n}$ gives the desired u.

Next, we assume that our Theorem with m replaced by some $m - 1 \in \{2,3,4,\cdots\}$ holds and will show that our Theorem with this m holds. Let $\{u_{j_1,\cdots,j_m}\}$ satisfy the assumptions of our Theorem. Applying Theorem 14.2 to each $\{u_{j_1,\cdots,j_{m-1},j}\}_{j=1,\cdots,n}$ $(j_1,\cdots,j_{m-1} \in \{1,\cdots,n\})$ gives $u_{j_1,\cdots,j_{m-1}}$ such that

$$\nabla u_{j_1,\cdots,j_{m-1}} \equiv (u_{j_1,\cdots,j_{m-1},j})_{j=1,\cdots,n} \quad on \ \Omega, \tag{14.55}$$

$$u_{j_1,\cdots,j_{m-1}}(x_0) = 0 \tag{14.56}$$

for some freezed $x_0 \in \Omega$. Then (14.55)–(14.56) and (14.50) imply

$$u_{j_1,\cdots,j_{m-1}} \equiv u_{j_{\tau(1)},\cdots,j_{\tau(m-1)}}$$

for any permutation τ of $\{1,\cdots,m-1\}$. Thus the hypothesis of the induction gives us u such that

$$\nabla^{m-1} u \equiv (u_{j_1,\cdots,j_{m-1}}) \quad on \ \Omega. \tag{14.57}$$

In particular,

$$\nabla^m u \equiv (u_{j_1,\cdots,j_m}) \quad on \ \Omega. \tag{14.58}$$

Furthermore, if (14.53) holds, then

$$\nabla \sum_{i=1}^{n} u_{i,i,j_3,\cdots,j_{m-1}} \equiv \left(\sum_{i=1}^{n} u_{i,i,j_3,\cdots,j_{m-1},j}\right)_{j=1,\cdots,n}$$
$$\equiv (0,\cdots,0) \ on \ \Omega,$$

which combined with (14.56) implies

$$\sum_{i=1}^{n} u_{i,i,j_3,\cdots,j_{m-1}} \equiv 0 \ on \ \Omega.$$

So, again the hypothesis of induction gives us a harmonic $u \in C^\infty(\Omega)$ that satisfies (14.57), in particular (14.58).   □

Notes. The case $m = 1$ of Theorem 14.1 was discovered by E. M. Stein–G. Weiss [60] and it was the beginning of the outburst of the theory of real Hardy spaces. The case $m \geq 2$ was discovered by A. P. Calderón–A. Zygmund [64]. The proof of Theorem 14.1 and Remark 14.1 are due to A. P. Calderón–A. Zygmund [64]. As for the properties of Gegenbauer polynomials see H. Bateman [53]. As for Theorems 14.2–3, see E. M. Stein–G. Weiss [68]. More difficult matters than Theorems 14.2–3 are discussed there.

XV. Subharmonicity, 2

Definition 15.1. Let

$$^t(v_1, \cdots, v_m) = \begin{bmatrix} v_1 \\ \vdots \\ v_m \end{bmatrix}.$$

In this section, for the sake of convenience we define $\mathbf{R}^m$ to be $\{^t(v_1 \cdots v_m) : v_1, \cdots, v_m \in \mathbf{R}\}$. For $\vec{a}_1, \cdots, \vec{a}_n \in \mathbf{R}^m$ let $(\vec{a}_1, \cdots, \vec{a}_n)$ be the $m \times n$ matrix whose j-th column vector is $\vec{a}_j$ for $j = 1, \cdots, n$.

Theorem 15.1. Let $n \geq 2$. Let $\{G_1, \cdots, G_n\}$ be a set of real constant $k \times m$ matrices such that

$$\text{rank}\,(G_1 \vec{a} \cdots, G_n \vec{a}) = n \qquad (15.1)$$

for any $\vec{a} \in \mathbf{R}^m \backslash \{\vec{0}\}$. Let $\vec{U}(x)$ be an $\mathbf{R}^m$-valued C^2-function defined on some neighbourhood of $x_0 \in \mathbf{R}^n$ such that

$$\vec{U}(x_0) \neq \vec{0}, \qquad (15.2)$$
$$\triangle \vec{U}(x_0) = \vec{0}, \qquad (15.3)$$
$$G_1 D_{x_1} \vec{U}(x_0) + \cdots + G_n D_{x_n} \vec{U}(x_0) = \vec{0}. \qquad (15.4)$$

Then there exists a positive constant $q < 1$, depending only on $\{G_1, \cdots, G_n\}$, such that

$$\left(\triangle(|\vec{U}|^q)\right)(x_0) \geq 0. \qquad (15.5)$$

Remark 15.1. This theorem can be regarded as a weak extension of the case $m = 1$ of Theorem 14.1. For the sake of simplicity let $n = 3$. Let $u(x) = u(x_1, x_2, x_3)$ be real and harmonic on an open set $\Omega \subset \mathbf{R}^3$ and let

$$\vec{U}(x) = {}^t(D_{x_1} u(x), D_{x_2} u(x), D_{x_3} u(x)).$$

Then, (15.3) is clear. Let

$$G_1 = \begin{bmatrix} 1 & 0 & 0 \\ 0 & 1 & 0 \\ 0 & 0 & 1 \\ 0 & 0 & 0 \end{bmatrix} \qquad G_2 = \begin{bmatrix} 0 & 1 & 0 \\ -1 & 0 & 0 \\ 0 & 0 & 0 \\ 0 & 0 & 1 \end{bmatrix} \qquad G_3 = \begin{bmatrix} 0 & 0 & 1 \\ 0 & 0 & 0 \\ -1 & 0 & 0 \\ 0 & -1 & 0 \end{bmatrix}$$

Then,

$$G_1 D_{x_1} \vec{U}(x) + G_2 D_{x_2} \vec{U}(x) + G_3 D_{x_3} \vec{U}(x) \equiv \vec{0}.$$

It is easy to see that if

$$\vec{a} = {}^t(a_1, a_2, a_3) \neq \vec{0}$$

and if

$$c_1 G_1 \vec{a} + c_2 G_2 \vec{a} + c_3 G_3 \vec{a} = \vec{0},$$

then

$$(c_1, c_2, c_3) = (0, 0, 0).$$

So, $\{G_1 \ G_2 \ G_3\}$ satisfies (15.1) with $n = 3$ for any $\vec{a} \in \mathbf{R}^3 \setminus \{\vec{0}\}$. Thus, Theorem 15.1 implies the existence of $q < 1$, which is independent of u and Ω such that

$$\triangle \left(|\nabla u(x)|^q \right) = \triangle \left(\left| \vec{U}(x) \right|^q \right) \geq 0 \ \text{if} \ \nabla u(x) \neq \vec{0}.$$

Remark 15.2. Theorem 15.1 cannot give an explicit value of q as in (14.3).

Lemma 15.1. *Let* $\{G_1, \cdots, G_n\}$ *be as in Theorem 15.1. Let* $\{\vec{a}_1, \cdots, \vec{a}_n\} \subset \mathbf{R}^m$ *and let*

$$G_1 \vec{a}_1 + \cdots + G_n \vec{a}_n = \vec{0}. \tag{15.6}$$

Then

$$\operatorname{rank}(\vec{a}_1, \cdots, \vec{a}_n) = 1. \tag{15.7}$$

Proof. Assume that $\{\vec{a}_1, \cdots, \vec{a}_n\}$ satisfies (15.6) and does not satisfy (15.7). Then there exist $\vec{a}_0 \in \mathbf{R}^m \setminus \{\vec{0}\}$ and $\{c_1, \cdots, c_n\} \subset \mathbf{R}$ such that

$$\vec{a}_j = c_j \vec{a}_0 \ (j = 1, \cdots, n), \tag{15.8}$$

$$(c_1, \cdots, c_n) = (0, \cdots, 0). \tag{15.9}$$

Substituting (15.8) into (15.6) gives

$$c_1 G_1 \vec{a}_0 + \cdots + c_n G_n \vec{a}_0 = \vec{0},$$

which combined with (15.9) gives a contradiction to (15.1). $\qquad \square$

Lemma 15.2. *Let* $\{G_1, \cdots, G_n\}$ *be as in Theorem 15.1. Let*

$$C_{15.1}(\{G_1, \cdots, G_n\})$$

$$= \sup \left\{ \frac{|{}^t A \vec{b}|^2}{\sum\limits_{j=1}^{n} |\vec{a}_j|^2 |\vec{b}|^2} : \vec{b} \in \mathbf{R}^m \setminus \{\vec{0}\}, \ A = (\vec{a}_1, \cdots, \vec{a}_n) \ and \right.$$

$$\left. \{\vec{a}_1, \cdots, \vec{a}_n\} \subset \mathbf{R}^m \ satisfies \ (15.6) \right\}$$

where ${}^t A$ *denotes the transposed matrix of* A. *Then*

$$C_{15.1} < 1. \tag{15.10}$$

Proof. Let

$$E = \left\{ (A, \vec{b}) = \left((\vec{a}_1, \cdots, \vec{a}_n), \vec{b} \right) : \{ \vec{a}_1, \cdots, \vec{a}_n, \vec{b} \} \subset \mathbf{R}^m \text{ satisfies (15.6)} \right.$$

$$\left. \text{and } \sum_{j=1}^{n} |\vec{a}_j|^2 = |\vec{b}| = 1 \right\}.$$

Then E can be identified with a compact subset of $\mathbf{R}^{mn+m}$ and

$$C_{15.1} = \max \left\{ |{}^t A \vec{b}|^2 : (A, \vec{b}) \in E \right\}. \qquad (15.11)$$

Let $(A, \vec{b}) \in E$. Then Lemma 15.1 implies (15.7). So,

$$|\vec{a}_i \cdot \vec{b}| \leq |\vec{a}_i||\vec{b}|$$

for some $i \in \{1, \cdots, n\}$. Thus,

$$|{}^t A \vec{b}| < 1,$$

which combined with (15.11) implies (15.10). $\qquad \square$

Proof of Theorem 15.1. By (15.2) and (15.3) we have

$$\Delta \left(|\vec{U}|^q \right) = q|\vec{U}|^{q-4} \left\{ \sum_{j=1}^{n} |D_{x_j}\vec{U}|^2 |\vec{U}|^2 - (2-q) \sum_{j=1}^{n} (D_{x_j}\vec{U}, \vec{U})^2 \right\} = (15.12) \text{ say,}$$

at $x = x_0$. Let

$$\vec{a}_j = (D_{x_j}\vec{U})(x_0), \quad A = (\vec{a}_1, \cdots, \vec{a}_n) \text{ and } \vec{b} = \vec{U}(x_0).$$

Since (15.4) implies (15.6), Lemma 15.2 implies

$$(15.12) \quad = \quad q|\vec{b}|^{q-4} \left\{ \sum_{j=1}^{n} |\vec{a}_j|^2 |\vec{b}|^2 - (2-q)|{}^t A \vec{b}|^2 \right\}$$

$$\geq \quad q|\vec{b}|^{q-2} \sum_{j=1}^{n} |\vec{a}_j|^2 \left\{ 1 - (2-q)C_{15.1} \right\}.$$

if $0 < q \leq 2$. Thus, if

$$2 - 1/C_{15.1} \leq q \leq 2 \text{ and } 0 < q,$$

then

$$(15.12) \geq 0.$$

Since (15.10) implies

$$2 - 1/C_{15.1} < 1,$$

we have obtained Theorem 15.1. $\qquad \square$

Definition 15.2. For a complex $m \times n$ matrix

$$
V = \begin{bmatrix}
v_{1,1} & v_{1,2} & \cdots & v_{1,n} \\
v_{2,1} & v_{2,2} & \cdots & v_{2,n} \\
\vdots & \vdots & \ddots & \vdots \\
v_{m,1} & v_{m.2} & \cdots & v_{m,n}
\end{bmatrix}
$$

let

$$
\tilde{V} = \begin{bmatrix}
\operatorname{Re} v_{1,1} & \operatorname{Re} v_{1,2} & \cdots & \operatorname{Re} v_{1,n} \\
\operatorname{Im} v_{1,1} & \operatorname{Im} v_{1,2} & \cdots & \operatorname{Im} v_{1,n} \\
\operatorname{Re} v_{2,1} & \operatorname{Re} v_{2,2} & \cdots & \operatorname{Re} v_{2,n} \\
\operatorname{Im} v_{2,1} & \operatorname{Im} v_{2,2} & \cdots & \operatorname{Im} v_{2,n} \\
\vdots & \vdots & \ddots & \vdots \\
\operatorname{Im} v_{m,1} & \operatorname{Im} v_{m,2} & \cdots & \operatorname{Im} v_{m,n}
\end{bmatrix} , \quad \operatorname{rank}_{\mathbf{R}} V = \operatorname{rank} \tilde{V},
$$

$$
\tilde{\tilde{V}} = \begin{bmatrix}
\operatorname{Re} v_{1,1} & -\operatorname{Im} v_{1,1} & \operatorname{Re} v_{1,2} & -\operatorname{Im} v_{1,2} & \cdots \\
\operatorname{Im} v_{1,1} & \operatorname{Re} v_{1,1} & \operatorname{Im} v_{1,2} & \operatorname{Re} v_{1,2} & \cdots \\
\operatorname{Re} v_{2,1} & -\operatorname{Im} v_{2,1} & \operatorname{Re} v_{2,2} & -\operatorname{Im} v_{2,2} & \cdots \\
\operatorname{Im} v_{2,1} & \operatorname{Re} v_{2,1} & \operatorname{Im} v_{2,2} & \operatorname{Re} v_{2,2} & \cdots \\
\cdots & \cdots & \cdots & \cdots & \cdots
\end{bmatrix} ,
$$

where $\tilde{V}$ is a real $2m \times n$ matrix and $\tilde{\tilde{V}}$ is a real $2m \times 2n$ matrix.

Corollary 15.1. *Let* $n \geq 2$. *Let* $\{G_1, \cdots, G_n\}$ *be a set of complex constant* $k \times m$ *matrices such that*

$$
\operatorname{rank}_{\mathbf{R}}(G_1 \vec{a}, \cdots, G_n \vec{a}) = n \tag{15.1}'
$$

for any $\vec{a} \in \mathbf{C}^m \backslash \{\vec{0}\}$. *Let* $\vec{U}(x)$ *be a* $\mathbf{C}^m$-*valued* C^2-*function defined on some neighbourhood of* $x_0 \in \mathbf{R}^n$ *such that* (15.2)–(15.4) *hold. Then the assertion of Theorem 15.1 holds.*

Proof. Since

$$
\operatorname{rank}(\tilde{\tilde{G}}_1 \tilde{\tilde{a}}, \cdots, \tilde{\tilde{G}}_n \tilde{\tilde{a}}) = \operatorname{rank}_{\mathbf{R}}(G_1 \vec{a}, \cdots, G_n \vec{a}) = n \quad \text{by } (15.1)'
$$

for any $\vec{a} \in \mathbf{C}^m \backslash \{\vec{0}\} = \mathbf{R}^{2m} \backslash \{\vec{0}\}$ and since (15.4) implies

$$
\tilde{\tilde{G}}_1 D_{x_1} \tilde{\tilde{U}}(x_0) + \cdots + \tilde{\tilde{G}}_n D_{x_n} \tilde{\tilde{U}}(x_0) = \vec{0},
$$

Thorem 15.1 implies the existence of $q < 1$ such that

$$
\left(\triangle (|\vec{U}|^q) \right)(x_0) = \left(\triangle \left(|\tilde{\tilde{U}}|^q \right) \right)(x_0) \geq 0.
$$

□

Notes. The theory in this section is due to R. Coifman–G. Weiss [70]. J.-A. Chao–M. Taibleson [73], J.-A. Chao [74], M. Taibleson [75] and S. Janson [77] extended the theory of R. Coifman–G. Weiss [70] to Hardy spaces on local fields and on some martingales.

XVI. Preliminaries for characterizations of H^p in terms of Fourier multipliers

First, we define $\|\cdot\|_{\Lambda_a}$ and $\|\cdot\|_{H^p}$ for $\mathbf{R}^m$-valued functions $\vec{f}(x) = (f_1(x), \cdots, f_m(x))$ defined on $\mathbf{R}^n$ by exactly the same way as in Sections 1 and 2. In the following, $|\vec{f}(x)|$ denotes $\left\{ \sum_{j=1}^{n} |f_j(x)|^2 \right\}^{1/2}$.

Definition 1.1′. For $a \geq 0$ and for $\vec{f} \in L^1_{\mathrm{loc}}(\mathbf{R}^n, \mathbf{R}^m)$ let

$$\|\vec{f}\|_{\Lambda_a} = \sup_B \inf_{\vec{P}:\deg \vec{P} \leq a} \int_B |\vec{f}(y) - \vec{P}(y)| dy |B|^{-1-a/n},$$

where B is taken over all balls B in $\mathbf{R}^n$ and where $\vec{P}(y)$ is taken over all $\mathbf{R}^m$-valued polynomials of degree $\leq a$. Let

$$\Lambda_a(\mathbf{R}^n, \mathbf{R}^m) = \left\{ \vec{f} \in L^1_{\mathrm{loc}}(\mathbf{R}^n, \mathbf{R}^m) : \|\vec{f}\|_{\Lambda_a} < \infty \right\},$$

$$\|\vec{f}\|_{\mathrm{BMO}} = \|\vec{f}\|_{\Lambda_0} \quad and \quad \mathrm{BMO}(\mathbf{R}^n, \mathbf{R}^m) = \Lambda_0(\mathbf{R}^n, \mathbf{R}^m).$$

Definition 1.2′. Let $p \in (0, 1]$. A function $\vec{a}(x) \in L^\infty(\mathbf{R}^n, \mathbf{R}^m)$ is called a (p, ∞)-*atom* if there exists a ball B such that

$$\mathrm{supp}\, \vec{a} \subset B, \quad \|\vec{a}\|_{L^\infty} \leq |B|^{-1/p},$$

$$\int \vec{a}(x) x^\alpha dx = \vec{0} \quad \text{provided } |\alpha| \leq n(1/p - 1).$$

By identifying a (p, ∞)-atom $\vec{a}(x)$ with a linear functional

$$\vec{f} \in \Lambda_{n(1/p-1)}(\mathbf{R}^n, \mathbf{R}^m) \longmapsto \int_{\mathbf{R}^n} \vec{f}(x) \cdot \vec{a}(x) dx \in \mathbf{R}$$

where $\vec{f}(x) \cdot \vec{a}(x)$ denotes the inner product in $\mathbf{R}^m$, $\vec{a}(x)$ can be regarded as an element in the unit ball of $\Lambda_{n(1/p-1)}(\mathbf{R}^n, \mathbf{R}^m)'$ ($=$ the dual space of $\Lambda_{n(1/p-1)}(\mathbf{R}^n, \mathbf{R}^m)/\{\mathbf{R}^m$-valued polynomials of degree $\leq n(1/p - 1)\}$). By the same way, $\vec{a}(x)$ can be regarded as an element of $\mathcal{D}(\mathbf{R}^n, \mathbf{R}^m)'$ ($=$ the set of continuous linear functionals on the topological linear space $\mathcal{D}(\mathbf{R}^n, \mathbf{R}^m)$).

Definition 2.1′. For $p \in (0, 1]$ and for $\vec{f} \in \mathcal{D}(\mathbf{R}^n, \mathbf{R}^m)'$ let

$$\|\vec{f}\|_{H^p} = \inf\left\{\|\{\lambda_j\}\|_{\ell^p} : \quad \text{there exists a sequence of } (p,\infty)\text{-atoms}\right.$$

$$\{\vec{a}_j(x)\}_j \subset L^\infty(\mathbf{R}^n, \mathbf{R}^m) \text{ such that}$$

$$\left. \vec{f} = \lim_{m\to\infty \text{ in } \mathcal{D}'} \sum_{j=1}^m \lambda_j \vec{a}_j(x)\right\}$$

where $\inf \emptyset = \infty$. Let

$$H^p(\mathbf{R}^n, \mathbf{R}^m) = \left\{\vec{f} \in \mathcal{D}(\mathbf{R}^n, \mathbf{R}^m)' : \|\vec{f}\|_{H^p} < \infty\right\}.$$

By the same argument as the scalar-valued case (see Remark 1.8), $H^p(\mathbf{R}^n, \mathbf{R}^m)$ can be regarded as a subspace of $\Lambda_{n(1/p-1)}(\mathbf{R}^n, \mathbf{R}^m)'$. Then, by the same argument as the scalar-valued case we have the following.

Lemma 2.2′ . Let $p \in (0,1]$ and $\vec{f} \in H^p(\mathbf{R}^n, \mathbf{R}^m)$. Then

$$\|\vec{f}\|_{\Lambda_{n(1/p-1)}(\mathbf{R}^n, \mathbf{R}^m)'} \leq \|\vec{f}\|_{H^p}.$$

Theorem 2.1′ . Let $p \in (0,1]$. Let T be a linear functional defined on $H^p(\mathbf{R}^n, \mathbf{R}^m)$ such that

$$\|T\| = \sup\left\{|T(\vec{f})|/\|\vec{f}\|_{H^p} : \vec{f} \in H^p(\mathbf{R}^n, \mathbf{R}^m)\setminus\{\vec{0}\}\right\} < \infty.$$

Then there exists an $\vec{h} \in \Lambda_{n(1/p-1)}(\mathbf{R}^n, \mathbf{R}^m)$ satisfying the following :

$$T(\vec{f}) = \langle \vec{h}, \vec{f}\rangle_{\Lambda_{n(1/p-1)}(\mathbf{R}^n, \mathbf{R}^m)} \text{ for any } \vec{f} \in H^p(\mathbf{R}^n, \mathbf{R}^m),$$

$$\|T\| = \|\vec{h}\|_{\Lambda_{n(1/p-1)}}.$$

Remark 16.1. Identifying $\mathbf{C}^m$ with $\mathbf{R}^{2m}$, we define $\Lambda_a(\mathbf{R}^n, \mathbf{C}^m)$, $\mathrm{BMO}(\mathbf{R}^n, \mathbf{C}^m)$ and $H^p(\mathbf{R}^n, \mathbf{C}^m)$ to be the same things with $\Lambda_a(\mathbf{R}^n, \mathbf{R}^{2m})$, $\mathrm{BMO}(\mathbf{R}^n, \mathbf{R}^{2m})$ and $H^p(\mathbf{R}^n, \mathbf{R}^{2m})$, respectively.

Remark 16.2. In the following, the notations $L^p(\mathbf{R}^n)$, $H^p(\mathbf{R}^n)$, $\Lambda_a(\mathbf{R}^n)$, $\mathcal{S}(\mathbf{R}^n)$, $\mathcal{S}_0(\mathbf{R}^n)$, $\mathcal{D}(\mathbf{R}^n), \cdots$ will denote $L^p(\mathbf{R}^n, \mathbf{R})$, $H^p(\mathbf{R}^n, \mathbf{R}), \cdots$, respectively.

Remark 16.3. It is clear that if $\vec{a}(x) = (a_1(x), \cdots, a_m(x))$ is an $\mathbf{R}^m$-valued (p,∞)-atom, then each $a_j(x)$ is an $\mathbf{R}$-valued (p,∞)-atom. On the other hand, if $a(x)$ is an $\mathbf{R}$-valued (p,∞)-atom, then $(0, \cdots, 0, a(x), 0, \cdots, 0)$ is an $\mathbf{R}^m$-valued (p,∞)-atom. Thus, we have

$$m^{-1}\sum_{j=1}^m \|f_j\|_{H^p}^p \leq \left\|\vec{f}\right\|_{H^p}^p \leq \sum_{j=1}^m \|f\|_{H^p}^p \tag{16.1}$$

for any $\vec{f} = (f_1, \cdots, f_m) \in \mathcal{D}(\mathbf{R}^n, \mathbf{R}^m)'$.

Next we introduce certain Fourier multipliers.

Definition 16.1. For $\theta \in L^\infty(S^{n-1}, \mathbf{C})$ and for $f \in L^2(\mathbf{R}^n, \mathbf{C})$ let

$$\mathsf{m}_\theta f = \mathcal{F}^{-1}\{\theta\,(\xi/|\xi|)\,\mathcal{F}f(\xi)\} \quad (\in L^2(\mathbf{R}^n, \mathbf{C})).$$

Remark 16.4. In the above we define $S^0 = \{-1, 1\}$. So, $L^\infty(S^0) = \mathbf{C}^\infty(S^0)$ denotes all functions defined on S^0.

Remark 16.5. Let $\theta \in L^\infty(S^{n-1}, \mathbf{C})$. It is clear that

$$\mathsf{m}_\theta f \in L^2(\mathbf{R}^n, \mathbf{R}) \text{ for all } f \in L^2(\mathbf{R}^n, \mathbf{R})$$

if and only if

$$\overline{\theta(\xi)} = \theta(-\xi) \text{ for a.e. } \xi \in S^{n-1}.$$

Definition 16.2. For $j = 1, 2, \cdots, n$ and for $f \in L^2(\mathbf{R}^n, \mathbf{C})$ let

$$R_j f = \mathcal{F}^{-1}\big(-i\xi_j|\xi|^{-1}\mathcal{F}f(\xi)\big).$$

These R_j's are called the *Riesz transforms*. For the sake of convenience let

$$R_0 f = f, \quad \vec{R}f = (R_0 f, R_1 f, \cdots, R_n f) \in L^2(\mathbf{R}^n, \mathbf{C}^{n+1}),$$
$$\vec{R}^m f = (R_{j_1} R_{j_2} \cdots R_{j_m} f)_{j_1, \cdots, j_m \in \{0, 1, \cdots, n\}} \in L^2(\mathbf{R}^n, \mathbf{C}^{(n+1)^m}).$$

If $n = 1$, then we write.

$$H = R_1.$$

The operator H is called the *Hilbert transform*.

Remark 16.6. By Remark 16.5, if $f \in L^2(\mathbf{R}^n, \mathbf{R})$, then

$$\vec{R}f \in L^2(\mathbf{R}^n, \mathbf{R}^{n+1}), \quad \vec{R}^m f \in L^2(\mathbf{R}^n, \mathbf{R}^{(n+1)^m}).$$

Theorem 16.1. *Let* $f \in L^2(\mathbf{R}^n, \mathbf{C})$, $\theta \in \mathbf{C}^\infty(S^{n-1}, \mathbf{C})$ *and* $p \in (0, 1]$. *Then*

$$\|\mathsf{m}_\theta f\|_{H^p} \le C(\theta, p)\|f\|_{H^p}.$$

Proof. Let $\psi \in \mathcal{S}_0(\mathbf{R}^n)$ be such that

$$\operatorname{supp} \mathcal{F}\phi \subset B(0, 1) \backslash B(0, 1/2),$$
$$\inf_{\xi \in S^{n-1}} \sup_{t>0} |\mathcal{F}\psi(t\xi)| > 0.$$

Put

$$\phi = \mathcal{F}^{-1}\{\theta\,(\xi/|\xi|)\,\mathcal{F}\psi(\xi)\}.$$

Then

$$\phi \in \mathcal{S}_0(\mathbf{R}^n, \mathbf{C}).$$

Thus

$$\|\mathsf{m}_\theta f\|_{H^p} \le C\|S_{\psi,1}\mathsf{m}_\theta f\|_{L^p} \quad \text{by Theorem 5.1}$$
$$= C\|S_{\phi,1}f\|_{L^p} \le C(\phi, p)\|f\|_{H^p} \quad \text{by Theorem 3.3.}$$

$\square$

Theorem 16.2. *Let* $p \in (0,1]$. *Let* $\{\theta_1, \cdots, \theta_m\} \subset C^\infty(S^{n-1}, \mathbf{C})$ *and let*

$$\inf_{\xi \in S^{n-1}} \sum_{j=1}^{m} |\theta_j(\xi)| > 0. \tag{16.2}$$

Let $f \in L^2(\mathbf{R}^n, \mathbf{C})$. *Then*

$$\|f\|_{H^p} \leq C(\{\theta_1, \cdots, \theta_m\}, p) \sum_{j=1}^{m} \|\mathrm{m}_{\theta_j} f\|_{H^p}. \tag{16.3}$$

Proof. For $\xi \in S^{n-1}$ let

$$\bar{\theta}_j(\xi) = \frac{\overline{\theta_j(\xi)}}{\displaystyle\sum_{k=1}^{m} |\theta_k(\xi)|^2}, \quad j = 1, \cdots, m. \tag{16.4}$$

Then

$$\|f\|_{H^p} = \left\| \sum_{j=1}^{m} \mathrm{m}_{\theta_j}\,\mathrm{m}_{\bar{\theta}_j} f \right\|_{H^p} \leq C(m,p) \sum \|\mathrm{m}_{\theta_j}\,\mathrm{m}_{\bar{\theta}_j} f\|_{H^p}$$

$$\leq C(\{\theta_1, \cdots, \theta_m\}, p) \sum \|\mathrm{m}_{\theta_j} f\|_{H^p} \text{ by Theorem 16.1.}$$

$$\square$$

Theorem 16.3. *Let* $p \in (0,1]$ *and* $\{\theta_1, \cdots, \theta_m\} \subset C^\infty(S^{n-1}, \mathbf{C})$ *be fixed. If* (16.3) *holds for any* $f \in L^2(\mathbf{R}^n, \mathbf{C})$, *then* (16.2) *holds.*

Proof. Assume

$$\sum_{j=1}^{m} |\theta_j(\xi_0)| = 0 \text{ for some } \xi_0 \in S^{n-1}. \tag{16.5}$$

Let

$$\phi \in \mathcal{S}(\mathbf{R}^n), \text{ supp } \mathcal{F}\phi \subset B(0, 1/2),$$
$$\mathcal{F}\phi(\xi) \equiv 1 \text{ on } B(0, 1/4),$$
$$\psi_j = \mathcal{F}^{-1}\left\{ \mathcal{F}\phi(\xi)\theta_j \left(\frac{\xi + \xi_0}{|\xi + \xi_0|} \right) \right\},$$
$$f = \mathcal{S}(\mathbf{R}^n, \mathbf{R}), \ \mathcal{F}f(0) \neq 0, \text{ supp } \mathcal{F}f \subset B(0, 1/4).$$

Since

$$\psi_j \in \mathcal{S}(\mathbf{R}^n, \mathbf{C}) \text{ and } \int \psi_j dx = 0,$$

we get

$$\sup_{t\in(0,1)} |f * (\psi_j)_t(x)| \leq C(f,\psi_j,p)\,(1+|x|)^{-1-n/p},$$

$\|f * (\psi_j)_t\|_{L^\infty} \to 0 \ (t \to +0).$

For $t \in (0,1)$ let

$$g_{j,t}(x) = e^{2\pi i \xi_0 \cdot x/t} \cdot f * (\psi_j)_t(x).$$

Then the above estimates and the condition on $\operatorname{supp} \mathcal{F} f$ imply

$$|g_{j,t}(x)| \leq C(f,\psi_j,p)\,(1+|x|)^{-1-n/p},$$

$$\|g_{j,t}\|_{L^\infty} \to 0 \ (t \to +0),$$

$$\operatorname{supp} \mathcal{F} g_{j,t} \subset B(\xi_0/t, 1/4) \subset \mathbf{R}^n \setminus \{0\}.$$

Thus, Remark 3.4 implies

$$\|g_{j,t}\|_{H^p} \to 0 \ (t \to +0). \tag{16.6}$$

Put

$$u_t(x) = t^{-n+n/p} e^{2\pi i \xi_0 \cdot x} (f)_{1/t}(x). \tag{16.7}$$

Then

$$\begin{aligned}
\|m_{\theta_j}(u_t)\|_{H^p}^p &= t^{-np+n} \left\| \mathcal{F}^{-1}\{\theta_j(\xi/|\xi|)\mathcal{F}f((\xi-\xi_0)/t)\} \right\|_{H^p}^p \\
&= t^{-np+n} \left\| \mathcal{F}^{-1}\{\mathcal{F}\phi(\xi-\xi_0)\theta_j(\xi/|\xi|)\mathcal{F}f((\xi-\xi_0)/t)\} \right\|_{H^p}^p \\
&= t^{-np+n} \left\| (g_{j,t})_{1/t} \right\|_{H^p}^p = \|g_{j,t}\|_{H^p}^p \to 0 \ (t \to +0) \text{ by } (16.6), \tag{16.8}
\end{aligned}$$

$$\|u_t\|_{H^p} \geq \|u_t\|_{L^p} = \|f\|_{L^p} = 0 \text{ by Theorem 2.5.} \tag{16.9}$$

Thus, (16.8) and (16.9) give a contradiction to (16.3). $\quad\square$

Definition 16.3. For $\theta \in L^\infty(S^{n-1}, \mathbf{C})$ let

$$\theta_\$ \xi = \operatorname{Re} \frac{\theta(\xi) + \theta(-\xi)}{2} + i \operatorname{Im} \frac{\theta(\xi) - \theta(-\xi)}{2},$$

$$\theta_\not\xi \xi = \operatorname{Im} \frac{\theta(\xi) + \theta(-\xi)}{2} - i \operatorname{Re} \frac{\theta(\xi) - \theta(-\xi)}{2}.$$

Remark 16.7. If $f \in L^2(\mathbf{R}^n, \mathbf{R})$, then

$$\operatorname{Re} m_\theta f = m_{\theta_\$} f, \quad \operatorname{Im} m_\theta f = m_{\theta_\not\xi} f.$$

Remark 16.8. If $f \in L^2(\mathbf{R}^n, \mathbf{C})$, then

$$\begin{aligned}
&\|m_{\theta_\$} f\|_{H^p} + \|m_{\theta_\not\xi} f\|_{H^p} \\
&= \|m_{\theta_\$} \operatorname{Re} f + i m_{\theta_\$} \operatorname{Im} f\|_{H^p} + \|m_{\theta_\not\xi} \operatorname{Re} f + i m_{\theta_\not\xi} \operatorname{Im} f\|_{H^p} \\
&\approx \|m_{\theta_\$} \operatorname{Re} f\|_{H^p} + \|m_{\theta_\$} \operatorname{Im} f\|_{H^p} + \|m_{\theta_\not\xi} \operatorname{Re} f\|_{H^p} + \|m_{\theta_\not\xi} \operatorname{Im} f\|_{H^p} \\
&\qquad\qquad\qquad\qquad\qquad\qquad\qquad \text{by Remark 16.5 and (16.1)} \\
&\approx \|m_{\theta_\$} \operatorname{Re} f + i m_{\theta_\not\xi} \operatorname{Re} f\|_{H^p} + \|m_{\theta_\$} \operatorname{Im} f + i m_{\theta_\not\xi} \operatorname{Im} f\|_{H^p} \\
&\qquad\qquad\qquad\qquad\qquad\qquad\qquad \text{by Remark 16.5 and (16.1)} \\
&= \|m_\theta \operatorname{Re} f\|_{H^p} + \|m_\theta \operatorname{Im} f\|_{H^p}.
\end{aligned}$$

Corollary 16.1. *Let $p \in (0,1]$ and $\{\theta_1, \cdots, \theta_m\} \subset C^\infty(S^{n-1}, \mathbf{C})$ be fixed. Then (16.3) holds for any*

$$f \in L^2(\mathbf{R}^n, \mathbf{R})$$

if and only if

$$\inf_{\xi \in S^{n-1}} \sum_{j=1}^{m} \{|\theta_j(\xi)| + |\theta_j(-\xi)|\} > 0. \tag{16.10}$$

Proof. Remark 16.8 implies

$$\sup \left\{ \frac{\|f\|_{H^p}}{\sum\limits_{j=1}^{m} \|m_{\theta_j} f\|_{H^p}} : f \in (L^2(\mathbf{R}^n, \mathbf{R}) \cap H^p(\mathbf{R}^n, \mathbf{R})) \setminus \{0\} \right\} \approx$$

$$\sup \left\{ \frac{\|f\|_{H^p}}{\sum\limits_{j=1}^{m} (\|m_{\theta_{j\$}} f\|_{H^p} + \|m_{\theta_{j\ell}} f\|_{H^p})} : f \in (L^2(\mathbf{R}^n, \mathbf{C}) \cap H^p(\mathbf{R}^n, \mathbf{C})) \setminus \{0\} \right\}. \tag{16.11}$$

On the other hand the condition (16.10) is equivalent with

$$\inf_{\xi \in S^{n-1}} \sum_{j=1}^{m} \{|\theta_{j\$}(\xi)| + |\theta_{j\ell}(\xi)|\} > 0.$$

Therefore, Corollary 16.1 follows from Theorems 16.2 and 16.3.   □

Finally we add the following whose proof we leave as an exercise.

Theorem 16.4. *Let $\Omega \subset \mathbf{R}^m$ be an open set. Let $v \in C(\Omega)$, $v(x) \geq 0$, $v \in C^2(\{x \in \Omega : v(x) > 0\})$ and*

$$\Delta v(x) \geq 0 \quad on \quad \{x \in \Omega : v(x) > 0\}.$$

Then, v is subharmonic on Ω, i.e.

$$\sigma_{m-1}(\partial B(x,r))^{-1} \int_{\partial B(x,r)} v(y) d\sigma_{m-1}(y) \geq v(x)$$

for any $\overline{B(x,r)} \subset \Omega$. (In other words, the distributional Laplacian Δv is a positive measure on Ω.)

XVII. Characrterization of H^p in terms of Riesz transforms

Theorem 17.1. *Let $f \in L^2(\mathbf{R}^n)$. Let*

$$m \begin{cases} \in \mathbf{N} & \text{if } n \in \{2, 3, 4, \cdots\}, \\ = 1 & \text{if } n = 1, \end{cases} \tag{17.1}$$

and

$$p \in \left(\frac{n-1}{n+m-1}, 1 \right]. \tag{17.2}$$

Then

$$\|\vec{R}^m f\|_{H^p} \le C(p, m, n) \|\vec{R}^m f\|_{L^p}. \tag{17.3}$$

Corollary 17.1. *Let f, m and p be as in Theorem 17.1. Then*

$$\|f\|_{H^p} \le C(p, m, n) \left\{ \|f\|_{L^p} + \sum_{k=1}^{m} \sum_{j_1=1}^{n} \cdots \sum_{j_k=1}^{n} \|R_{j_1} R_{j_2} \cdots R_{j_k} f\|_{L^p} \right\}.$$

Lemma 17.1. *Let m and f be as in Theorem 17.1. Then there exists a harmonic function $u(x, t)$ defined on $\mathbf{R}^{n+1}_+ = \{(x, t) : x \in \mathbf{R}^n, \ t > 0\}$ such that*

$$P(\cdot, t) * (\vec{R}^m f)(x) \equiv \nabla^m_{t,x} u(x, t),$$

where $P(x, t)$ is the Poisson kernel of R^{n+1}_+ and

$$\nabla_{t,x} = (D_t, D_{x_1}, \cdots, D_{x_n}).$$

Proof. For $j_1, \cdots, j_m \in \{0, 1, \cdots, n\}$ let

$$u_{j_1, \cdots, j_m}(x, t) = P(\cdot, t) * (R_{j_1} \cdots R_{j_m} f)(x).$$

Identifying t with x_0 let

$$u_{j_1, \cdots, j_m, j_{m+1}} = D_{x_{j_{m+1}}} u_{j_1, \cdots, j_m}, \ j_{m+1} \in \{0, 1, \cdots, n\}.$$

It is easy to see that

$$u_{j_1, \cdots, j_m} = u_{j_{\tau(1)}, \cdots, j_{\tau(m)}}$$

for any permutation τ of $\{1, \cdots, m\}$. By taking Fourier transforms it is easy to see that

$$u_{j_1, \cdots, j_{m+1}} = u_{j_{\rho(1)}, \cdots, j_{\rho(m+1)}}$$

for any permutation ρ of $\{1, \cdots, m+1\}$, that if $m = 1$, then

$$\sum_{j=0}^{n} u_{j,j} \equiv 0$$

and that if $m \in \{2, 3, 4, \cdots\}$, then

$$\sum_{j=0}^{n} u_{j,j,j_3,\cdots,j_m} \equiv 0 \text{ for any } j_3, \cdots, j_m \in \{0, 1, \cdots, n\}.$$

So, the existence of the desired harmonic function $u(x,t)$ follows from Thorems 14.2 and 14.3.

Lemma 17.2. *Let m and f be as in Theorem 17.1. Let*

$$q \in \left[\frac{n-1}{n+m-1}, +\infty\right) \cap (0, +\infty).$$

Then

$$\left|P(\cdot,t) * (\vec{R}^m f)(x)\right|^q \leq P(\cdot,t) * \left(|\vec{R}^m f|^q\right)(x) \tag{17.4}$$

for any $(x,t) \in \mathbf{R}_+^{n+1}$ In particular

$$N_{P(\cdot,1),0}(\vec{R}^m f)(x) \leq M_q\left(|\vec{R}^m f|\right)(x),$$
$$N_{P(\cdot,1),1}(\vec{R}^m f)(x) \leq C(q,n)M_q\left(|\vec{R}^m f|\right)(x), \tag{17.5}$$

where

$$N_{P(\cdot,1),0}(\vec{R}^m f)(x) = \sup_{t>0}\left|P(\cdot,t) * \vec{R}^m f(x)\right|,$$
$$N_{P(\cdot,1),1}(\vec{R}^m f)(x) = \sup_{(y,t)\in\Gamma(x,1)}\left|P(\cdot,t) * \vec{R}^m f(y)\right|.$$

Proof. Lemma17.1, Theorems 14.1 and 16.4 imply the subharmonicity of

$$\left|P(\cdot,t) * (\vec{R}^m f)(x)\right|^q$$

as a function of $(x,t) \in \mathbf{R}_+^{n+1}$. Let $\varepsilon > 0$. Then, the condition $f \in L^2$ implies

$$\left|P(\cdot,t+\varepsilon) * (\vec{R}^m f)(x)\right| \to 0 \quad (|x|+t \to \infty, \ (x,t) \in \mathbf{R}_+^{n+1}),$$

which combined with the above subharmonicity implies

$$\left|P(\cdot,t+\varepsilon) * (\vec{R}^m f)(x)\right|^q \leq P(\cdot,t) * \left(|P(\cdot,\varepsilon) * (\vec{R}^m f)|^q\right)(x).$$

So, letting $\varepsilon \to +0$ and using Lebesgue's dominated convergence theorem gives (17.4) under the additional condition $q \leq 2$. The case $q > 2$ follows from the case $q = 2$ and Hölder's inequality. $\square$

Proof of Theorem 17.1. Take

$$q \in \left[\frac{n-1}{n+m-1}, \, p \right) \cap (0,p).$$

Then

$$\|\vec{R}^m f\|_{H^p} \le C(p,m,n) \left\| N_{P(\cdot,1),0}(\vec{R}^m f) \right\|_{L^p}$$

by the latter half of Remark 9.2

$$\le C \left\| M_q(|\vec{R}^m f|) \right\|_{L^p} \qquad \text{by Lemma 17.2}$$

$$\le C\|\vec{R}^m f\|_{L^p} \qquad \text{by } p > q \text{ and by Corollary 0.4.}$$

Corollary 17.1 follows from Theorem 17.1 and (16.1). $\qquad\qquad\square$

Theorem 17.2. *Let* $\{\theta_1, \cdots, \theta_m\} \subset C^\infty(S^{n-1}, \mathbf{C})$ *be such that there exists a set of complex constant* $k \times m$ *matrices* $\{G_0, G_1, \cdots, G_n\}$ *satisfying*

$$\mathrm{rank}_{\mathbf{R}}(G_0\vec{a}, G_1\vec{a}, \cdots, G_n\vec{a}) = n+1 \text{ for any } \vec{a} \in \mathbf{C}^m\backslash\{\vec{0}\}, \qquad (17.6)$$

$$\{G_0 - i\xi_1 G_1 - \cdots - i\xi_n G_n\} \begin{pmatrix} \theta_1(\xi) \\ \vdots \\ \theta_m(\xi) \end{pmatrix} \equiv \begin{pmatrix} 0 \\ \vdots \\ 0 \end{pmatrix} \text{ for any } \xi \in S^{n-1}.$$

$$(17.7)$$

Let $f \in L^2(\mathbf{R}^n, \mathbf{C})$ *and let*

$$\vec{K}f(x) = \left(\mathsf{m}_{\theta_j} f(x) \right)_{j=1,\cdots,m}. \qquad (17.8)$$

Then there exists $q_{17.1}(\{\theta_1, \cdots, \theta_m\}) < 1$ *such that if*

$$p \in (q_{17.1}, 1], \qquad (17.9)$$

then

$$\|\vec{K}f\|_{H^p} \le C(p, \{\theta_1, \cdots, \theta_m\})\|\vec{K}f\|_{L^p}. \qquad (17.10)$$

Corollary 17.2. *Let* $\{\theta_1, \cdots, \theta_m\}$, $\{G_0, \cdots, G_n\}$, f *and* $q_{17.1}$ *be as in Theorem 17.2. Furthermore, assume* (16.2) *and* (17.9). *Let* $f \in L^2(\mathbf{R}^n, \mathbf{C})$. *Then*

$$\|f\|_{H^p} \le C(p, \{\theta_1, \cdots, \theta_m\}) \sum_{j=1}^m \|\mathsf{m}_{\theta_j} f\|_{L^p}.$$

Proof of Theorem 17.2. By taking Fouirer transforms, it is easy to see that the condition (17.7) implies

$$G_0 D_t \big(P(\cdot,t) * (\vec{K}f)(x)\big) + G_1 D_{x_1}\big(P(\cdot,t) * (\vec{K}f)(x)\big)$$

$$+ \cdots + G_n D_{x_n}\big(P(\cdot,t) * (\vec{K}f)(x)\big) \equiv \vec{0}$$

$$(x,t) \in \mathbf{R}_+^{n+1}. \qquad (17.11)$$

(17.6), (17.11), Corollary 15.1 and Theorem 16.4 imply the subharmonicity of

$$\left| P(\cdot, t) * (\vec{K}f)(x) \right|^q$$

for some $q < 1$, which is independent of f. Then the same argument as in the proof of Lemma 17.2 implies

$$\left| P(\cdot, t) * (\vec{K}f)(x) \right|^q \leq P(\cdot, t) * \left(|\vec{K}f|^q \right)(x)$$

for any $(x, t) \in \mathbf{R}_+^{n+1}$. So, the same argument as the proof of Theorem 17.1 implies (17.10). Corollary 17.2 follows from Theorems 17.2 and 16.2.   □

Notes. Theorem 17.1 for the case "$m = 1$, $n \geq 2$" is essentially due to E. M. Stein–G. Weiss [60]. Theorem 17.1 for the case "$m \geq 2$" is essentially due to A. P. Calderón–A. Zygmund [64]. Theorem 17.2 is an easy consequence of R. Coifman–G. Weiss [70].

XVIII. Other results on the characterization of H^p in terms of Fourier multipliers

Theorem 18.1. *Let $p \in (0,1]$, $f \in L^2(\mathbf{R}^n, \mathbf{C})$, $\nu \in S^{n-1}$, $a \in (0,1)$ and*

$$\operatorname{supp} \mathcal{F}f \subset \{\xi \in \mathbf{R}^n : |\xi \cdot \nu| \geq a|\xi|\}. \tag{18.1}$$

Then

$$\begin{aligned}
c(a,p,n)\|f\|_{H^p(\mathbf{R}^n)}^p &\leq \int_{\{x' \in \mathbf{R}^n : x' \cdot \nu = 0\}} \|f(x' + \cdot \nu)\|_{H^p(\mathbf{R}^1)}^p d\sigma_{n-1}(x') \\
&\leq C(a,p,n)\|f\|_{H^p(\mathbf{R}^n)}^p, \tag{18.2}
\end{aligned}$$

where $\|f(x' + \cdot \nu)\|_{H^p(\mathbf{R}^1)}$ denotes the $H^p(\mathbf{R}^1)$-norm of $f(x' + \cdot \nu)$ as a funtion of $\cdot \in \mathbf{R}^1$ and $d\sigma_{n-1}$ is the $(n-1)$-dimensional Lebesgue measure.

Corollary 18.1. *Let $p \in (0,1]$, $f \in L^2(\mathbf{R}^n, \mathbf{C})$, $\nu \in S^{n-1}$, $a \in (0,1)$ and*

$$\operatorname{supp} \mathcal{F}f \subset \{\xi \in \mathbf{R}^n : \xi \cdot \nu \geq a|\xi|\}. \tag{18.3}$$

Then

$$\|f\|_{H^p} \leq C(a,p,n)\|f\|_{L^p}. \tag{18.4}$$

Corollary 18.2. *Let $p \in (0,1]$, $\{\theta_1, \cdots, \theta_m\} \subset C^\infty(S^{n-1}, \mathbf{C})$, $\{\nu_1, \cdots, \nu_m\} \subset S^{n-1}$, $a \in (0,1)$ and let*

$$\operatorname{supp} \theta_j \subset \{\xi \in S^{n-1} : \xi \cdot \nu_j \geq a\}, \quad (j = 1, \cdots, m). \tag{18.5}$$

Let $f \in L^2(\mathbf{R}^n, \mathbf{C})$ and let (16.2) hold (or let $f \in L^2(\mathbf{R}^n, \mathbf{R})$ and let (16.10) hold). Then

$$\|f\|_{H^p} \leq C(p, \{\theta_1, \cdots, \theta_m\}) \sum_{j=1}^{m} \|m_{\theta_j} f\|_{L^p}. \tag{18.6}$$

Proof of Theorem 18.1 We may assume

$$\nu = (1, 0, \cdots, 0). \tag{18.7}$$

Let $\psi_1 \in \mathcal{S}(\mathbf{R}^1)$ and $\psi_2 \in \mathcal{S}(\mathbf{R}^{n-1})$ be such that

$$\mathcal{F}\psi_1(0) = 1, \quad \operatorname{supp} \mathcal{F}\psi_1 \subset (-a, a), \tag{18.8}$$

$$\mathcal{F}\psi_2(\xi_1, \cdots, \xi_{n-1}) = 1 \text{ if } \xi_1^2 + \cdots + \xi_{n-1}^2 \leq \sqrt{1-a^2}. \tag{18.9}$$

Let

$$\psi(x) = \psi_1(x_1)\psi_2(x_2, \cdots, x_n).$$

Then (18.1), (18.8) and (18.9) imply

$$\mathcal{F}f(\xi)\mathcal{F}\psi(t\xi) = \mathcal{F}f(\xi)\mathcal{F}\psi_1(t\xi_1)\mathcal{F}\psi_2(t\xi_2, \cdots, t\xi_n) = \mathcal{F}f(\xi)\mathcal{F}\psi_1(t\xi_1) \quad (18.10)$$

for all $t > 0$. Since both

$$f * (\psi)_t(x) \quad (18.11)$$

and

$$\int_{\mathbf{R}^1} f(y_1, x_2, \cdots, x_n)\frac{1}{t}\psi_1\left(\frac{x_1 - y_1}{t}\right) dy_1 \quad (18.12)$$

are measurable functions of $(x, t) \in \mathbf{R}_+^{n+1}$, (18.10) and the condtion $f \in L^2$ imply

$$(18.11) = (18.12) \text{ for a.e. } (x, t) \in \mathbf{R}_+^{n+1}.$$

Furthermore, since (18.11) is continuous on $\mathbf{R}_+^{n+1}$ and since (18.12) is continuous as a function of $(x_1, t) \in \mathbf{R}_+^2$ for a.e. $(x_2, \cdots, x_n) \in \mathbf{R}_+^{n-1}$, we have that

$$(18.11) = (18.12) \text{ for all } (x_1, t) \in \mathbf{R}_+^2$$

for a.e. $(x_2, \cdots, x_n) \in \mathbf{R}^{n-1}$. Therefore,

$$\int_{\mathbf{R}^n} N_{\psi,0}f(x)^p dx = \int_{\mathbf{R}^{n-1}} dx_2 \cdots dx_n \int_{\mathbf{R}^1} \sup_{t>0} |(18.12)|^p dx_1$$

which combined with Remark 9.2 implies (18.2).   □

Lemma 18.1. *Let $p \in (0, 1]$, $h \in L^2(\mathbf{R}^1, \mathbf{C})$ and supp $\mathcal{F}h \subset [0, +\infty)$. Then*

$$\|h\|_{H^p(\mathbf{R}^1)} \leq C(p)\|h\|_{L^p(\mathbf{R}^1)}.$$

This is clear from the case $n = 1$ of Theorem 17.1, because $\operatorname{Im} h = H \operatorname{Re} h$ by supp $\mathcal{F}h \subset [0, +\infty)$.

Proof of Corollary 18.1. We may assume (18.7). Then the condition $f \in L^2$ and (18.3) (with (18.7)) imply that for a.e. $(x_2, \cdots, x_n) \in \mathbf{R}^{n-1}$ the function

$$f(y_1, x_2, \cdots, x_n), \quad (18.13)$$

which we regard as a function of one variable y_1, belongs to $L^2(\mathbf{R}^1, \mathbf{C})$ and its Fourier transform is supported by $[0, +\infty)$. So, Lemma 18.1 implies

$$\|(18.13)\|_{H^p(\mathbf{R}^1)} \approx \|(18.13)\|_{L^p(\mathbf{R}^1)}. \quad (18.14)$$

Then, (18.4) follows from the first inequality of (18.2) and (18.14).   □

Proof of Corollary 18.2.

$$\|f\|_{H^p} \leq C(p, \{\theta_1, \cdots, \theta_m\}) \sum_{j=1}^{m} \|m_{\theta_j} f\|_{H^p} \text{ by Theorem 16.2}$$

$$\leq C \sum \|m_{\theta_j} f\|_{L^p} \text{ by (18.5) and Corollary 18.1.}$$

$\square$

Theorem 18.2. *Let $p \in (0, 1]$, and $f \in L^2(\mathbf{R}^n, \mathbf{C})$. Let*

$$\text{supp} \, \mathcal{F} f \subset \{\xi = (\xi_1, \cdots, \xi_n) \in \mathbf{R}^n : \xi_1, \cdots, \xi_n \geq 0\}. \tag{18.15}$$

Let $a > 1/p - 1$ and

$$G_a^* f(x) = \sup \left\{ \left| \int_{\mathbf{R}^n} \frac{1}{t_1} \phi_1 \left(\frac{x_1 - y_1}{t_1} \right) \cdots \frac{1}{t_n} \phi_n \left(\frac{x_n - y_n}{t_n} \right) f(y) dy \right| \right.$$

$$\left. : \phi_1, \cdots, \phi_n \in \mathcal{B}_a(\mathbf{R}^1), \ t_1, \cdots, t_n > 0 \right\}, \tag{18.16}$$

where $x = (x_1, \cdots, x_n)$ and $y = (y_1, \cdots, y_n)$. Then

$$\|G_a^* f\|_{L^p} \leq C(a, p, n) \|f\|_{L^p}. \tag{18.17}$$

Remark 18.1. The inequality (18.17) implies

$$\|f\|_{H^p} \leq C(a, p, n) \|f\|_{L^p}.$$

Then, by dilations into the direction ν and by rotations, Corollary 18.1 follows from Theorem 18.2.

Definition 18.1. For $(\lambda, t) \in \mathbf{R}_+^2$ let

$$\rho(\lambda, t) = \frac{t}{\pi(\lambda^2 + t^2)},$$

which is the Poisson kernel of $\mathbf{R}_+^2$.

Definition 18.2. For $f \in L_{\text{loc}}^1(\mathbf{R}^n)$, $j \in \{1, 2, \cdots, n\}$ and for $x = (x_1, \cdots, x_n) \in \mathbf{R}^n$ let

$$M_j f(x) = \sup_{s>0} \frac{1}{2s} \int_{x_j - s}^{x_j + s} |x_1, \cdots, x_{j-1}, y_j, x_{j+1}, \cdots, x_n)| \, dy_j.$$

Lemma 18.2. *Let $q > 0$ and $f \in L^2(\mathbf{R}^n, \mathbf{C})$. Assume (18.15). Then*

$$\sup_{t_1, \cdots, t_n > 0} \left| \int_{\mathbf{R}^n} \rho(x_1 - y_1, t_1) \cdots \rho(x_n - y_n, t_n) f(y) dy \right|^q$$

$$\leq M_1 M_2 \cdots M_n \left(|f|^q \right)(x). \tag{18.18}$$

Proof. The condition (18.15) implies that

$$\int_{\mathbf{R}^n} \rho(x_1 - y_1, t_1) \cdots \rho(x_n - y_n, t_n) f(y) dy =$$

$$\int_{\mathbf{R}^1} \rho(x_1 - y_1, t_1) dy_1 \int_{\mathbf{R}^{n-1}} \rho(x_2 - y_2, t_2) \cdots \rho(x_n - y_n, t_n) f(y) dy_2 \cdots dy_n$$

$$(18.19)$$

is analytic as a function of $x_1 + it_1 \in \mathbf{R}^2_+$ for each fixed $((x_2, t_2), \cdots, (x_n, t_n)) \in (\mathbf{R}^2_+)^{n-1}$. So, $|(18.19)|^q$ is subharmonic as a function of $(x_1, t_1) \in \mathbf{R}^2_+$. Since

$$\int_{\mathbf{R}^1} dy_1 \left| \int_{\mathbf{R}^{n-1}} \rho(x_2 - y_2, t_2) \cdots \rho(x_n - y_n, t_n) f(y) dy_2 \cdots dy_n \right|^2$$
$$\leq C \cdot (t_2 \cdots t_n)^{-1} \|f\|_{L^2(\mathbf{R}^n)}^2 < \infty,$$

the subharmonicity of $|(18.19)|^q$ implies

$$|(18.19)|^q \quad \leq \quad \int_{\mathbf{R}^1} \rho(x_1 - y_1, t_1) dy_1$$

$$\times \quad \left| \int_{\mathbf{R}^{n-1}} \rho(x_2 - y_2, t_2) \cdots \rho(x_n - y_n, t_n) f(y) dy_2 \cdots dy_n \right|^q.$$

By the same argument we can show

$$\left| \int_{\mathbf{R}^{n-1}} \rho(x_2 - y_2, t_2) \cdots \rho(x_n - y_n, t_n) f(y) dy_2 \cdots dy_n \right|^q$$

$$\leq \int_{\mathbf{R}^1} \rho(x_2 - y_2, t_2) dy_2 \left| \int_{\mathbf{R}^{n-2}} \rho(x_3 - y_3, t_3) \cdots \rho(x_n - y_n, t_n) f(y) dy_3 \cdots dy_n \right|^q$$

for all $((x_2, t_2), \cdots, (x_n, t_n)) \in (\mathbf{R}^2_+)^{n-1}$ for a.e. $y_1 \in \mathbf{R}^1$. Thus

$$|(18.19)|^q \quad \leq \quad \int_{\mathbf{R}^1} \rho(x_1 - y_1, t_1) dy_1 \int_{\mathbf{R}^1} \rho(x_2 - y_2, t_2) dy_2$$

$$\times \quad \left| \int_{\mathbf{R}^{n-2}} \rho(x_3 - y_3, t_3) \cdots \rho(x_n - y_n, t_n) f(y) dy_3 \cdots dy_n \right|^q.$$

Repeating this argument n times gives

$$|(18.19)|^q \quad \leq \quad \int_{\mathbf{R}^1} \rho(x_1 - y_1, t_1) dy_1 \int_{\mathbf{R}^1} \rho(x_2 - y_2, t_2) dy_2$$

$$\cdots \int_{\mathbf{R}^1} \rho(x_n - y_n, t_n) |f(y)|^q dy_n$$

wich implies (18.18).   □

Proof of Theorem 18.2. Let $q \in (1/(a+1), p)$. Let

$$\phi_1, \cdots, \phi_n \in \mathcal{B}_a(\mathbf{R}^1).$$

Fix $((x_2, t_2), \cdots, (x_n, t_n)) \in (\mathbf{R}_+^2)^{n-1}$ and put

$$g_1(y_1) = \int_{\mathbf{R}^{n-1}} \frac{1}{t_2} \phi_2\left(\frac{x_2 - y_2}{t_2}\right) \cdots \frac{1}{t_n} \phi_n\left(\frac{x_n - y_n}{t_n}\right) f(y) dy_2 \cdots dy_n.$$

Since

$$\int_{\mathbf{R}^1} |g_1(y_1)|^2 \, dy_1 \le C \frac{1}{t_2 \cdots t_n} \|f\|_{L^2(\mathbf{R}^n)}^2 < \infty,$$

applying Theorem 9.3 to $g_1 \in L^2(\mathbf{R}^1)$ gives

$$\sup_{t_1>0} \left| \int_{\mathbf{R}^n} \frac{1}{t_1} \phi_1\left(\frac{x_1 - y_1}{t_1}\right) \frac{1}{t_2} \phi_2\left(\frac{x_2 - y_2}{t_2}\right) \cdots \frac{1}{t_n} \phi_n\left(\frac{x_n - y_n}{t_n}\right) f(y) dy \right|^q$$

$$= \sup_{t_1>0} \left| \int_{\mathbf{R}^1} \frac{1}{t_1} \phi_1\left(\frac{x_1 - y_1}{t_1}\right) g_1(y_1) dy_1 \right|^q \le G_a g_1(x_1)$$

$$\le C(a,q) \sup_{s_1>0} \frac{1}{2s_1} \int_{x_1-s_1}^{x_1+s_1} dz_1 \sup_{t_1>0} \left| \int_{\mathbf{R}^1} \rho(z_1 - y_1, t_1) g(y_1) dy_1 \right|^q$$

$$= C(a,q) \sup_{s_1>0} \frac{1}{2s_1} \int_{x_1-s_1}^{x_1+s_1} dz_1$$

$$\times \sup_{t_1>0} \left| \int_{\mathbf{R}^n} \rho(z_1 - y_1, t_1) \frac{1}{t_2} \phi_2\left(\frac{x_2 - y_2}{t_2}\right) \cdots \frac{1}{t_n} \phi_n\left(\frac{x_n - y_n}{t_n}\right) f(y) dy \right|^q.$$

$$(18.20)$$

Next repeating the same argument with respect to the second variable we get

$$\sup_{t_2>0} \left| \int_{\mathbf{R}^n} \rho(z_1 - y_1, t_1) \frac{1}{t_2} \phi_2\left(\frac{x_2 - y_2}{t_2}\right) \cdots \frac{1}{t_n} \phi_n\left(\frac{x_n - y_n}{t_n}\right) f(y) dy \right|^q$$

$$= \sup_{t_2>0} \left| \int_{\mathbf{R}^1} \frac{1}{t_2} \phi_2\left(\frac{x_2 - y_2}{t_2}\right) dy_2 \int_{\mathbf{R}^{n-1}} \rho(z_1 - y_1, t_1) \frac{1}{t_3} \phi_3\left(\frac{x_3 - y_3}{t_3}\right) \right.$$

$$\left. \cdots \frac{1}{t_n} \phi_n\left(\frac{x_n - y_n}{t_n}\right) f(y) dy_1 dy_3 \cdots dy_n \right|^q$$

$$\le C(a,q) \sup_{s_2>0} \frac{1}{2s_2} \int_{x_2-s_2}^{x_2+s_2} dz_2 \sup_{t_2>0} \left| \int_{\mathbf{R}^n} \rho(z_1 - y_1, t_1) \rho(z_2 - y_2, t_2) \right.$$

$$\left. \times \frac{1}{t_3} \phi_3\left(\frac{x_3 - y_3}{t_3}\right) \cdots \frac{1}{t_n} \phi_n\left(\frac{x_n - y_n}{t_n}\right) f(y) dy \right|^q. \quad (18.21)$$

Combining (18.20) and (18.21) gives

$$\sup_{t_1,t_2>0}\left|\int_{\mathbf{R}^n}\frac{1}{t_1}\phi_1\left(\frac{x_1-y_1}{t_1}\right)\frac{1}{t_2}\phi_2\left(\frac{x_2-y_2}{t_2}\right)\cdots\frac{1}{t_n}\phi_n\left(\frac{x_n-y_n}{t_n}\right)f(y)dy\right|^q$$

$$\le C(a,q)^2\sup_{s_1>0}\frac{1}{2s_1}\int_{x_1-s_1}^{x_1+s_1}dz_1\sup_{s_2>0}\frac{1}{2s_2}\int_{x_2-s_2}^{x_2+s_2}dz_2$$

$$\times\sup_{t_1,t_2>0}\left|\int_{\mathbf{R}^n}\rho(z_1-y_1,t_1)\rho(z_2-y_2,t_2)\right.$$

$$\left.\times\frac{1}{t_3}\phi_3\left(\frac{x_3-y_3}{t_3}\right)\cdots\frac{1}{t_n}\phi_n\left(\frac{x_n-y_n}{t_n}\right)f(y)dy\right|^q.$$

Repeating this process n times gives

$$G_a^*f(x)^q\ \le\ C(a,q)^n\sup_{s_1>0}\frac{1}{2s_1}\int_{x_1-s_1}^{x_1+s_1}dz_1\cdots\sup_{s_n>0}\frac{1}{2s_n}\int_{x_n-s_n}^{x_n+s_n}dz_n$$

$$\times\sup_{t_1,\cdots,t_n>0}\left|\int_{\mathbf{R}^n}\rho(z_1-y_1,t_1)\cdots\rho(z_n-y_n,t_n)f(y)dy\right|^q.$$

Combining this with Lemma 18.2 gives

$$G_a^*f(x)^p\le\left\{C(a,q)^nM_1M_2\cdots M_nM_1M_2\cdots M_n\big(|f|^q\big)(x)\right\}^{p/q}.$$

Therefore, the condition $p/q>1$ and the repeated use of the 1-dimensional Hardy-Littlewood maximal theorem imply (18.17). $\square$

Notes. Corollary 18.2 is due to L. Carleson [76]. He showed it from Lemma 18.2. Theorem 18.1 is due to R. Coifman–B. Dahlberg [79]. They showed it on nonisotropic Hardy spaces and extended Corollary 18.2 there.

XIX. Fefferman's original proof of

$$\left| \int_{\mathbf{R}^n} \vec{R}f(x) \cdot \vec{g}(x)\,dx \right| \leq C(n)\|\vec{R}f\|_{L^1}\|\vec{g}\|_{\text{BMO}} \qquad (19.1)$$

where

$$f \in L^2(\mathbf{R}^n, \mathbf{R}), \ \vec{g} \in L^2(\mathbf{R}^n, \mathbf{R}^{n+1}) \qquad (19.2)$$

In Sections 19–20, we let

$$\Delta = D_t^2 + D_{x_1}^2 + \cdots + D_{x_n}^2, \ \nabla = \nabla_{t,x} = (D_t, D_{x_1}, \cdots, D_{x_n}),$$

$$|\nabla u| = \left\{ \sum_{j=0}^{n}(D_{x_j}u)^2 \right\}^{1/2},$$

$$|\nabla^2 u| = \left\{ \sum_{j=0}^{n}\sum_{i=0}^{n}(D_{x_j}D_{x_i}u)^2 \right\}^{1/2}, \ \text{where } D_{x_0} = D_t.$$

Lemma 17.1 implies the existence of a harmonic function $u(x,t)$ on $\mathbf{R}_+^{n+1}$ such that

$$P(\cdot,t) * \vec{R}f(x) = \nabla u(x,t) \ \text{on } \mathbf{R}_+^{n+1}.$$

We may assume $\nabla u \not\equiv \vec{0}$. Let

$$E = \left\{ (x,t) \in \mathbf{R}_+^{n+1} : \nabla u(x,t) = \vec{0} \right\}.$$

Then, the harmonicity of u implies

$$|E| = 0. \qquad (19.3)$$

Theorem 14.1 with $m = 1$, $q = 1$ and with $n+1$ in place of n implies

$$\frac{|\nabla^2 u(x,t)|^2}{|\nabla u(x,t)|} \leq (n+1)\Delta|\nabla u(x,t)| \ \text{on } \mathbf{R}_+^{n+1}\backslash E. \qquad (19.4)$$

So, Theorem 16.4 implies that the distributional Laplacian $\Delta|\nabla u|$ is a positive measure on $\mathbf{R}_+^{n+1}$. Then, (19.3)–(19.4) imply

$$\frac{|\nabla^2 u|^2}{|\nabla u|} \leq (n+1)\Delta|\nabla u| \ \text{on } \mathbf{R}_+^{n+1} \qquad (19.5)$$

in the sense of measures.

Take $\phi \in L^1(\mathbf{R}^n)$, depending only on the dimension n, such that

$$c(\phi)\phi \in \mathcal{B}_1^0 \text{ for some } c(\phi) > 0,$$

$$\int_0^{+\infty} \mathcal{F}(D_t P(\cdot,t))(\xi)\mathcal{F}\phi(t\xi)dt = 1 \text{ for any } \xi \in \mathbf{R}^n\backslash\{0\}.$$

Let

$$\vec{v}(x,t) = \vec{g} * (\phi)_t(x).$$

Then

$$\left| \int \vec{R}f(x) \cdot \vec{g}(x)dx \right| = \left| \iint_{\mathbf{R}_+^{n+1}} D_t P(\cdot,t) * \vec{R}f(x) \cdot (\phi)_t * \vec{g}(x)dxdt \right|$$

$$\text{by Plancherel's theorem}$$

$$= \left| \iint D_t \nabla u(x,t) \cdot \vec{v}(x,t)dxdt \right|$$

$$\leq \left\{ \iint \frac{|\nabla^2 u(x,t)|^2}{|\nabla u(x,t)|}tdxdt \right\}^{1/2} \left\{ \iint |\nabla u(x,t)||\vec{v}(x,t)|^2 dxdt/t \right\}^{1/2}$$

$$= (19.6)^{1/2} \cdot (19.7)^{1/2}, \text{ say.}$$

For $r > 1$ let

$$D_r = \{(x,t) \in \mathbf{R}_+^{n+1} : |x|^2 + t^2 < r^2, \ t > 1/r\}.$$

Then

$$(19.6) \leq (n+1) \iint_{\mathbf{R}_+^{n+1}} t\Delta|\nabla u|dxdt \text{ by } (19.5)$$

$$= (n+1) \lim_{r \to +\infty} \iint_{D_r} t\Delta|\nabla u|dxdt$$

$$= (n+1) \lim_{r \to +\infty} \liminf_{s \to +0} \iint_{D_r} t\Delta((\psi)_s * |\nabla u|)dxdt$$

$$\text{where } \psi \in \mathcal{D}(\mathbf{R}^{n+1}), \ \psi \geq 0, \ \iint_{\mathbf{R}^{n+1}} \psi(x,t)dxdt = 1$$

$$(\psi)_s(x,t) = s^{-n-1}\psi(x/s,t/s) \text{ and where the convolution}$$

$$* \text{ is taken on } \mathbf{R}^{n+1}$$

$$\leq (n+1) \liminf_{r \to +\infty} \liminf_{s \to +0} \int_{(x,t)\in\partial D_r} \{(\psi)_s * |\nabla u| + t|\nabla((\psi)_s * |\nabla u|)|\} d\sigma_n$$

$$\text{by Green's theorem, where } d\sigma_n \text{ denotes the area of } \partial D_r$$

$$\leq \cdots\cdots\cdots\cdots \quad \{(\psi)_s * |\nabla u| + t \cdot (\psi)_s * |\nabla|\nabla u||\}d\sigma_n$$

$$\leq \cdots\cdots\cdots\cdots \quad \{(\psi)_s * |\nabla u| + t \cdot (\psi)_s * |\nabla^2 u|\}d\sigma_n$$

$$\text{because since } \nabla u \text{ is a } C^1\text{-function, the}$$

$$\text{distributional derivative } \nabla|\nabla u| \text{ is locally bounded}$$

$$\text{and } \left|\nabla|\nabla u|\right| \le |\nabla^2 u|,$$

$$= (n+1)\liminf_{r\to+\infty}\int_{\partial D_r}\{|\nabla u|+t|\nabla^2 u|\}\, d\sigma_n$$

$$\le C(n)\liminf_{r\to+\infty}\|N_1(|\nabla u|+t|\nabla^2 u|)\|_{L^1}\|\{d\sigma_n \text{ on } \partial D_r\}\|_{\mathcal{C}} \quad \text{by Lemma 13.5}$$

$$\le C(n)\left\|N_1(|\nabla u|+t|\nabla^2 u|)\right\|_{L^1}$$

$$\le C(n)\left\|N_2(|\nabla u|)\right\|_{L^1} \text{ by } N_1(t|\nabla^2 u|)\le C(n)N_2(|\nabla u|) \text{ which follows}$$
$$\text{from the same reason as (11.10).} \tag{19.8}$$

$$(19.7)\le C(n)\left\|N_1(|\nabla u|)\right\|_{L^1}\left\||\vec v(x,t)|^2\, dxdt/t\right\|_{\mathcal{C}} \quad \text{by Lemma 13.5}$$

$$\le C(n)\left\|N_1(|\nabla u|)\right\|_{L^1}\|\vec g\|_{\mathrm{BMO}}^2 \qquad \text{by Lemma 12.1.} \tag{19.9}$$

Therefore, combining (19.8)–(19.9) implies

$$(19.6)^{1/2}\cdot(19.7)^{1/2}\le C(n)\left\|N_2(|\nabla u|)\right\|_{L^1}\|\vec g\|_{\mathrm{BMO}}$$

$$\le C(q,n)\big\|M_q(|\vec R f|)\big\|_{L^1}\|\vec g\|_{\mathrm{BMO}} \text{ by (17.5) where } q\in[(n-1)/n,1)\cap(0,1)$$

$$\le C(q,n)\big\|\vec R f\big\|_{L^1}\|\vec g\|_{\mathrm{BMO}} \text{ by Corollary 0.4.}$$

Notes. The argument in this section is a modification of C. Fefferman's proof of his epoch-making H^1-BMO duality theorem, which is in C. Fefferman–E. Stein [72] p.p.147–148.

XX. Varopoulos's proof of (19.1), where

$$\vec{R}f \in L^1(\mathbf{R}^n, \mathbf{R}^{n+1}) \cap L^\infty(\mathbf{R}^n, \mathbf{R}^{n+1}), \ \vec{g} \in \mathrm{BMO}(\mathbf{R}^n, \mathbf{R}^{n+1})$$
$$(20.1)$$

and supp $\vec{g}$ is compact

Lemma 20.1. *Let $g \in \mathrm{BMO}(\mathbf{R}^n, \mathbf{R})$ and supp g be compact. Then there exists $v(x,t) \in C^\infty(\mathbf{R}^{n+1}_+, \mathbf{R})$ such that supp v is bounded,*

$$\||\nabla v(x,t)|\, dxdt\|_{\mathcal{C}} \leq C(n)\|g\|_{\mathrm{BMO}}, \ \ where \ \nabla = \nabla_{t,x}, \quad (20.2)$$

$$\left|g(x) - \lim_{t \to +0} v(x,t)\right| \leq C(n)\|g\|_{\mathrm{BMO}} \ \ a.e. \ x \in \mathbf{R}^n. \quad (20.3)$$

Proof. We may assume

$$\|g\|_{\mathrm{BMO}} \leq 1 \ \ and \ \ \mathrm{supp}\, g \subset [1/4,\ 3/4]^n. \quad (20.4)$$

Let $\{h_j\}_{j=-\infty}^0$ be as in Lemma 13.3. Let

$$\psi \in C^\infty\big((0,+\infty)\big), \ \psi(t) \equiv 1 \ \text{on} \ (0,1/2], \ \psi(t) \equiv 0 \ \text{on} \ [1,+\infty),$$

$$\eta \in \mathcal{D}(\mathbf{R}^n), \ \eta(x) \geq 0, \ \int \eta(x)dx = 1, \ \mathrm{supp}\, \eta \subset B(0,1).$$

Let

$$v(x,t) = \sum_{j=-\infty}^0 h_j * (\eta)_{2^{-j^2}}(x)\psi(t/2^j).$$

Then

$$v \in C^\infty(\mathbf{R}^{n+1}_+),$$

$$|\nabla v(x,t)| \leq C(\psi, \eta) \sum_{j=-\infty}^0 \Big\{ 2^{-j}|h_j| * (\eta)_{2^{-j^2}}(x)\chi_{[2^{j-1},\ 2^j]}(t)$$

$$+ |\nabla_x h_j| * (\eta)_{2^{-j^2}}(x)\chi_{(0,2^j]}(t)\Big\}.$$

Thus, for any ball $B \subset \mathbf{R}^n$

$$\iint_{Q(B)} |\nabla v(x,t)|\, dxdt$$

$$\leq \sum_{2^j \leq 2\ell(B)} \int_{2B} \big(|h_j(x)| + 2^j|\nabla_x h_j(x)|\big)\, dx + \sum_{2^j > \ell(B)} \|\nabla_x h_j\|_{L^\infty}\ell(B)|B|$$

$$\leq C|B| \quad \text{by } (13.11)', \ (13.13)' \ \text{and} \ (13.16),$$

which implies (20.2). Since

$$\sum_{j=-\infty}^{0} \|h_j - h_j * (\eta)_{2^{-j^2}}\|_{L^\infty} \leq C \sum \|\nabla_x h_j\|_{L^\infty} 2^{-j^2} \leq C \text{ by (13.16)}$$

and since

$$\lim_{t\to+0} v(x,t) = \sum_{j=-\infty}^{0} h_j * (\eta)_{2^{-j^2}}(x) \text{ a.e. } x \in \mathbf{R}^n$$

$$\text{by } \sum \|h_j * (\eta)_{2^{-j^2}}\|_{L^1} \leq \sum \|h_j\|_{L^1} < \infty,$$

(20.3) follows from (13.12)'.   □

Now, we begin the proof of (19.1). Applying Lemma 20.1 to each component of $\vec{g} = (g_0, \cdots, g_n)$ gives $\vec{v}(x,t) = (v_0(x,t), \cdots, v_n(x,t)) \in \mathbf{C}^\infty(\mathbf{R}_+^{n+1}, \mathbf{R}^{n+1})$ such that supp $\vec{v}$ is bounded,

$$\| |\nabla \vec{v}(x,t)| dx dt \|_{\mathcal{C}} \leq C \|\vec{g}\|_{\text{BMO}}, \qquad (20.2)'$$

$$\left| \vec{g}(x) - \lim_{t\to+0} \vec{v}(x,t) \right| \leq C \|\vec{g}\|_{\text{BMO}}, \text{ a.e. } x \in \mathbf{R}^n. \qquad (20.3)'$$

Then

$$\int \vec{R}f(x) \cdot \vec{g}(x) dx$$

$$= \int \vec{R}f(x) \cdot \lim_{t\to+0} \vec{v}(x,t) dx + \int \vec{R}f(x) \cdot \left\{ \vec{g}(x) - \lim_{t\to+0} \vec{v}(x,t) \right\} dx$$

$$= (20.5) + (20.6), \text{ say.}$$

By (20.3)' we have

$$|(20.6)| \leq C \|\vec{R}f\|_{L^1} \|\vec{g}\|_{\text{BMO}}.$$

The hard part is (20.5). Let

$$\vec{u}(x,t) = (u_0(x,t), \cdots, u_n(x,t)) = P(\cdot,t) * \vec{R}f(x).$$

Then

$$-(20.5) = -\int \lim_{\varepsilon\to+0} \vec{u}(x,\varepsilon) \cdot \vec{v}(x,\varepsilon) dx$$

$$\text{by } \vec{u}(x,\varepsilon) \to \vec{R}f(x) \ (\varepsilon \to +0) \text{ a.e. } x \in \mathbf{R}^n$$

$$= -\lim_{\varepsilon\to+0} \int \vec{u}(x,\varepsilon) \cdot \vec{v}(x,\varepsilon) dx$$

$$\text{by } \int \sup_{\varepsilon>0} |\vec{u}(x,\varepsilon) \cdot \vec{v}(x,\varepsilon)| dx \leq \|\vec{R}f\|_{L^\infty} \iint |D_t\vec{v}| dx dt < \infty$$

$$= \lim_{\varepsilon\to+0} \left\{ \iint_{\mathbf{R}^n \times (\varepsilon,+\infty)} \vec{u}(x,t) \cdot D_t\vec{v}(x,t) dx dt + \iint_{\mathbf{R}^n \times (\varepsilon,+\infty)} D_t\vec{u}(x,t) \cdot \vec{v}(x,t) dx dt \right\}$$

$$= \lim\{(20.7)_\varepsilon + (20.8)_\varepsilon\}, \text{ say.}$$

Since
$$D_t \vec{u} = \left(-\sum_{j=1}^n D_{x_j} u_j, D_{x_1} u_0, \cdots, D_{x_n} u_0 \right),$$

integrations by parts imply

$$|(20.8)_\varepsilon| = \left| \iint_{\mathbf{R}^n \times (\varepsilon, +\infty)} \left\{ \left(-\sum_{j=1}^n D_{x_j} u_j \right) v_0 + \sum_{j=1}^n (D_{x_j} u_0) v_j \right\} dx dt \right|$$

$$= \left| \iint_{\mathbf{R}^n \times (\varepsilon, +\infty)} \left\{ \sum u_j D_{x_j} v_0 - \sum u_0 D_{x_j} v_j \right\} dx dt \right|$$

$$\le C \iint_{\mathbf{R}^{n+1}_+} |\vec{u}(x,t)| |\nabla \vec{v}(x,t)| dx dt.$$

Then,

$$|(20.7)_\varepsilon| + |(20.8)_\varepsilon| \le C \iint |\vec{u}(x,t)| |\nabla \vec{v}(x,t)| dx dt$$

$$\le C \left\| N_1(\vec{R}f) \right\|_{L^1} \||\nabla \vec{v}(x,t)| dx dt\|_C \quad \text{by Lemma 13.5}$$

$$\le C \| M_q(|\vec{R}f|) \|_{L^1} \|\vec{g}\|_{\mathrm{BMO}} \quad \text{by (17.5) and (20.2)}', \text{ where}$$
$$q \in [(n-1)/n,\ 1) \cap (0,1)$$

$$\le C \|\vec{R}f\|_{L^1} \|\vec{g}\|_{\mathrm{BMO}} \quad \text{by Corollary 0.4.}$$

Combining the above estimates gives (19.1).

Remark 20.1. Theorem 5.1 of P. W. Jones [78] shows that we can replace (20.3) in Lemma 20.1 by

$$g(x) = \lim_{t \to +0} v(x,t) \quad \text{a.e. } x \in \mathbf{R}^n. \tag{20.3}'$$

This is an easy consequence of Theorem 2.2 of P. W. Jones [78] which is listed in Remark 13.2.

Notes. The argument in this section is a modification of N. Th. Varopoulos [77].

The comparison between Lemma 20.1 and Remark 12.3 is very interesting. As for the relation between $\|t|\nabla v|^2 dx dt\|_C$ and $\||\nabla v| dx dt\|_C$, important results have been obtained by J. B. Garnett [81] p. 348, B. Dahlberg [80a] and N. Th. Varopouls [78].

A similar situation occurs in the proof of the corona theorem. L. Carleson's proof corresponds to Lemma 20.1 and T. Wolff's proof corresponds to Remark 12.3. (For Carleson's argument, see L. Carleson [62], [70] and J. B. Garnett [81] p. 342. For Wolff's argument, see T. Gamelin [80], P. Koosis [80] p. 369 and J. B. Garnett [81] p. 325.)

Therse matters are very precisely discussed in J. B. Garnett [81].

XXI. The Fefferman-Stein decomposition of BMO

In sections 17–20 we investigated our theory from the viewpoint of H^p, especially H^1. In the following part of this book, we will investigate our theory from the viewpoint of the dual spaces of H^p, especially BMO (= the dual space of H^1).

Theorem 16.1 implies that if $\theta \in C^\infty(S^{n-1}, \mathbf{C})$, then m_θ can be extended uniquely as a bounded operator $m_\theta : H^1(\mathbf{R}^n, \mathbf{C}) \to H^1(\mathbf{R}^n, \mathbf{C})$. (It is easy to see that it is the same thing with $\mathcal{F}^{-1}\big(\theta(\xi/|\xi|)\mathcal{F}f(\xi)\big)$ defined in the sense of tempered distributions.) So, we can consider its adjoint operator.

Remark 21.1. (Important.) As we noticed in Remark 16.1, we identify the point $(p_1, \cdots, p_m) \in \mathbf{C}^m$ with $(\mathrm{Re}\,p_1, \mathrm{Im}\,p_1, \cdots, \mathrm{Re}\,p_m, \mathrm{Im}\,p_m) \in \mathbf{R}^{2m}$. Neglecting the complex structure of $\mathbf{C}^m$, for $\vec{p} = (p_1, \cdots, p_m)$, $\vec{q} = (q_1, \cdots, q_m) \in \mathbf{C}^m$ and for $\vec{\nu} = (\nu_1, \cdots, \nu_{2m}) \in \mathbf{R}^{2m}$ we define the inner products $\vec{p} \cdot \vec{q}$ and $\vec{p} \cdot \vec{\nu}$ by

$$\vec{p} \cdot \vec{q} = \sum_{j=1}^m \{(\mathrm{Re}\,p_j)(\mathrm{Re}\,q_j) + (\mathrm{Im}\,p_j)(\mathrm{Im}\,q_j)\},$$

$$\vec{p} \cdot \vec{\nu} = \sum_{j=1}^m \{(\mathrm{Re}\,p_j)\nu_{2j-1} + (\mathrm{Im}\,p_j)\nu_{2j}\}.$$

As we noticed in Remark 16.1, $H^1(\mathbf{R}^n, \mathbf{C}^m)$ and $\mathrm{BMO}(\mathbf{R}^n, \mathbf{C}^m)$ are defined to be the same things with $H^1(\mathbf{R}^n, \mathbf{R}^{2m})$ and $\mathrm{BMO}(\mathbf{R}^n, \mathbf{R}^{2m})$, respectively. For $\vec{f} = (f_1, \cdots, f_m) \in H^1(\mathbf{R}^n, \mathbf{C}^m)$ and $\vec{g} = (g_1, \cdots, g_m) \in \mathrm{BMO}(\mathbf{R}^n, \mathbf{C}^m)$ let

$$\Big\langle \vec{g}, \vec{f} \Big\rangle_{\mathrm{BMO}(\mathbf{R}^n, \mathbf{R}^{2m})}$$
$$= \Big\langle (\mathrm{Re}\,g_1, \mathrm{Im}\,g_1, \cdots, \mathrm{Im}\,g_m), (\mathrm{Re}\,f_1, \mathrm{Im}\,f_1, \cdots, \mathrm{Im}\,f_m) \Big\rangle_{\mathrm{BMO}(\mathbf{R}^n, \mathbf{R}^{2m})}.$$

In particular, if $f \in H^1(\mathbf{R}^n, \mathbf{C})$ and $g \in \mathrm{BMO}(\mathbf{R}^n, \mathbf{C})$, then

$$\langle g, f \rangle_{\mathrm{BMO}(\mathbf{R}^n, \mathbf{R}^2)} = \langle \mathrm{Re}\,g, \mathrm{Re}\,f \rangle_{\mathrm{BMO}(\mathbf{R}^n, \mathbf{R})} + \langle \mathrm{Im}\,g, \mathrm{Im}\,f \rangle_{\mathrm{BMO}(\mathbf{R}^n, \mathbf{R})}.$$

Definition 21.1. For $g \in \mathrm{BMO}(\mathbf{R}^n, \mathbf{C})$, and $\theta \in C^\infty(S^{n-1}, \mathbf{C})$ let $\tilde{m}_\theta g \in \mathrm{BMO}(\mathbf{R}^n, \mathbf{C})$, be such that

$$\langle \tilde{m}_\theta g, f \rangle_{\mathrm{BMO}(\mathbf{R}^n, \mathbf{R}^2)} = \langle g, m_{\bar\theta} f \rangle_{\mathrm{BMO}(\mathbf{R}^n, \mathbf{R}^2)}$$

for all $f \in H^1(\mathbf{R}^n, \mathbf{C})$. ($\tilde{m}_\theta g$ is determined modulo constants. So, to be precise, $\tilde{m}_\theta$ is an operator from $\mathrm{BMO}(\mathbf{R}^n, \mathbf{C})/\mathbf{C}$ into itself. It is easy to see $\tilde{m}_\theta(ag + bh) = a\tilde{m}_\theta g + b\tilde{m}_\theta h$ for any $g, h \in \mathrm{BMO}(\mathbf{R}^n, \mathbf{C})$ and $a, b \in \mathbf{C}$.)

Definition 21.2. For $g \in \mathrm{BMO}(\mathbf{R}^n, \mathbf{C})$ let

$$\tilde{R}_j g = \tilde{m}_{-\xi_j/|\xi|} g \quad (j = 1, \cdots, n), \quad \tilde{R}_0 g = \tilde{m}_1 g.$$

For $g \in \mathrm{BMO}(\mathbf{R}^1, \mathbf{C})$ let $\tilde{H}g = \tilde{R}_1 g$.

Remark 21.2. Theorem 16.1 implies

$$\|\tilde{m}_\theta g\|_{\mathrm{BMO}} \le C(\theta)\|g\|_{\mathrm{BMO}}. \tag{21.1}$$

Remark 21.3. If $g \in \mathrm{BMO}(\mathbf{R}^n, \mathbf{C}) \cap L^2(\mathbf{R}^n, \mathbf{C})$, $f \in \mathcal{S}(\mathbf{R}^n, \mathbf{C})$ and $\int f \, dx = 0$, then

$$\int \tilde{m}_\theta g(x) \overline{f(x)} \, dx = \langle \tilde{m}_\theta g, f \rangle_{\mathrm{BMO}(\mathbf{R}^n, \mathbf{R}^2)} + i\langle \tilde{m}_\theta g, if \rangle_{\mathrm{BMO}(\mathbf{R}^n, \mathbf{R}^2)}$$

by Lemma 1.10 and by Theorem 2.2

$$= \langle g, m_{\bar\theta} f \rangle_{\mathrm{BMO}(\mathbf{R}^n, \mathbf{R}^2)} + i\langle g, im_{\bar\theta} f \rangle_{\mathrm{BMO}(\mathbf{R}^n, \mathbf{R}^2)}$$

$$= \int g(x) \overline{m_{\bar\theta} f(x)} \, dx = \int m_\theta g(x) \overline{f(x)} \, dx$$

by $f, g \in L^2$ and by Theorem 2.2.

Therefore, if $g \in \mathrm{BMO}(\mathbf{R}^n, \mathbf{C}) \cap L^2(\mathbf{R}^n, \mathbf{C})$, then

$$\tilde{m}_\theta g = m_\theta g \quad \text{in } \mathcal{S}(\mathbf{R}^n \mathbf{C})'_c/\mathbf{C}.$$

Remark 21.4. If $\overline{\theta(\xi)} \equiv \theta(-\xi)$, then $\tilde{m}_\theta$ can be regarded as a bounded operator from $\mathrm{BMO}(\mathbf{R}^n, \mathbf{R})/\mathbf{R}$ into itself. In particular, $\tilde{R}_j$'s can be regarded so.

Theorem 21.1. *Let $\{\theta_1, \cdots, \theta_m\} \subset C^\infty(S^{n-1}, \mathbf{C})$. Then the following two conditions are equivalent:*

$$\sup\left\{ \frac{\|f\|_{H^1}}{\sum\limits_{j=1}^m \|m_{\bar\theta_j} f\|_{L^1}} : f \in H^1(\mathbf{R}^n, \mathbf{C}) \backslash \{0\} \right\} < \infty, \tag{21.2}$$

$$\mathrm{BMO}(\mathbf{R}^n, \mathbf{C})/\mathbf{C} = \sum_{j=1}^m \tilde{m}_{\theta_j} L^\infty(\mathbf{R}^n, \mathbf{C}). \tag{21.3}$$

Proof. The operator

$$f \in H^1(\mathbf{R}^n, \mathbf{C}) \;\to\; (m_{\bar\theta_j} f)_{j=1,\cdots,m} \in L^1(\mathbf{R}^n, \mathbf{C}^m)$$

has a bounded inverse if and only if its adjoint operator

$$(k_j)_{j=1,\cdots,m} \in L^\infty(\mathbf{R}^n, \mathbf{C}^m) \;\to\; \sum_{j=1}^m \tilde{m}_{\theta_j} k_j \in \mathrm{BMO}(\mathbf{R}^n, \mathbf{C})/\mathbf{C}$$

is surjective. $\qquad\qquad\qquad\qquad\qquad\qquad\qquad\qquad\qquad\qquad\qquad\square$

Noticing Remarks 21.4, 16.5 and 16.7 we have

Corollary 21.1. *Let* $\{\theta_1, \cdots, \theta_m\} \subset C^\infty(S^{n-1}, \mathbf{C})$. *Let*

$$\bar\theta_j(\xi) \equiv \theta_j(-\xi) \quad (j = 1, \cdots, m).$$

Then the following are equivalent :

$$\sup\left\{ \frac{\|f\|_{H^1}}{\displaystyle\sum_{j=1}^m \|m_{\bar\theta_j} f\|_{L^1}} : f \in H^1(\mathbf{R}^n, \mathbf{R})\backslash\{0\} \right\} < \infty,$$

$$\mathrm{BMO}(\mathbf{R}^n, \mathbf{R})/\mathbf{R} = \sum_{j=1}^m \tilde{m}_{\theta_j} L^\infty(\mathbf{R}^n, \mathbf{R}).$$

Corollary 21.2. *Let* $\{\theta_1, \cdots, \theta_m\} \subset C^\infty(S^{n-1}, \mathbf{C})$. *Then the following two conditions are equivalent :*

$$\sup\left\{ \frac{\|f\|_{H^1}}{\displaystyle\sum_{j=1}^m \|m_{\bar\theta_j} f\|_{L^1}} : f \in H^1(\mathbf{R}^n, \mathbf{R})\backslash\{0\} \right\} < \infty,$$

$$\mathrm{BMO}(\mathbf{R}^n, \mathbf{R})/\mathbf{R} = \sum_{j=1}^m \left\{ \tilde{m}_{\theta_{j\$}} L^\infty(\mathbf{R}^n, \mathbf{R}) + \tilde{m}_{\theta_{j\ell}} L^\infty(\mathbf{R}^n, \mathbf{R}) \right\}$$

$$= \mathrm{Re} \sum_{j=1}^m \tilde{m}_\theta L^\infty(\mathbf{R}^n, \mathbf{C}).$$

Corollary 21.1 combined with the case $p = m = 1$ of Theorem 17.1 implies the following which is called the Fefferman-Stein decomposition of $\mathrm{BMO}(\mathbf{R}^n)$.

Corollary 21.3. $\mathrm{BMO}(\mathbf{R}^n, \mathbf{R})/\mathbf{R} = \displaystyle\sum_{j=1}^m \tilde{R}_j L^\infty(\mathbf{R}^n, \mathbf{R}).$

Next, we look at the above theorem from a little bit general point of view.

Definition 21.3. For a subspace $S \subset H^1(\mathbf{R}^n, \mathbf{R}^m)$ let

$$S_* = \left\{ \vec{g} \in \mathrm{BMO}(\mathbf{R}^n, \mathbf{R}^m) : \langle \vec{g}, \vec{f} \rangle_{\mathrm{BMO}(\mathbf{R}^n, \mathbf{R}^m)} = 0 \text{ for all } \vec{f} \in S \right\}.$$

Theorem 21.2. *Let $S \subset H^1(\mathbf{R}^n, \mathbf{R}^m)$ be a subspace. Then the following two values coincide :*

$$\sup \left\{ \frac{\|\vec{f}\|_{H^1}}{\|\vec{f}\|_{L^1}} : \vec{f} \in S \backslash \{0\} \right\}, \tag{21.4}$$

$$\sup \left\{ \frac{\inf \left\{ \|\vec{g} - \vec{h}\|_{L_\infty} : \vec{h} \in S_* \right\}}{\|\vec{g}\|_{\mathrm{BMO}}} : \vec{g} \in \mathrm{BMO}(\mathbf{R}^n, \mathbf{R}^m) \backslash \mathbf{R}^m \right\} \tag{21.5}$$

Proof. The dual space of "S as a subspace of $H^1(\mathbf{R}^n, \mathbf{R}^m)$" is

$$\mathrm{BMO}(\mathbf{R}^n, \mathbf{R}^m)/S_*.$$

The dual space of "S as a subspace of $L^1(\mathbf{R}^n, \mathbf{R}^m)$" is

$$(L^\infty(\mathbf{R}^n, \mathbf{R}^m) + S_*)/S_*$$

endowed with the norm

$$\|\vec{g} + S_*\| = \inf \left\{ \|\vec{g} - \vec{h}\|_{L^\infty} : \vec{h} \in S_* \right\}. \tag{21.6}$$

So,

$$(21.4) = \left\{ \text{the operator norm of} \right.$$
$$\vec{f} \in S \ (\subset L^1(\mathbf{R}^n, \mathbf{R}^m)) \ \rightarrow \ \vec{f} \in S \ (\subset H^1(\mathbf{R}^n, \mathbf{R}^m)) \left. \right\}$$
$$= \left\{ \text{the operator norm of} \right.$$
$$\vec{g} + S_* \in \mathrm{BMO}(\mathbf{R}^n, \mathbf{R}^m)/S_* \ \rightarrow \ \vec{g} + S_* \in (L^\infty(\mathbf{R}^n, \mathbf{R}^m) + S_*)/S_* \left. \right\}$$

(where if $\mathrm{BMO}(\mathbf{R}^n, \mathbf{R}^m) \not\subset L^\infty(\mathbf{R}^n, \mathbf{R}^m) + S_*$, then this operator norm is defined to be $+\infty$,)

$$= \sup \left\{ \frac{\inf \left\{ \|\vec{g} - \vec{h}\|_{L^\infty} : \vec{h} \in S_* \right\}}{\inf \left\{ \|\vec{g} - \vec{h}\|_{\mathrm{BMO}} : \vec{h} \in S_* \right\}} : \vec{g} \in \mathrm{BMO}(\mathbf{R}^n, \mathbf{R}^m) \backslash S_* \right\} = (21.5).$$

$\square$

Corollary 21.4. *Let* $S \subset H^1(\mathbf{R}^n, \mathbf{R}^m)$ *be a subspace. Then the following three conditions are equivalent.*

$$(21.4) < \infty,$$
$$(21.5) < \infty,$$
$$\mathrm{BMO}(\mathbf{R}^n, \mathbf{R}^m) = L^\infty(\mathbf{R}^n, \mathbf{R}^m) + S_*. \qquad (21.7)$$

Proof. $(21.4) < \infty \Leftrightarrow (21.5) < \infty$ by Theorem 21.2

$\Leftrightarrow$ "$(L^\infty(\mathbf{R}^n, \mathbf{R}^m) + S_*)/S_* \hookrightarrow \mathrm{BMO}(\mathbf{R}^n, \mathbf{R}^m)/S_*$" is surjective
$\Leftrightarrow (21.7)$. $\qquad\qquad\qquad\qquad\qquad\qquad\qquad\qquad\qquad\qquad\qquad\square$

Corollary 21.5. *Let* $\{\theta_1, \cdots, \theta_m\} \subset C^\infty(S^{n-1}, \mathbf{C})$,

$$S_{\{\theta_1, \cdots, \theta_m\}} = \left\{ (\mathsf{m}_{\bar{\theta}_j} f)_{j=1,\cdots,m} : f \in H^1(\mathbf{R}^n, \mathbf{C}) \right\} \subset H^1(\mathbf{R}^n, \mathbf{C}^m),$$
$$(21.8)$$

$$S_{\{\theta_1, \cdots, \theta_m\}*} = \Big\{ \vec{g} = (g_j)_{j=1,\cdots,m} \in \mathrm{BMO}(\mathbf{R}^n, \mathbf{C}^m) :$$
$$\sum_{j=1}^m \tilde{\mathsf{m}}_{\theta_j} g_j = 0 \ \ in \ \mathrm{BMO}(\mathbf{R}^n, \mathbf{C})/\mathbf{C} \Big\}. \quad (21.9)$$

Then the following three conditions are equivalent :

$$\sup \left\{ \frac{\displaystyle\sum_{j=1}^m \|\mathsf{m}_{\bar{\theta}_j} f\|_{H^1}}{\displaystyle\sum_{j=1}^m \|\mathsf{m}_{\bar{\theta}_j} f\|_{L^1}} : f \in H^1(\mathbf{R}^n, \mathbf{C}), \ \sum_{j=1}^m \|\mathsf{m}_{\bar{\theta}_j} f\|_{L^1} = 0 \right\} < \infty,$$
$$(21.10)$$

$$\sup \left\{ \frac{\inf\left\{ \|\vec{g} - \vec{h}\|_{L^\infty} : \vec{h} \in S_{\{\theta_1, \cdots, \theta_m\}*} \right\}}{\|\vec{g}\|_{\mathrm{BMO}}} : \vec{g} \in \mathrm{BMO}(\mathbf{R}^n, \mathbf{C}^m)\backslash\mathbf{C}^m \right\} < \infty,$$
$$(21.11)$$

$$\sum_{j=1}^m \tilde{\mathsf{m}}_{\theta_j} \mathrm{BMO}(\mathbf{R}^n, \mathbf{C}) = \sum_{j=1}^m \tilde{\mathsf{m}}_{\theta_j} L^\infty(\mathbf{R}^n, \mathbf{C}). \qquad (21.12)$$

Furthermore if

$$\sum_{j=1}^m |\theta_j(\xi)| = 0 \ \ for \ any \ \xi \in S^{n-1}, \qquad (21.13)$$

then the following three conditions are equivalent :

$$\sup \left\{ \frac{\|f\|_{H^1}}{\sum_{j=1}^{m} \|\mathsf{m}_{\bar\theta_j} f\|_{L^1}} : f \in H^1(\mathbf{R}^n, \mathbf{C}) \backslash \{0\} \right\} < \infty, \qquad (21.10)'$$

(21.11),

$$\mathrm{BMO}(\mathbf{R}^n, \mathbf{C})/\mathbf{C} = \sum_{j=1}^{m} \tilde{\mathsf{m}}_{\theta_j} L^\infty(\mathbf{R}^n, \mathbf{C}). \qquad (21.12)'$$

Proof. The equivalence between $\mathrm{BMO}(\mathbf{R}^n, \mathbf{C}^m) = L^\infty(\mathbf{R}^n, \mathbf{C}^m) + S_{\{\theta_1, \cdots, \theta_m\}*}$ and (21.12) is clear. So, the former half of Corollary 21.5 is a direct consequence of Corollary 21.4 (with its $\mathbf{R}^m$ and S replaced by $\mathbf{R}^{2m} (= \mathbf{C}^m)$ and $S_{\{\theta_1, \cdots, \theta_m\}}$, respectively.)

Next, assume (21.13). Theorems 16.1–16.2 imply

$$\|f\|_{H^1} \approx \sum_{j=1}^{m} \|\mathsf{m}_{\bar\theta_j} f\|_{H^1} \text{ for any } f \in H^1(\mathbf{R}^n, \mathbf{C}).$$

If we define θ_j by (16.4), then

$$\mathrm{BMO}(\mathbf{R}^n, \mathbf{C})/\mathbf{C} \supset \sum_{j=1}^{m} \tilde{\mathsf{m}}_{\theta_j} \mathrm{BMO}(\mathbf{R}^n, \mathbf{C})$$

$$\supset \sum_{j=1}^{m} \tilde{\mathsf{m}}_{\theta_j} \tilde{\mathsf{m}}_{\theta_j} \mathrm{BMO}(\mathbf{R}^n, \mathbf{C}) = \mathrm{BMO}(\mathbf{R}^n, \mathbf{C})/\mathbf{C}.$$

Thus, the latter half is a direct consequence of the former half. $\square$

Corollary 21.6. *Let* $\{\theta_1, \cdots, \theta_m\}$ *be as in Corollary 21.1. Let*

$$S_{\{\theta_1, \cdots, \theta_m\}(r)} = \left\{ (\mathsf{m}_{\bar\theta_j} f)_{j=1, \cdots, m} : f \in H^1(\mathbf{R}^n, \mathbf{R}) \right\} \subset H^1(\mathbf{R}^n, \mathbf{R}^m), \quad (21.14)$$

$$(21.15)$$

$$S_{\{\theta_1, \cdots, \theta_m\}*(r)} = \left\{ \vec{g} = (g_j)_{j=1, \cdots, m} \in \mathrm{BMO}(\mathbf{R}^n, \mathbf{R}^m) : \right.$$

$$\left. \sum_{j=1}^{m} \tilde{\mathsf{m}}_{\theta_j} g_j = 0 \text{ in } \mathrm{BMO}(\mathbf{R}^n, \mathbf{R})/\mathbf{R} \right\}. \qquad (21.16)$$

Then the following three conditions are equivalent :

$$\sup \left\{ \frac{\sum_{j=1}^{m} \|\mathsf{m}_{\bar\theta_j} f\|_{H^1}}{\sum_{j=1}^{m} \|\mathsf{m}_{\bar\theta_j} f\|_{L^1}} : f \in H^1(\mathbf{R}^n, \mathbf{R}), \sum_{j=1}^{m} \|\mathsf{m}_{\bar\theta_j} f\|_{L^1} \neq 0 \right\} < \infty, \qquad (21.17)$$

$$\sup\left\{\frac{\inf\left\{\|\vec{g}-\vec{h}\|_{L^{\infty}}:\vec{h}\in S_{\{\theta_1,\cdots,\theta_m\}*(r)}\right\}}{\|\vec{g}\|_{\mathrm{BMO}}}:\vec{g}\in\mathrm{BMO}(\mathbf{R}^n,\mathbf{R}^m)\backslash\mathbf{R}^m\right\}<\infty,$$

$$(21.18)$$

$$\sum_{j=1}^{m}\tilde{\mathfrak{m}}_{\theta_j}\mathrm{BMO}(\mathbf{R}^n,\mathbf{R})=\sum_{j=1}^{m}\tilde{\mathfrak{m}}_{\theta_j}L^{\infty}(\mathbf{R}^n,\mathbf{R}). \qquad (21.19)$$

Furthermore if (21.13) *holds, then the following three conditions are equivalent :*

$$\sup\left\{\frac{\|f\|_{H^1}}{\sum_{j=1}^{m}\|\mathfrak{m}_{\bar{\theta}_j}f\|_{L^1}}:f\in H^1(\mathbf{R}^n,\mathbf{R})\backslash\{0\}\right\}<\infty, \qquad (21.16)'$$

(21.18),

$$\mathrm{BMO}(\mathbf{R}^n,\mathbf{R})/\mathbf{R}=\sum_{j=1}^{m}\tilde{\mathfrak{m}}_{\theta_j}L^{\infty}(\mathbf{R}^n,\mathbf{R}). \qquad (21.18)'$$

Finally we add one more small theorem that we will refer to in the next section.

Theorem 21.3. *Let* $S\subset H^1(\mathbf{R}^n,\mathbf{R}^m)$ *be a subspace. Then*

$$(21.5)\le C(n)\sup\left\{\frac{\inf\left\{\|\vec{g}-\vec{h}\|_{L^{\infty}}:\vec{h}\in S_*\right\}}{\|\vec{g}\|_{\mathrm{BMO}}}:\vec{g}\in\mathcal{D}(\mathbf{R}^n,\mathbf{R}^m)\backslash\{\vec{0}\}\right\}. (21.20)$$

Proof. For $\vec{g}\in\mathrm{BMO}(\mathbf{R}^n,\mathbf{R}^m)$ and $r>0$ let

$$\mathrm{tr}(\vec{g},r)(x)=\frac{\vec{g}(x)}{\max\left\{1,|\vec{g}(x)|/r\right\}}.$$

Then the same reasonning as Theorem 2.2 shows

$$\left.\begin{array}{l}\mathrm{tr}(\vec{g},r)\in L^{\infty}(\mathbf{R}^n,\mathbf{R}^m),\\ \|\mathrm{tr}(\vec{g},r)\|_{\mathrm{BMO}}\le\|\vec{g}\|_{\mathrm{BMO}},\\ \int\mathrm{tr}(\vec{g},r)(x)\cdot\vec{f}(x)dx\ \rightarrow\ \langle\vec{g},\vec{f}\rangle_{\mathrm{BMO}}\ (r\rightarrow+\infty)\end{array}\right\} \qquad (21.21)$$

for any $\vec{f}\in H^1(\mathbf{R}^n,\mathbf{R}^m)$. Thus

$$(21.5)=\sup\left\{\frac{|\langle\vec{g},\vec{f}\rangle_{\mathrm{BMO}}|}{\|\vec{g}\|_{\mathrm{BMO}}\|\vec{f}\|_{L^1}}:\vec{f}\in S\backslash\{\vec{0}\},\ \vec{g}\in\mathrm{BMO}(\mathbf{R}^n,\mathbf{R}^m)\backslash\mathbf{R}^m\right\}$$

$$=\sup\left\{\frac{|\langle\vec{g},\vec{f}\rangle_{\mathrm{BMO}}|}{\|\vec{g}\|_{\mathrm{BMO}}\|\vec{f}\|_{L^1}}:\vec{f}\in S\backslash\{\vec{0}\},\ \vec{g}\in L^{\infty}(\mathbf{R}^n,\mathbf{R}^m)\backslash\mathbf{R}^m\right\}\quad\text{by (21.21)}$$

$$=(21.21),\text{ say.}$$

Next, let
$$\vec{g} \in L^\infty(\mathbf{R}^n, \mathbf{R}^m) \text{ and } \vec{f} \in H^1(\mathbf{R}^n, \mathbf{R}^m).$$
Let
$$\phi \in \mathcal{D}(\mathbf{R}^n) \text{ and } \phi(x) \equiv 1 \text{ near } x = 0.$$
Then

$$\int \vec{f}(x) \cdot \{\phi(x/r)(\vec{g}(x) - \text{av}(\vec{g}, B(0,r)))\} dx$$

$$= \int \phi(x/r)\vec{f}(x) \cdot \vec{g}(x)dx + \text{av}(\vec{g}, B(0,r)) \cdot \int (1 - \phi(x/r)) \vec{f}(x)dx$$

$$\text{by } \int \vec{f}dx = \vec{0}$$

$$\rightarrow \int \vec{f}(x) \cdot \vec{g}(x)dx \quad (r \to \infty) \tag{21.22}$$

and
$$\|\phi(x/r)(\vec{g}(x) - \text{av}(\vec{g}, B(0,r)))\|_{\text{BMO}} \leq C(\phi)\|\vec{g}\|_{\text{BMO}} \tag{21.23}$$
by the $\mathbf{R}^m$-valued version of (1.9). Thus

$$(21.21)$$

$$\leq C(n) \sup \left\{ \frac{|\langle \vec{g}, \vec{f}\rangle_{\text{BMO}}|}{\|\vec{g}\|_{\text{BMO}}\|\vec{f}\|_{L^1}} : \vec{f} \in S\setminus\{\vec{0}\}, \ \vec{g} \in L^\infty(\mathbf{R}^n, \mathbf{R}^m)\setminus\{\vec{0}\}, \right.$$

$$\left. \text{supp } \vec{g} \text{ is compact} \right\} \text{ by } (21.22)\text{--}(21.23)$$

$$= C(n) \sup \left\{ \frac{|\langle \vec{g}, \vec{f}\rangle_{\text{BMO}}|}{\|\vec{g}\|_{\text{BMO}}\|\vec{f}\|_{L^1}} : \vec{f} \in S\setminus\{\vec{0}\}, \ \vec{g} \in \mathcal{D}(\mathbf{R}^n, \mathbf{R}^m) \right\}$$

$$\text{by mollifying } \vec{g}$$

$$= \{\text{the right-hand side of } (21.20)\}.$$

$$\square$$

Remark 21.5. Since
$$m_{\check{\theta}}f = (m_\theta \check{f})\check{\ },$$
where $\check{f}(x) = f(-x)$, we can replace

$$\sum_{j=1}^m \|m_{\check{\theta}_j}f\|_{L^1}, \ \sum_{j=1}^m \|m_{\check{\theta}_j}f\|_{H^1}$$

in Theorem 21.1, Corollaries 21.1–21.2 and 21.5–21.6, by

$$\sum_{j=1}^m \|m_{\theta_j}f\|_{L^1}, \ \sum_{j=1}^m \|m_{\theta_j}f\|_{H^1}.$$

Notes. The Fefferman-Stein decomposition of $BMO(\mathbf{R}^n)$ is due to C. Fefferman [71] and C. Fefferman–E. M. Stein [72].

H. Helson–G. Szegö [60], combined with R. Hunt–B. Muckenhoupt– R. Wheeden [73], gives another proof of the Fefferman-Stein decomposition of $BMO(\mathbf{R}^1)$. Their result is much finer than that of C. Fefferman–E. M. Stein. It says that

$$f = g_0 + Hg_1 \text{ with } g_0 \in L^\infty, \ \|g_1\|_{L^\infty} < \pi/2,$$

if and only if e^f satisfies the A_2-condition. (As for the result of Helson–Szegö see J. B. Garnett [81] p. 147 and R. Coifman–P. Jones–Rubio de Francia [83]. (See also T. Wolff [pre].) As for the A_p-condition see B. Muckenhoupt [72, 79], R. Coifman–C. Fefferman [74], A. Córdoba–C. Fefferman [76], P. Jones [80b], Rubio de Francia [84] and García-Cuerva–Rubio de Francia [85].) D. Sarason [75] is related to the Fefferman-Stein decomposition and important. (See also D. Sarason [73].)

XXII. A constructive proof of the Fefferman-Stein decomposition of BMO

Let

$$S_{\vec{R}} = \{(f, -R_1 f, \cdots, -R_n f) : f \in H^1(\mathbf{R}^n, \mathbf{R})\} \subset H^1(\mathbf{R}^n, \mathbf{R}^{n+1}), \quad (22.1)$$

$$S_{\vec{R}*} = \Big\{ \vec{g} = (g_j)_{j=0,\cdots,n} \in \mathrm{BMO}(\mathbf{R}^n, \mathbf{R}^{n+1}) :$$

$$\sum_{j=0}^{n} \tilde{R}_j g_j = 0 \text{ in } \mathrm{BMO}(\mathbf{R}^n, \mathbf{R})/\mathbf{R} \Big\}. \quad (22.2)$$

In this section we will show the following.

Theorem 22.1. *Let* $\vec{g} \in \mathrm{BMO}(\mathbf{R}^n, \mathbf{R}^{n+1})$ *and*

$$\mathrm{supp}\, \vec{g} \subset B(0,1). \quad (22.3)$$

Then there exists $\vec{h} \in S_{\vec{R}*}$ *such that*

$$\left| \vec{g}(x) - \vec{h}(x) \right| \leq C_{22.1}(n) \|\vec{g}\|_{\mathrm{BMO}} (1 + |x|)^{-n-1}. \quad (22.4)$$

Recall Theorem 21.3. Then Theorem 22.1 implies

$$\sup \left\{ \frac{\inf \left\{ \|\vec{g} - \vec{h}\|_{L^\infty} : \vec{h} \in S_{\vec{R}*} \right\}}{\|\vec{g}\|_{\mathrm{BMO}}} : \vec{g} \in \mathrm{BMO}(\mathbf{R}^n, \mathbf{R}^{n+1}) \backslash \mathbf{R}^{n+1} \right\}$$

$$\leq C(n) C_{22.1}(n) < \infty.$$

So, by Corollary 21.6, Theorem 22.1 gives another proof of

$$\sup \left\{ \frac{\|f\|_{H^1}}{\|\vec{R}f\|_{L^1}} : f \in H^1(\mathbf{R}^n, \mathbf{R}) \backslash \{0\} \right\} < \infty$$

namely the case $p = 1$ of Theorem 17.1. The advantage of the argument in this section is that it can be applied to operators other than Riesz transforms. (See Sections 24–27.)

In the argument of Section 17 what is essential is the subharmonicity of

$$\left| \vec{R}f * P(\cdot, t)(x) \right|^q$$

on $\mathbf{R}_+^{n+1}$ for $q \in [(n-1)/n, 1) \bigcap (0, 1)$. In this section we do not use subharmonicity. Instead of subharmonicity we use the following trivial lemma.

Lemma 22.1. *Let $b \in L^2(\mathbf{R}^n, \mathbf{R}) \cap \text{BMO}(\mathbf{R}^n, \mathbf{R})$. Let $\vec{\nu} = (\nu_0, \cdots, \nu_n) \in S^n$. Let*

$$\vec{p}(x) = \begin{bmatrix} \nu_0 b(x) - \sum_{j=1}^n \nu_j R_j b(x) \\ \nu_1 b(x) + \nu_0 R_1 b(x) \\ \vdots \qquad \vdots \\ \nu_n b(x) + \nu_0 R_n b(x) \end{bmatrix} \in L^2(\mathbf{R}^n, \mathbf{R}^{n+1}) \cap \text{BMO}(\mathbf{R}^n, \mathbf{R}^{n+1}).$$

(22.5)

Then

$$\vec{p} \in S_{\vec{R}_*}, \tag{22.6}$$
$$\vec{p}(x) \cdot \vec{\nu} \equiv b(x). \tag{22.7}$$

For the proof of Theorem 22.1 it is enough to show the following.

Lemma 22.2. *Let $\vec{g}$ be as in Theorem 22.1. Then there exists $\vec{h} \in S_{\vec{R}_*}$ and $\vec{v} \in \text{BMO}(\mathbf{R}^n, \mathbf{R}^{n+1})$ such that*

$$\|\vec{h}\|_{\text{BMO}} \le C_{22.2}(n)\|\vec{g}\|_{\text{BMO}}, \tag{22.8}$$
$$\|\vec{v}\|_{\text{BMO}} \le C_{22.2}(n)\|\vec{g}\|_{\text{BMO}}^2, \tag{22.9}$$
$$\text{supp}\,\vec{v} \subset B(0,3), \tag{22.10}$$
$$\left|\vec{g}(x) - \vec{h}(x) - \vec{v}(x)\right| \le \chi_{B(0,3)}(x) + |x|^{-n-1}\chi_{B(0,3)^c}(x). \tag{22.11}$$

Proof of "Lemma 22.2 → Theorem 22.1". We may assume

$$\|\vec{g}\|_{\text{BMO}} = \varepsilon/C_{22.2} \quad \text{where} \quad \varepsilon = 1/(2 \cdot 3^{n+1}).$$

First, applying Lemma 22.2 to $\vec{g}$ gives

$$\vec{h}_1 \in S_{\vec{R}_*} \quad \text{and} \quad \vec{v}_1 \in \text{BMO}(\mathbf{R}^n, \mathbf{R}^{n+1})$$

such that

$$\|\vec{h}_1\|_{\text{BMO}} \le C_{22.2}\|\vec{g}\|_{\text{BMO}} = \varepsilon, \tag{22.8$_1$}$$
$$\|\vec{v}_1\|_{\text{BMO}} \le \varepsilon\|\vec{g}\|_{\text{BMO}} = \varepsilon^2/C_{22.2}, \tag{22.9$_1$}$$
$$\text{supp}\,\vec{v}_1 \subset B(0,3), \tag{22.10$_1$}$$
$$\left|\vec{g}(x) - \vec{h}_1(x) - \vec{v}_1(x)\right| \le \chi_{B(0,3)}(x) + |x|^{-n-1}\chi_{B(0,3)^c}(x). \tag{22.11$_1$}$$

The conditions (22.10)$_1$–(22.11)$_1$ imply

$$\left|\vec{h}_1(x)\right| \le |x|^{-n-1} \quad \text{on } B(0,3)^c. \tag{22.12$_1$}$$

Next, applying the above argument to $\varepsilon^{-1}\vec{v}_1$ with dilation and multiplying ε to the obtained $\vec{h}$ and $\vec{v}$, we get

$$\vec{h}_2 \in S_{\tilde{R}^*} \text{ and } \vec{v}_2 \in \text{BMO}(\mathbf{R}^n, \mathbf{R}^{n+1})$$

such that

$$\|\vec{h}_2\|_{\text{BMO}} \leq \varepsilon C_{22.2}\|\varepsilon^{-1}\vec{v}_1\|_{\text{BMO}} \leq \varepsilon^2, \qquad (22.8)_2$$

$$|\vec{h}_2(x)| \leq \varepsilon(|x|/3)^{-n-1} \text{ on } B(0,3^2)^c, \qquad (22.12)_2$$

$$\|\vec{v}_2\|_{\text{BMO}} \leq \varepsilon\varepsilon\|\varepsilon^{-1}\vec{v}_1\|_{\text{BMO}} \leq \varepsilon^3/C_{22.2}, \qquad (22.9)_2$$

$$\text{supp}\,\vec{v}_2 \subset B(0,3^2), \qquad (22.10)_2$$

$$\left|\vec{v}_1(x) - \vec{h}_2(x) - \vec{v}_2(x)\right| \leq \varepsilon\left\{\chi_{B(0,3^2)}(x) + |x/3|^{-n-1}\chi_{B(0,3^2)^c}(x)\right\}. \qquad (22.11)_2$$

Substituting $(22.11)_2$ into $(22.11)_1$ gives

$$\left|\vec{g}(x) - \sum_{k=1}^{2}\vec{h}_k(x) - \vec{v}_2(x)\right| \leq \sum_{k=0}^{1}\varepsilon^k 4^{n+1}\left(1+|x|/3^k\right)^{-n-1}. \qquad (22.13)_2$$

Repeating this process gives

$$\{\vec{h}_j\} \subset S_{\tilde{R}^*} \text{ and } \{\vec{v}_j\} \subset \text{BMO}(\mathbf{R}^n, \mathbf{R}^{n+1}) \ (j=1,2,\cdots)$$

such that

$$\|\vec{h}_j\|_{\text{BMO}} \leq \varepsilon^j, \qquad (22.8)_j$$

$$|\vec{h}_j(x)| \leq \varepsilon^{j-1}(|x|/3^{j-1})^{-n-1} \text{ on } B(0,3^j)^c, \qquad (22.12)_j$$

$$\|\vec{v}_j\|_{\text{BMO}} \leq \varepsilon^{j+1}/C_{22.2}, \qquad (22.9)_j$$

$$\text{supp}\,\vec{v}_j \subset B(0,3^j), \qquad (22.10)_j$$

$$\left|\vec{g}(x) - \sum_{k=1}^{j}\vec{h}_k(x) - \vec{v}_j(x)\right| \leq \sum_{k=0}^{j-1}\varepsilon^k 4^{n+1}\left(1+|x|/3^k\right)^{-n-1} \leq C(1+|x|)^{-n-1}. \qquad (22.13)_j$$

The conditions $(22.9)_j$ and $(22.10)_j$ imply

$$\|\vec{v}_j\|_{L^1} \leq C\varepsilon^j 3^{jn} \to 0 \ (j \to \infty).$$

The conditions $(22.8)_j$ and $(22.12)_j$ imply

$$\|\vec{h}_j\|_{L^1} \leq C\varepsilon^j 3^{jn} \leq C0.5^j.$$

Let

$$\vec{h} = \sum_{k=1}^{\infty}\vec{h}_k \in L^1.$$

Then letting $j \to \infty$ in $(22.13)_j$ gives

$$\left|\vec{g}(x) - \vec{h}(x)\right| \leq C(1+|x|)^{-n-1}.$$

Since $\vec{h} \in S_{\tilde{R}^*}$ by $\vec{h}_j \in S_{\tilde{R}^*}$, we get (22.4). $\qquad\square$

For the proof of Lemma 22.2 we prepare several lemmas.

Lemma 22.3. *Let* $\theta \in C^\infty(S^{n-1}, \mathbf{C})$. *Let* $I \subset \mathbf{R}^n$ *be a cube. Let* $b \in C^2(\mathbf{R}^n, \mathbf{R})$,

$$\operatorname{supp} b \subset I, \tag{22.14}$$

$$\|\nabla^2 b\|_{L^\infty} \leq \ell(I)^{-2}, \tag{22.15}$$

$$\int_{\mathbf{R}^n} b(x)dx = 0. \tag{22.16}$$

Let $p = m_\theta b$. *Then,*

$$p \in C^1(\mathbf{R}^n, \mathbf{C}),$$

$$|p(x)| + \ell(I)|\nabla p(x)| \leq C_{22.3}(\theta)\,(1 + |x - x_I|/\ell(I))^{-n-1}, \tag{22.17}$$

$$\int_{\mathbf{R}^n} p(x)dx = 0, \tag{22.18}$$

where x_I *is the center of* I *and where* $C_{22.3}(\theta)$ *depends only on* n *and*

$$\left\{ \|\nabla^k \theta\|_{L^1(S^{n-1})} : k = 0, 1, 2, \cdots, n+2 \right\}. \tag{22.19}$$

Proof. Let $\psi \in \mathcal{S}_0(\mathbf{R}^n)$ be such that

$$\int_0^{+\infty} \mathcal{F}\psi(t\xi)dt/t = 1 \text{ for any } \xi \in \mathbf{R}^n \backslash \{0\}.$$

Let

$$\eta = \mathcal{F}^{-1}\left\{ \mathcal{F}\psi(\xi)\theta(\xi/|\xi|) \right\}.$$

Then

$$|\nabla^j \eta(x)| \leq C(\psi, j) \sum_{k=0}^{n+2} \|\nabla^k \theta\|_{L^1(S^{n-1})}(1 + |x|)^{-n-2} \tag{22.20}$$

and

$$\int_\varepsilon^{1/\varepsilon} (\eta)_t \frac{dt}{t} * b \to p \text{ in } L^2 \ (\varepsilon \to +0).$$

Let $\varepsilon \in (0, 1)$. If

$$x \notin 2I,$$

then

$$\left| \int_\varepsilon^{1/\varepsilon} (\eta)_t \frac{dt}{t} * b(x) \right| = \left| \int_\varepsilon^{1/\varepsilon} dt/t \int_I \left\{ (\eta)_t(x - y) - (\eta)_t(x - x_I) \right\} b(y)dy \right|$$

by (22.16) and (22.14)

$$\leq C(\psi, \theta) \int_\varepsilon^{1/\varepsilon} dt/t \int_I \ell(I)t^{-n-1}\,(1 + |x - x_I|/t)^{-n-2}\,|b(y)|dy \text{ by (22.20)}$$

$$\leq C\ell(I)^{n+1} \int_\varepsilon^{1/\varepsilon} t^{-n-1}\,(1 + |x - x_I|/t)^{-n-2}\,dt/t$$

$$\text{by } \|b\|_{L^\infty} \le C\ell(I)^2 \|\nabla^2 b\|_{L^\infty} \le C$$

$$\le C\ell(I)^{n+1} |x - x_I|^{-n-1}, \tag{22.21}$$

$$\left| \nabla \left\{ \int_\varepsilon^{1/\varepsilon} (\eta)_t \frac{dt}{t} * b(x) \right\} \right| \le \int_\varepsilon^{1/\varepsilon} dt/t \int_I |\nabla(\eta)_t(x - y)| \, |b(y)| dy$$

$$\le \int_\varepsilon^{1/\varepsilon} dt/t \int_I t^{-n-1} (1 + |x - x_I|/t)^{-n-2} |b(y)| dy \text{ by } (22.20)$$

$$\le C\ell(I)^n |x - x_I|^{-n-1} \text{ by } \|b\|_{L^\infty} \le C. \tag{22.22}$$

If
$$x \in 2I,$$

then

$$\left| \nabla \left\{ \int_{\min\{\varepsilon, \ell(I)\}}^{\min\{1/\varepsilon, \ell(I)\}} (\eta)_t \frac{dt}{t} * b(x) \right\} \right| = \left| \int_{..}^{..} dt/t \int (\eta)_t(x - y) \nabla b(y) dy \right|$$

$$\le \int_{..}^{..} dt/t \int |(\eta)_t(x - y)| \, |\nabla b(y) - \nabla b(x)| \, dy \text{ by } \int \eta dx = 0$$

$$\le \int_{..}^{..} dt/t \int C t^{-n} (1 + |x - y|/t)^{-n-2} \ell(I)^{-2} |y - x| dy$$

$$\text{by } (22.20) \text{ and } (22.15)$$

$$\le \int_0^{\ell(I)} C\ell(I)^{-2} dt \le C\ell(I)^{-1}, \tag{22.23}$$

$$\left| \nabla \left\{ \int_{\max\{\varepsilon, \ell(I)\}}^{\max\{1/\varepsilon, \ell(I)\}} (\eta)_t \frac{dt}{t} * b(x) \right\} \right| \le \int_{\max\{\varepsilon, \ell(I)\}}^{\max\{1/\varepsilon, \ell(I)\}} \|\nabla(\eta)_t\|_{L^\infty} \|b\|_{L^1} dt/t$$

$$\le \int_{\ell(I)}^\infty C t^{-n-1} |I| dt/t = C\ell(I)^{-1}, \tag{22.24}$$

$$\left| \int_\varepsilon^{1/\varepsilon} (\eta)_t \frac{dt}{t} * b(x) \right| \le C \text{ by a similar argument.}$$

Since the convergences in (22.22)–(22.24) as $\varepsilon \to +0$ are uniform with respect to $x \in \mathbf{R}^n$, the distributional derivative ∇p is continuous. So, p belongs to $C^1(\mathbf{R}^n)$ and satisfies (22.17). (22.18) is clear from $\mathcal{F}p(0) = 0$ (because $\mathcal{F}b(0) = 0$). $\square$

Lemma 22.4. *Let I and J be cubes in $\mathbf{R}^n$. Let $\ell(I) \ge \ell(J)$. Let p_I, $p_J \in C^1(\mathbf{R}^n, \mathbf{R})$ and*

$$|p_I(x)| + \ell(I) |\nabla p_I(x)| \le (1 + |x - x_I|/\ell(I))^{-n-1}, \tag{22.25}$$

$$|p_J(x)| + \ell(J) |\nabla p_J(x)| \le (1 + |x - x_J|/\ell(J))^{-n-1}, \tag{22.26}$$

$$\int_{\mathbf{R}^n} p_I dx = \int_{\mathbf{R}^n} p_J dx = 0, \tag{22.27}$$

where x_I is the center of I. Then

$$\left| \int_{\mathbf{R}^n} p_I p_J dx \right| \leq C(n)|I||J|\ell(J)^{1/(n+2)} \left(\ell(I) + |x_I - x_J| \right)^{-n-1/(n+2)}.$$

$$(22.28)$$

Proof. Let

$$r_0 = \ell(J)^{1/(n+2)} \left(\ell(I) + |x_I - x_J| \right)^{1-1/(n+2)},$$
$$B_0 = B(x_J, r_0),$$
$$p_{J,1}(x) = p_J(x)\chi_{B_0}(x) + |B_0|^{-1} \int_{B_0^c} p_J(y)dy \chi_{B_0}(x),$$
$$p_{J,2}(x) = p_J(x)\chi_{B_0^c}(x) - |B_0|^{-1} \int_{B_0^c} p_J(y)dy \chi_{B_0}(x).$$

Then

$$p_J = p_{J,1} + p_{J,2},$$

$$\text{supp } p_{J,1} \subset \overline{B_0}, \quad \|p_{J,1}\|_{L^1} \leq C|J|, \quad \int p_{J,1}dx = 0 \text{ by (22.27)}, \quad (22.29)$$

$$\|p_{J,2}\|_{L^\infty} \leq C \left((\ell(I) + |x_I - x_J|) / \ell(J) \right)^{-n-1/(n+2)}. \qquad (22.30)$$

So, (22.29) and (22.25) imply

$$\left| \int p_I p_{J,1} dx \right| \leq r_0 \sup_{y \in B_0} |\nabla p_I(y)| \|p_{J,1}\|_{L^1}$$
$$\leq Cr_0\ell(I)^{-1} (1 + |x_I - x_J|/\ell(I))^{-n-1} |J|$$
$$\leq C\ell(J)^{1/(n+2)} \left(\ell(I) + |x_I - x_J| \right)^{-n-1/(n+2)} |I||J|.$$

(22.30) and (22.25) imply

$$\left| \int p_I p_{J,2} dx \right| \leq \|p_I\|_{L^1} \|p_{J,2}\|_{L^\infty}$$
$$\leq C|I| \left((\ell(I) + |x_I - x_J|) / \ell(J) \right)^{-n-1/(n+2)}.$$

$\square$

Lemma 22.5. *Let* $\{\lambda_I\}_I \subset [0, +\infty)$ *and* $\{p_I\}_I \subset C^1(\mathbf{R}^n, \mathbf{R})$*, where* I *is taken over all dyadic cubes in* $\mathbf{R}^n$*. Let*

$$\sum_I \lambda_I^2 |I| \leq \infty,$$

$$|p_I(x)| + \ell(I)|\nabla p_I(x)| \leq (1 + |x_I - x_J|/\ell(I))^{-n-1},$$

$$\int p_I dx = 0.$$

Then $\sum_I \lambda_I p_I$ converges in L^2 independently of the order of summation and

$$\left\|\sum_I \lambda_I p_I\right\|_{L^2}^2 \le C(n) \sum_I \lambda_I^2 |I|, \tag{22.31}$$

$$\left\|\sum_I \lambda_I p_I\right\|_{BMO}^2 \le C(n) \left\|\sum_I \lambda_I^2 |I| \delta_{(x_I, \ell(I))}\right\|_C, \tag{22.32}$$

where $\delta_{(x,t)}$ is the point mass concentrated at the point $(x,t) \in \mathbf{R}_+^{n+1}$.

Proof of (22.31).

$$\left\|\sum \lambda_I p_I\right\|_{L^2}^4 \le \left\{2 \sum_{I,J: \ell(J) \le \ell(I)} \lambda_I \lambda_J \left|\int p_I p_J \, dx\right|\right\}^2$$

$$\le C \left\{\sum \lambda_I \lambda_J |I| |J| \ell(J)^{1/(n+2)} (\ell(I) + |x_I - x_J|)^{-n-1/(n+2)}\right\}^2$$
$$\text{by Lemma 22.4}$$

$$\le C \left\{\sum \lambda_I^2 |I| |J| \ell(J)^{1/(n+2)} (\ell(I) + |x_I - x_J|)^{-n-1/(n+2)}\right\}$$
$$\times \left\{\sum \lambda_J^2 |I| |J| \ell(J)^{1/(n+2)} (\ell(I) + |x_I - x_J|)^{-n-1/(n+2)}\right\}$$
$$\text{by Schwarz's inequality}$$

$$= C \left\{\sum_I \lambda_I^2 |I| \sum_{J: \ell(J) \le \ell(I)} |J| \ell(J)^{1/(n+2)} (\ell(I) + |x_I - x_J|)^{-n-1/(n+2)}\right\}$$

$$\times \left\{\sum_J \lambda_J^2 |J| \sum_{I: \ell(I) \ge \ell(J)} |I| \ell(J)^{1/(n+2)} (\ell(I) + |x_I - x_J|)^{-n-1/(n+2)}\right\}$$

$$\le C \left\{\sum_I \lambda_I^2 |I| \sum_{k: 2^k \le \ell(I)} 2^{k/(n+2)} \int (\ell(I) + |x_I - y|)^{-n-1/(n+2)} \, dy\right\}$$

$$\times \left\{\sum_J \lambda_J^2 |J| \sum_{k: 2^k \ge \ell(J)} \ell(J)^{1/(n+2)} \int (2^k + |y - x_J|)^{-n-1/(n+2)} \, dy\right\}$$

$$= C \left\{\sum_I \lambda_I^2 |I|\right\} \left\{\sum_J \lambda_J^2 |J|\right\}.$$

$\square$

Proof of (22.32). We may assume

$$\left\|\sum_I \lambda_I^2 |I| \delta_{(x_I, \ell(I))}\right\|_C = 1. \tag{22.33}$$

Note that

$$\lambda_I \leq C \left\| \sum \lambda_I^2 |I| \delta_{(x_I, \ell(I))} \right\|_C^{1/2}. \tag{22.34}$$

Take any ball $B = B(x_B, r_B)$. Then

$$\sum_I \lambda_I p_I = \sum_{I:x_I \in 2B, \ell(I) \leq r_B} \lambda_I p_I + \sum_{I:x_I \notin 2B, \ell(I) \leq r_B} + \sum_{I:\ell(I) > r_B}$$

$$= q_1 + q_2 + q_3, \text{ say,}$$

where x_I is the center of a dyadic cube I. Then

$$\|q_2\|_{L^1(B)} \leq C \sum_{k=-\infty}^{[\log_2 r_B]} \sum_{I:x_I \notin 2B, \ell(I)=2^k} \int_B |p_I(x)| dx \text{ by (22.33)–(22.34)}$$

$$\leq C \sum \sum |B| \left(1 + |x_B - x_I|/2^k\right)^{-n-1}$$

$$\leq C \sum_{k=-\infty}^{[\log_2 r_B]} |B| 2^{-kn} \int_{(2B)^c} \left(1 + |x_B - y|/2^k\right)^{-n-1} dy \leq C|B|, \tag{22.35}$$

$$\|\nabla q_3\|_{L^\infty(B)} \leq C \sum_{k=[\log_2 r_B]+1}^{\infty} \sum_{I:\ell(I)=2^k} \|\nabla p_I\|_{L^\infty(B)} \text{ by (22.34)}$$

$$\leq C \sum \sum 2^{-k} \left(1 + |x_B - x_I|/2^k\right)^{-n-1}$$

$$\leq C \sum_{k=[\log_2 r_B]+1}^{\infty} 2^{-kn} \int_{\mathbf{R}^n} 2^{-k} \left(1 + |x_B - y|/2^k\right)^{-n-1} dy \leq C/r_B. \tag{22.36}$$

Therefore

$$\int_B \left| \sum_I \lambda_I p_I(x) - q_3(x_B) \right| dx/|B|$$

$$\leq \|q_1\|_{L^1(B)}/|B| + \|q_2\|_{L^1(B)}/|B| + C r_B \|\nabla q_3\|_{L^\infty(B)}$$

$$\leq \|q_1\|_{L^2(\mathbf{R}^n)}/|B|^{1/2} + C + C \text{ by (22.35)–(22.36)}$$

$$\leq C \left\{ \sum_{I:x_I \in 2B, \ell(I) \leq r_B} \lambda_I^2 |I| \right\}^{1/2} /|B|^{1/2} + C \text{ by (22.31)}$$

$$\leq C \text{ by (22.33).}$$

$$\square$$

Definition 22.1. For a cube $I \subset \mathbf{R}^n$ let

$$T(I) = I \times (\ell(I)/2, \ell(I)].$$

Lemma 22.6. *Let $\vec{g}$ be as in Theorem 22.1. Then there exist $\{\lambda_I\}_I \subset [0, +\infty)$ and $\{\vec{b}_I\}_I \subset \mathcal{D}(\mathbf{R}^n, \mathbf{R}^{n+1})$, where I is taken over all dyadic cubes in $\mathbf{R}^n$, such that*

$$\lambda_I = 0 \text{ and } \vec{b}_I(x) \equiv \vec{0}$$

if $I \notin \{I : \text{dyadic cube in } \mathbf{R}^n \text{ such that } I \subset B(0, 1.1),$
$$\ell(I) \leq 2^{j(n)}\} = (22.37), \tag{22.38}$$

$$\left\| \sum_I \lambda_I^2 |I| \delta_{(x_I, \ell(I))} \right\|_{\mathcal{C}} \leq C(n) \|\vec{g}\|_{\text{BMO}}^2, \tag{22.39}$$

$$\text{supp } \vec{b}_I \subset 3I, \tag{22.40}$$

$$\int \vec{b}_I \, dx = \vec{0}, \tag{22.41}$$

$$\|\nabla^2 \vec{b}_I\|_{L^\infty} \leq \ell(I)^{-2}, \tag{22.42}$$

$$\text{supp}(\vec{g} - \sum_I \lambda_I \vec{b}_I) \subset B(0, 1.2), \tag{22.43}$$

$$\left\| \vec{g} - \sum_I \lambda_I \vec{b}_I \right\|_{L^\infty} \leq C(n) \|\vec{g}\|_{\text{BMO}}, \tag{22.44}$$

where

$$j(n) = [-\log_2(20n)] \tag{22.45}$$

and where the convergence in L^2 of $\sum_I \lambda_I \vec{b}_I$ follows from (22.38)–(22.39) and Lemma 22.5.

Proof. Let

$$\phi \in \mathcal{D}(\mathbf{R}^n, \mathbf{R}), \quad \text{supp } \phi \subset B(0, 1), \quad \int \phi \, dx = 0,$$

$$\int_0^\infty \mathcal{F}\phi(t\xi)^2 \, dt/t = 1 \text{ if } \xi \in \mathbf{R}^n \backslash \{0\}.$$

For a dyadic cube I let

$$\lambda_I' = \begin{cases} \left\{ |I|^{-1} \displaystyle\iint_{T(I)} |\vec{g} * (\phi)_t(y)|^2 \, dy \, dt/t \right\}^{1/2} & \text{if } \ell(I) \leq 2^{j(n)}, \\ 0 & \text{otherwise,} \end{cases}$$

$$\vec{b}_I'(x) = \begin{cases} \displaystyle\iint_{T(I)} \vec{g} * (\phi)_t(y)(\phi)_t(x - y) \, dy \, dt/t/\lambda_I' & \text{if } \lambda_I' \neq 0, \\ 0 & \text{otherwise.} \end{cases}$$

Then

$$\begin{aligned} |\nabla^2 \vec{b}_I'(x)| &= \left| \iint_{T(I)} \vec{g} * (\phi)_t(y) \nabla^2 (\phi)_t(x - y) \, dy \, dt/t \right| \lambda_I' \\ &\leq C_{22.4}(\phi) \ell(I)^{-2}. \end{aligned}$$

Put

$$\lambda_I = C_{22.4}\lambda_I' \text{ and } \vec{b}_I = \vec{b}_I'/C_{22.4}.$$

If $\ell(I) \leq 2^{j(n)}$ and if $I \not\subset B(0, 1.1)$, then

$$\vec{g} * (\phi)_t(y) = \vec{0} \text{ for } (y, t) \in T(I)$$

by (22.3) and by $\operatorname{supp} \phi \subset B(0, 1)$. Thus, we get (22.38). (22.39) follows from Lemma 12.1. (22.40)–(22.42) are clear. (22.43) follows from (22.3), (22.40) and (22.38). Since

$$|\vec{g} * (\phi)_t(y)| \leq C(\phi)t^{-n}\|\vec{g}\|_{L^1} \leq C(\phi)t^{-n}\|\vec{g}\|_{\text{BMO}} \text{ by (22.3)},$$

(22.44) follows from substituting this into

$$\vec{g}(x) - \sum_{I \in (22.37)} \lambda_I \vec{b}_I = \iint_{\mathbf{R}^n \times (2^{j(n)}, +\infty)} \vec{g} * (\phi)_t(y)(\phi)_t(x - y) dy dt/t.$$

$\square$

Definition 22.2. For $\{\lambda_I\}_I \subset [0, +\infty)$, where I is taken over all dyadic cubes in $\mathbf{R}^n$, and for $j \in \mathbf{Z}$ let

$$\eta_j^{(1)}(x) = \sum_{I:\ell(I)=2^j} \lambda_I \left(1 + |x - x_I|/2^j\right)^{-n-1/2},$$

$$\eta_j^{(2)}(x) = \sum_{k=j}^{\infty} 0.99^{k-j} \eta_k^{(1)}(x),$$

$$\eta_j^{(3)}(x) = \left\{ \sum_{I:\ell(I)=2^j} \lambda_I^2 \left(1 + |x - x_I|/2^j\right)^{-n-1/2} \right\}^{1/2},$$

$$\eta_j^{(4)}(x) = \left\{ \sum_{k=j}^{\infty} 0.99^{k-j} \eta_k^{(3)}(x)^2 \right\}^{1/2},$$

$$\eta_j^{(0)}(x, y) = \begin{cases} \eta_j^{(2)}(x)|x - y|/2^j & \text{if } |x - y| < 2^j, \\ \displaystyle\sum_{k=j}^{[\log_2 |x-y|]} \left(\eta_k^{(2)}(x) + \eta_k^{(2)}(y)\right) & \text{if } |x - y| \geq 2^j. \end{cases}$$

Lemma 22.7. *In this lemma, for the sake of simplicity we write*

$$\| \cdots \|_{\mathcal{C}} \text{ for } \left\| \sum_I \lambda_I^2 |I| \delta_{(x_I, \ell(I))} \right\|_{\mathcal{C}}.$$

Let $\{\lambda_I\}_I$ be as in Definition 22.2. Then

$$\eta_j^{(1)}(x) \le C(n)\eta_j^{(1)}(y) \ \text{ if } \ |x-y| \le 2^j, \tag{22.46}$$

$$\eta_j^{(4)}(x) \le C(n)\eta_j^{(4)}(y) \ \text{ if } \ |x-y| \le 2^j, \tag{22.47}$$

$$\sum_{k=j}^{\infty} 0.9^{k-j}\eta_k^{(4)}(x)^2 \le 11\eta_J^{(4)}(x)^2, \tag{22.48}$$

$$2^{-jn}\int \eta_j^{(4)}(y)^2 \left(1+|x-y|/2^j\right)^{-n-1/2} dy \le C(n)\eta_j^{(4)}(x)^2, \tag{22.49}$$

$$\eta_j^{(1)}(x) \le \eta_j^{(2)}(x) \le C(n)\eta_j^{(4)}(x) \le C(n)\|\cdots\|_C^{1/2}, \tag{22.50}$$

$$\eta_j^{(0)}(x,y) \le C(n)\|\cdots\|_C^{1/2}\log_2\left(2+|x-y|/2^j\right), \tag{22.51}$$

$$\sum_{I:\ell(I)=2^j} \lambda_I \left(1+|x-x_I|/2^j\right)^{-n-1}\eta_{j+1}^{(0)}(x,x_I)$$

$$\le C(n)\min\left\{\eta_j^{(4)}(x)^2, \ \|\cdots\|_C^{1/2}\eta_j^{(1)}(x)\right\}, \tag{22.52}$$

$$\left\|\sum_{j=-\infty}^{\infty}\eta_j^{(4)}(x)^2\delta_{t=2^j}\right\|_C \le C(n)\|\cdots\|_C. \tag{22.53}$$

(22.46)–(22.48) are easy. (22.49) follows from

$$2^{-jn}\int \left(1+|z-y|/2^k\right)^{-n-1/2}\left(1+|x-y|/2^j\right)^{-n-1/2} dy$$

$$\le C(n)\left(1+|z-x|/2^k\right)^{-n-1/2} \ \text{ if } \ k \ge j. \tag{22.54}$$

The last inequality of (22.50) follows from (22.34). (22.51) is easy from $\eta_j^{(2)}(x) \le C(n)\|\cdots\|_C^{1/2}$ which follows from (22.50).

Proof of (22.52).

the left-hand side of (22.52)

$$\le \sum_{I:\ell(I)=2^j} \lambda_I \left(1+|x-x_I|/2^j\right)^{-n-1} C\|\cdots\|_C^{1/2}\log_2\left(2+|x-x_I|/2^{j+1}\right)$$

by (22.51)

$$\le C\|\cdots\|_C^{1/2}\eta_j^{(1)}(x).$$

the left-hand side of (22.52)

$$\le \sum_{I:\ell(I)=2^j} \lambda_I \left(1+|x-x_I|/2^j\right)^{-n-1} \sum_{k=j+1}^{[\log_2(2^{j+1}+|x-x_I|)]} \left(\eta_k^{(2)}(x)+\eta_k^{(2)}(x_I)\right)$$

$$\le \sum_{k>j}\sum_{\substack{I:\ell(I)=2^j,\\ x_I\notin B(x,2^k-2^{j+1})}} \lambda_I \left(1+|x-x_I|/2^j\right)^{-n-1}\left(\eta_k^{(2)}(x)+\eta_k^{(2)}(x_I)\right)$$

$$\leq \left\{ \sum \sum \lambda_I^2 \left(1 + |x - x_I|/2^j\right)^{-n-1} \right\}^{1/2}$$

$$\times \left\{ \sum \sum \left(\eta_k^{(2)}(x) + \eta_k^{(2)}(x_I) \right)^2 \left(1 + |x - x_I|/2^j\right)^{-n-1} \right\}^{1/2}$$

$$\leq C \left\{ \sum \sum 2^{(j-k)/2} \lambda_I^2 \left(1 + |x - x_I|/2^j\right)^{-n-1/2} \right\}^{1/2}$$

$$\times \left\{ \sum \sum \left(\eta_k^{(2)}(x)^2 + \eta_k^{(2)}(x_I)^2 \right) 2^{(j-k)(n+1)} (1 + |x - x_I|/2^k)^{-n-1} \right\}^{1/2}$$

$$\leq C \left\{ \sum_{k>j} 2^{(j-k)/2} \eta_j^{(3)}(x)^2 \right\}^{1/2}$$

$$\times \left\{ \sum_{k>j} 2^{j-k} \int \left(\eta_k^{(4)}(x)^2 + \eta_k^{(4)}(y)^2 \right) 2^{-kn} \left(1 + |x - y|/2^k\right)^{-n-1} dy \right\}^{1/2}$$

$$\text{by } \eta_k^{(2)} \leq C \eta_k^{(4)} \text{ and } (22.47)$$

$$\leq C \eta_j^{(3)}(x) \left\{ \sum_{k>j} 2^{j-k} \eta_k^{(4)}(x)^2 \right\}^{1/2} \text{ by } (22.49)$$

$$\leq C \eta_j^{(4)}(x)^2 \text{ by } (22.48).$$

$$\square$$

Proof of (22.53). Take any ball $B = B(x_B, r_B) \subset \mathbf{R}^n$. Then

$$\int_B \sum_{j=-\infty}^{[\log_2 r_B]} \eta_j^{(3)}(x)^2 dx = \sum_{j=-\infty}^{[\log_2 r_B]} \sum_{I:\ell(I)=2^j} \lambda_I^2 \int_B \left(1 + |x - x_I|/2^j\right)^{-n-1/2} dx$$

$$\leq C \sum \sum \lambda_I^2 \min \left\{ |I|, |B| \left(\text{dist}(x_I, B)/\ell(I) \right)^{-n-1/2} \right\}$$

$$\leq C \sum \sum \lambda_I^2 |I| \left(1 + |x_I - x_B|/r_B\right)^{-n-1/2}$$

$$= C \iint_{\mathbf{R}^n \times (0, r_B]} \left(1 + |x - x_B|/r_B\right)^{-n-1/2} \sum_I \lambda_I^2 |I| \delta_{(x_I, \ell(I))}$$

$$\leq C \| \cdots \|_c |B|. \tag{22.55}$$

So, if $m \in \mathbf{N}_0$, then

$$\iint_{Q(B)} \sum_{j=-\infty}^{\infty} \eta_{j+m}^{(3)}(x)^2 \delta_{t=2^j} = \sum_{j=-\infty}^{[\log_2 r_B]+m} \int_B \eta_j^{(3)}(x)^2 dx$$

$$= \sum_{j=-\infty}^{[\log_2 r_B]} \cdots + \sum_{j=[\log_2 r_B]+1}^{[\log_2 r_B]+m} \cdots$$

$$\leq C \| \cdots \|_c |B| + m C \| \cdots \|_c |B| \text{ by } (22.55) \text{ and } (22.50).$$

Thus

$$\left\| \sum_{j=-\infty}^{\infty} \eta_{j+m}^{(3)}(x)^2 \delta_{t=2^j} \right\|_c \leq C(m+1) \| \cdots \|_c.$$

Thus,

$$\left\| \sum_{j=-\infty}^{\infty} \eta_j^{(4)}(x)^2 \delta_{t=2^j} \right\|_c \leq \sum_{m=0}^{\infty} 0.99^m \left\| \sum_{j=-\infty}^{\infty} \eta_{j+m}^{(3)}(x)^2 \delta_{t=2^j} \right\|_c$$

$$\leq C \sum 0.99^m (m+1) \| \cdots \|_c \leq C \| \cdots \|_c.$$

$\square$

Definition 22.3. For $\vec{v} \in \mathbf{R}^m$ let

$$U(\vec{v}) = \begin{cases} \vec{v}/|\vec{v}| & \text{if } \vec{v} \neq \vec{0}, \\ (1, 0, \cdots, 0) & \text{if } \vec{v} = \vec{0}. \end{cases}$$

Definition 22.4. For $t > -1$ let

$$v(t) = (1+t)^{1/2} - 1.$$

Proof of Lemma 22.2. The letter I denotes dyadic cubes in $\mathbf{R}^n$. The letter j denotes integers. Let $A > 0$ be a sufficiently large number depending only on the dimension n. We may assume

$$\|\vec{g}\|_{\mathrm{BMO}} \leq A^{-100}, \tag{22.56}$$

because otherwise we can take

$$\vec{v} \equiv \vec{g}, \quad \vec{h} \equiv \vec{0}.$$

Applying Lemma 22.6 gives $\{\lambda_I, \vec{b}_I\}_I$ that satisfy (22.38)–(22.44). By (22.43)–(22.44), for the proof of Lemma 22.2 we can replace the claim (22.11) by

$$\left| \sum_{I \in (22.37)} \lambda_I \vec{b}_I(x) - \vec{h}(x) - \vec{v}(x) \right| \leq \chi_{B(0,3)}(x) + |x|^{-n-1} \chi_{B(0,3)^c}(x). \tag{22.11*}$$

Let $\left\{ \eta_j^{(1)}, \eta_j^{(2)}, \eta_j^{(4)}, \eta_j^{(0)} \right\}$ be as in Definition 22.2. Then

$$\lambda_I \leq A \|\vec{g}\|_{\mathrm{BMO}} \text{ (recall (22.39) and (22.34))}, \tag{22.57}$$

$$\|\eta_j^{(1)}\|_{L^\infty} \leq \|\eta_j^{(2)}\|_{L^\infty} \leq A \|\vec{g}\|_{\mathrm{BMO}} \ (\leq A^{-99}) \text{ by (22.50) and (22.39)}, \tag{22.58}$$

$$\left\| \sum_{j=-\infty}^{\infty} \eta_j^{(4)}(x)^2 \delta_{t=2^j} \right\|_c \leq A \|\vec{g}\|_{\mathrm{BMO}}^2 \text{ by (22.53) and (22.39)}. \tag{22.59}$$

Let

$$\vec{p}_I(x) \equiv \vec{0} \ \text{if} \ \ell(I) > 2^{j(n)}, \tag{22.60}$$

$$\vec{\phi}_j(x) \equiv \vec{0} \ \text{if} \ j > j(n), \tag{22.61}$$

where $j(n)$ is defined by (22.45). We will construct

$$\{\vec{p}_I\}_{\ell(I) \le 2^{j(n)}}, \ \{\vec{\phi}_j\}_{j \le j(n)} \subset C^1(\mathbf{R}^n, \mathbf{R}^{n+1}),$$

so that the following hold :

$$|\vec{p}_I(x)| + \ell(I)|\nabla \vec{p}_I(x)| \le A \left(1 + |x - x_I|/\ell(I)\right)^{-n-1}, \tag{22.62}$$

$$\int \vec{p}_I(x)dx = \vec{0}, \tag{22.63}$$

$$\vec{p}_I \in S_{\vec{R}^*}, \tag{22.64}$$

$$|\vec{\kappa}_j(x)| \le 1 \quad (\text{see (22.69)}), \tag{22.65}$$

$$|\vec{\phi}_j(x)| \le A^{10} \min \left\{ \eta_j^{(4)}(x)^2, \|\vec{g}\|_{\text{BMO}}\eta_j^{(1)}(x) \right\}, \tag{22.66}$$

$$|\nabla \vec{\phi}_j(x)| \le 2^{-j}A^{10}\|\vec{g}\|_{\text{BMO}}\eta_j^{(1)}(x), \tag{22.67}$$

$$\text{supp} \, \vec{\phi}_j \subset \left\{ x \in \mathbf{R}^n : \sum_{I:\ell(I)>2^j} \lambda_I |\vec{b}_I(x) + \vec{p}_I(x)| \ge 0.9 \right\}, \tag{22.68}$$

where

$$\vec{\kappa}_j = \sum_{I:\ell(I)>2^j} \lambda_I(\vec{b}_I + \vec{p}_I) - \sum_{k \ge j} \vec{\phi}_k. \tag{22.69}$$

We grant this construction temporarily and finish the proof of Lemma 22.2.

Put

$$\vec{h} = -\sum_I \lambda_I \vec{p}_I, \quad \vec{v} = \sum_{k=-\infty}^{\infty} \vec{\phi}_k,$$

where I is taken over all dyadic cubes. (Since (22.38)–(22.39) imply

$$\sum_I \lambda_I^2 |I| < \infty,$$

$\sum_I \lambda_I \vec{p}_I$ converges in L^2 by (22.62), (22.63) and by (22.31) and $\sum_k \vec{\phi}_k$ converges in L^1 by (22.66) and by $\sum \|\eta_k^{(4)}\|_{L^2}^2 \le C \sum_I \lambda_I^2 |I|$.) Then (22.8) follows from (22.32) and (22.39). "$\vec{h} \in S_{\vec{R}^*}$" follows from (22.64). So, we will show (22.9), (22.10) and (22.11)*.

Proof of (22.10). Note that

$$\text{supp} \sum_{I:(22.37)} \lambda_I \vec{b}_I \subset B(0, 1.2).$$

So, if $x \in B(0,3)^c$, then

$$\sum_{I:(22.37)} \lambda_I \left(|\vec{b}_I(x)| + |\vec{p}_I(x)| \right) = \sum \lambda_I |\vec{p}_I(x)|$$

$$\leq \sum A \|\vec{g}\|_{\text{BMO}} \cdot A \left(1 + |x - x_I|/\ell(I)\right)^{-n-1} \quad \text{by (22.57) and (22.62)}$$

$$\leq A^{-97} |x|^{-n-1} \quad \text{by (22.56).} \tag{22.70}$$

Thus, (22.68) implies supp $\vec{\phi}_j \subset B(0,3)$ from which (22.10) follows.

Proof of (22.11)*. Letting $j \to \infty$ in (22.65) implies

$$\left| \sum_{I:(22.37)} \lambda_I \vec{b}_I(x) - \vec{h}(x) - \vec{v}(x) \right| \leq 1.$$

Furthermore, if $x \in B(0,3)^c$, then (22.10) and (22.70) imply

$$\left| \sum_{I:(22.37)} \lambda_I \vec{b}_I(x) - \vec{h}(x) - \vec{v}(x) \right| = \left| \sum \lambda_I \vec{b}_I(x) - \vec{h}(x) \right| \quad \leq \quad A^{-97} |x|^{-n-1}$$

$$\leq \quad |x|^{-n-1}.$$

Proof of (22.9). Take any ball $B = (x_B, r_B) \subset \mathbf{R}^n$. Then

$$\int_B \left| \vec{v}(x) - \sum_{k=[\log_2 r_B]+1}^{\infty} \vec{\phi}_k(x_B) \right| dy/|B|$$

$$\leq \int_B \sum_{k=-\infty}^{[\log_2 r_B]} |\vec{\phi}_k(x)| dx/|B| + \int_B \sum_{k=[\cdots]+1}^{\infty} |\vec{\phi}_k(x) - \vec{\phi}_k(x_B)| dx/|B|$$

$$\leq \int_B \sum_{k=-\infty}^{[\log_2 r_B]} A^{10} \eta_k^{(4)}(x)^2 dx/|B| + \sum_{k=[\cdots]+1}^{\infty} 2^{-k} A^{10} \|\vec{g}\|_{\text{BMO}} \|\eta_k^{(1)}\|_{L^\infty} r_B$$

$$\text{by (22.66)–(22.67)}$$

$$\leq A^{10} \| \sum \eta_k^{(4)}(x)^2 \delta_{t=2^k} \|_C + 2A^{10} \|\vec{g}\|_{\text{BMO}} A \|\vec{g}\|_{\text{BMO}} \quad \text{by (22.58)}$$

$$\leq A^{12} \|\vec{g}\|_{\text{BMO}}^2 \quad \text{by (22.59).}$$

This concludes the proof of Lemma 22.2. $\qquad \square$

The construction of $\{\vec{p}_I\}_I$ and $\{\vec{\phi}_j\}$. We will show this by induction. Let $j \leq j(n) + 1$. Suppose that $\{\vec{p}_I\}_{\ell(I) \geq 2^j}$ and $\{\vec{\phi}_k\}_{k \geq j}$ have been constructed so that (22.62)–(22.68) holds. (Recall (22.60)–(22.61). Note that (22.65) for the case $j = j(n) + 1$ is clear from $\vec{k}_{j(n)+1} \equiv \vec{0}$.) We will give the construction of

$$\{\vec{p}_I\}_{\ell(I)=2^{j-1}} \quad \text{and} \quad \vec{\phi}_{j-1}. \tag{22.71}$$

Let $\vec{\kappa}_j$ be as in (22.69).

Claim 1. There exist $\psi_j \in C^1(\mathbf{R}^n)$ such that

$$\psi_J(x) \begin{cases} = 1 & \text{if } |\vec{\kappa}_j(x)| \geq 0.99 \\ = 0 & \text{if } |\vec{\kappa}_j(x)| \leq 0.9 \\ \in [0,1] & \text{otherwise,} \end{cases} \tag{22.72}$$

$$\|\nabla\psi_j\|_{L^\infty} \leq 2^{-j}. \tag{22.73}$$

Let I be a dyadic cube with $\ell(I) = 2^{j-1}$. Applying Lemma 22.1 with $b(x) = -\vec{b}_I(x) \cdot U(\vec{\kappa}_j(x_I))$ and with $\vec{v} = U(\vec{\kappa}_j(x_I))$ gives $\vec{p}_I \in S_{\vec{R}^*}$ such that

$$\left(\vec{p}_I(x) + \vec{b}_I(x)\right) \cdot U(\vec{\kappa}_j(x_I)) \equiv 0. - \tag{22.74}$$

The formula (22.5) and Lemma 22.3 imply (22.62)–(22.63).

Put

$$\vec{\rho}(x) = \sum_{I:\ell(I)=2^{j-1}} \lambda_I \left(\vec{b}_I(x) + \vec{p}_I(x)\right), \tag{22.75}$$

$$\vec{\tau}(x) = \vec{\kappa}_j(x) + \vec{\rho}(x), \tag{22.76}$$

$$\vec{\phi}_{j-1}(x) = \psi_j(x)\left(|\vec{\tau}(x)| - |\vec{\kappa}_j(x)|\right)U(\vec{\tau}(x)). \tag{22.77}$$

(See Fig. 22.1.) Then

$$\vec{\kappa}_{j-1}(x) = \vec{\tau}(x) - \vec{\phi}_{j-1}(x). \tag{22.78}$$

Claim 2. $\quad |\vec{\kappa}_{j-1}(x)| \leq 1. \tag{22.79}$

Claim 3. $\quad |\vec{\phi}_{j-1}(x)| \leq A^{10} \min\left\{\eta_{j-1}^{(4)}(x)^2, \|\vec{g}\|_{\mathrm{BMO}}\eta_{j-1}^{(1)}(x)\right\}. \tag{22.80}$

Claim 4. $\quad |\nabla\vec{\phi}_{j-1}(x)| \leq 2^{-j+1}A^{10}\|\vec{g}\|_{\mathrm{BMO}}\eta_{j-1}^{(1)}(x). \tag{22.81}$

Claim 5. $\quad \mathrm{supp}\,\vec{\phi}_{j-1} \subset \left\{x \in \mathbf{R}^n : \sum_{I:\ell(I)=2^j} \lambda_I \left|\vec{b}_I(x) + \vec{p}_I(x)\right| \geq 0.9\right\}. \tag{22.82}$

This concludes the construction of (22.71). $\qquad\qquad\qquad\qquad\square$

Finally, we prove Claims 1–5.

First note that (22.40), (22.42) and (22.62) imply

$$|\vec{b}_I(x)| + |\vec{p}_I(x)| + \ell(I)\left(|\nabla\vec{b}_I(x)| + |\nabla\vec{p}_I(x)|\right)$$

$$\leq 2A\left(1 + |x - x_I|/\ell(I)\right)^{-n-1}. \tag{22.83}$$

Claim 6. $|\vec{\kappa}_j(x) - \vec{\kappa}_j(y)| \leq A^2\eta_j^{(0)}(x,y)$ for any $x, y \in \mathbf{R}^n$, $\tag{22.84}$

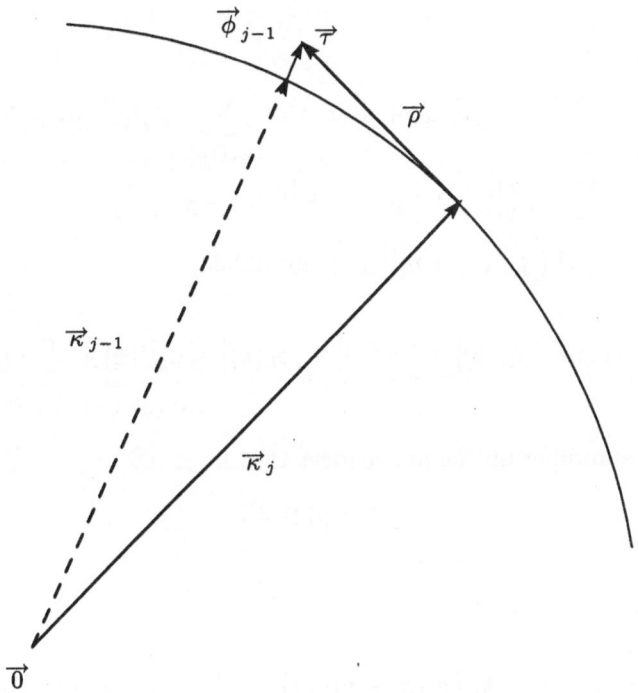

Fig. 22.1: The construction of $\vec{\phi}_{j-1}$

$$|\nabla \vec{\kappa}_j(x)| \leq 2^{-j} A^2 \eta_j^{(2)}(x). \tag{22.84}*$$

Proof. If

$$2^k > |x - y|,$$

then

$$\left| \sum_{I:\ell(I)=2^k} \lambda_I \left(\vec{b}_I(x) + \vec{p}_I(x) \right) - \sum_{I:\ell(I)=2^k} \lambda_I \left(\vec{b}_I(y) + \vec{p}_I(y) \right) \right|$$

$$\leq \sum \lambda_I \left(\left| \vec{b}_I(x) - \vec{b}_I(y) \right| + |\vec{p}_I(x) - \vec{p}_I(y)| \right)$$

$$\leq C \sum \lambda_I A 2^{-k} \left(1 + |x_I - x|/2^k \right)^{-n-1} |x - y| \text{ by (22.83)}$$

$$\leq C A 2^{-k} \eta_k^{(1)}(x)|x - y|,$$

$$\left| \vec{\phi}_k(x) - \vec{\phi}_k(y) \right| \leq C 2^{-k} A^{10} \|\vec{g}\|_{\mathrm{BMO}} \eta_k^{(1)}(x)|x - y| \text{ by (22.67) and (22.46)}$$

$$\leq 2^{-k} \eta_k^{(1)}(x)|x - y| \text{ by (22.56)}.$$

If

$$2^k \leq |x - y|,$$

then

$$\left| \sum_{I:\ell(I)=2^k} \lambda_I \left(\vec{b}_I(x) + \vec{p}_I(x) \right) - \sum_{I:\ell(I)=2^k} \lambda_I \left(\vec{b}_I(y) + \vec{p}_I(y) \right) \right|$$

$$\leq \sum \lambda_I \left(\left| \vec{b}_I(x) + \vec{p}_I(x) \right| + \left| \vec{b}_I(y) + \vec{p}_I(y) \right| \right)$$

$$\leq CA \left(\eta_k^{(1)}(x) + \eta_k^{(1)}(y) \right) \quad \text{by (22.83)},$$

$$\left| \vec{\phi}_k(x) - \vec{\phi}_k(y) \right| \leq \left| \vec{\phi}_k(x) \right| + \left| \vec{\phi}_k(y) \right| \leq \eta_k^{(1)}(x) + \eta_k^{(1)}(y)$$

$$\text{by (22.66) and (22.56).}$$

Therefore, summing up the above gives that if

$$|x - y| \leq 2^j,$$

then

$$|\vec{\kappa}_j(x) - \vec{\kappa}_j(y)|$$

$$\leq \sum_{k=j}^{\infty} \left\{ \left| \sum_{\ell(I)=2^k} \lambda_I \left(\vec{b}_I(x) + \vec{p}_I(x) \right) - \sum_{\ell(I)=2^k} \lambda_I \left(\vec{b}_I(y) + \vec{p}_I(y) \right) \right| \right.$$

$$\left. + \left| \vec{\phi}_k(x) - \vec{\phi}_k(y) \right| \right\}$$

$$\leq \sum_{k=j}^{\infty} CA2^{-k} \eta_k^{(1)}(x) |x - y| \leq A^2 2^{-j} \eta_j^{(2)}(x) |x - y|$$

and that if

$$|x - y| > 2^j,$$

then

$$|\vec{\kappa}_j(x) - \vec{\kappa}_j(y)|$$

$$\leq \sum_{k=j}^{\infty} \left\{ \left| \sum_{\ell(I)=2^k} \lambda_I \left(\vec{b}_I(x) + \vec{p}_I(x) \right) - \sum_{\ell(I)=2^k} \lambda_I \left(\vec{b}_I(y) + \vec{p}_I(y) \right) \right| \right.$$

$$\left. + \left| \vec{\phi}_k(x) - \vec{\phi}_k(y) \right| \right\}$$

$$\leq \sum_{k=j}^{[\log_2 |x-y|]} CA \left(\eta_k^{(1)}(x) + \eta_k^{(1)}(y) \right) + \sum_{k=[\log_2 |x-y|]+1}^{\infty} CA2^{-k} \eta_k^{(1)}(x) |x - y|$$

$$\leq A^2 \sum_{k=j}^{[\log_2 |x-y|]} \left(\eta_k^{(2)}(x) + \eta_k^{(2)}(y) \right).$$

This concludes the proof of (22.84). (22.84)* is clear from (22.84). □

Proof of Claim 1. (22.84)* and (22.58) imply

$$\text{dist}\left(\left\{x \in \mathbf{R}^n : |\vec{\kappa}_j(x)| \geq 0.99\right\}, \left\{x \in \mathbf{R}^n : |\vec{\kappa}_j(x)| \leq 0.9\right\}\right) \geq A2^j.$$

This guarantees the existence of the desired ψ_j. □

Claim 7. $\qquad |\vec{\rho}(x)| + 2^{j-1}|\nabla\vec{\rho}(x)| \leq A^2 \eta_{j-1}^{(1)}(x).$ $\qquad\qquad$ (22.85)

$\qquad\qquad\quad |\vec{\rho}(x)| < 0.01.$ $\qquad\qquad\qquad\qquad\qquad\qquad\qquad$ (22.86)

Claim 8. $\qquad 2^{j-1}|\nabla\vec{\tau}(x)| \leq A^3 \eta_{j-1}^{(2)}(x).$ $\qquad\qquad\qquad\qquad$ (22.87)

$\qquad\qquad\quad$ If $|\vec{\kappa}_j(x)| \geq 0.9,$ then $|\vec{\tau}(x)| \geq 0.89.$ $\qquad\qquad$ (22.88)

Claim 9. $\qquad \vec{\kappa}_{j-1}(x) = \left\{\psi_j(x)|\vec{\kappa}_j(x)| + (1 - \psi_j(x))\,|\vec{\tau}(x)|\right\} U(\vec{\tau}(x)).$ (22.89)

$\qquad\qquad\quad$ If $|\vec{\kappa}_j(x)| \geq 0.99,$ then $|\vec{\kappa}_{j-1}(x)| = |\vec{\kappa}_j(x)|.$ $\qquad\qquad$ (22.90)

(22.85) is clear from (22.83). (22.86) is clear from (22.85) and (22.58). (22.87) is clear from (22.85) and (22.84)*. (22.88) is clear from (22.86). (22.89) is clear from, (22.77)–(22.78). (22.90) is clear from (22.89) and (22.72).

Proof of Claim 2. Let $x \in \mathbf{R}^n$. If $|\vec{\kappa}_j(x)| \geq 0.99,$ then

$$|\vec{\kappa}_{j-1}(x)| = |\vec{\kappa}_j(x)| \text{ by (22.90)}$$
$$\leq 1 \text{ by the hypothesis of induction.}$$

If

$$|\vec{\kappa}_j(x)| < 0.99,$$

then

$$|\vec{\kappa}_{j-1}(x)| \leq \max\left\{|\vec{\kappa}_j(x)|, |\vec{\tau}(x)|\right\} \text{ by (22.89)}$$
$$\leq |\vec{\kappa}_j(x)| + |\vec{\rho}(x)| \leq 0.99 + 0.01 \text{ by (22.86).}$$

□

Claim 10. If $\vec{\kappa}_j(x) \neq \vec{0},$ then

$$|\vec{\tau}(x)| - |\vec{\kappa}_j(x)|$$
$$= |\vec{\kappa}_j(x)|\upsilon\left(2|\vec{\kappa}_j(x)|^{-1}\vec{\rho}(x) \cdot U(\vec{\kappa}_j(x)) + |\vec{\kappa}_j(x)|^{-2}|\vec{\rho}(x)|^2\right). \text{(22.91)}$$

(Recall Definition 22.4.)

Proof. $|\vec{\tau}(x)| = |\vec{\kappa}_j(x) + \vec{\rho}(x)|$

$$= |\vec{\kappa}_j(x)|\left\{1 + 2|\vec{\kappa}_j(x)|^{-1}\left(\vec{\rho}(x) \cdot U(\vec{\kappa}_j(x))\right) + |\vec{\kappa}_j(x)|^{-2}|\vec{\rho}(x)|^2\right\}^{1/2}$$
$$= |\vec{\kappa}_j(x)| + \{\text{the right-hand side of (22.91)}\}.$$

□

Claim 11. If $|\vec{\kappa}_j(x)| \geq 0.1$, then

$$|\vec{\rho}(x) \cdot U(\vec{\kappa}_j(x))| \leq A^4 \min\left\{\eta_{j-1}^{(4)}(x)^2, \|\vec{g}\|_{\mathrm{BMO}}\eta_{j-1}^{(1)}(x)\right\}, \quad (22.92)$$

$$2^{j-1}|(\nabla\vec{\rho}(x)) \cdot U(\vec{\kappa}_j(x))| \leq A^4\|\vec{g}\|_{\mathrm{BMO}}\eta_{j-1}^{(1)}(x). \quad (22.93)$$

Proof. {the left-hand side of (22.92)}

$$= \left|\sum_{I:\ell(I)=2^{j-1}} \lambda_I \left(\vec{b}_I(x) + \vec{p}_I(x)\right) \cdot U(\vec{\kappa}_j(x))\right|$$

$$= \left|\sum \lambda_I \left(\vec{b}_I(x) + \vec{p}_I(x)\right) \cdot (U(\vec{\kappa}_j(x)) - U(\vec{\kappa}_j(x_I)))\right| \text{ by (22.74)}$$

$$\leq \sum \lambda_I \left|\vec{b}_I(x) + \vec{p}_I(x)\right| \cdot 20|\vec{\kappa}_j(x) - \vec{\kappa}_j(x_I)| \text{ by } |\vec{\kappa}_j(x)| \geq 0.1$$

$$\leq \sum \lambda_I 2A(1 + |x - x_I|/2^{j-1})^{-n-1} \cdot 20A^2\eta_j^{(0)}(x, x_I)$$

$$\text{by (22.83) and (22.84)}$$

$$\leq \{\text{the right-hand side of (22.92)}\} \text{ by (22.52) and (22.39)}.$$

(22.93) can be proved by the same way. $\qquad\square$

If $\vec{\kappa}_j(x) \neq \vec{0}$, then put

$$\mu(x) = 2|\vec{\kappa}_j(x)|^{-1}\vec{\rho}(x) \cdot U(\vec{\kappa}_j(x)) + |\vec{\kappa}_j(x)|^{-2}|\vec{\rho}(x)|^2. \quad (22.94)$$

Then, by (22.77), (22.91) and (22.72)

$$\vec{\phi}_{j-1}(x) = \begin{cases} \psi_j(x)|\vec{\kappa}_j(x)|v(\mu(x))U(\vec{\tau}(x)) & \text{if } |\vec{\kappa}_j(x)| \geq 0.9, \\ \vec{0} & \text{if } |\vec{\kappa}_j(x)| < 0.9. \end{cases} \quad (22.95)$$

Claim 12. If $|\vec{\kappa}_j(x)| \geq 0.1$, then

$$|\mu(x)| \leq A^5 \min\left\{\eta_{j-1}^{(4)}(x)^2, \|\vec{g}\|_{\mathrm{BMO}}\eta_{j-1}^{(1)}(x)\right\} (\leq \|\vec{g}\|_{\mathrm{BMO}} < 0.5), (22.96)$$

$$2^{j-1}|\nabla\mu(x)| \leq A^5\|\vec{g}\|_{\mathrm{BMO}}\eta_{j-1}^{(1)}(x). \quad (22.97)$$

Proof.

$$|\mu(x)| \leq 2|\vec{\kappa}_j(x)|^{-1}A^4 \min\left\{\eta_{j-1}^{(4)}(x)^2, \|\vec{g}\|_{\mathrm{BMO}}\eta_{j-1}^{(1)}(x)\right\}$$

$$+|\vec{\kappa}_j(x)|^{-2}A^4\eta_{j-1}^{(1)}(x)^2 \quad \text{by (22.92) and (22.85)}$$

$$\leq \{\text{the right-hand side of (22.96)}\}$$

$$\text{by } |\vec{\kappa}_j(x)| \geq 0.1, (22.58) \text{ and by } \eta_{j-1}^{(1)} \leq C\eta_{j-1}^{(4)}.$$

$$2^{j-1}|\nabla\mu(x)|$$

$$\leq C2^{j-1}\left\{|\vec{\kappa}_j(x)|^{-2}|\nabla\vec{\kappa}_j(x)||\vec{\rho}(x)| + |\vec{\kappa}_j(x)|^{-1}|\nabla\vec{\rho}(x) \cdot U(\vec{\kappa}_j(x))|\right.$$

$$+|\vec{\kappa}_j(x)|^{-3}|\nabla\vec{\kappa}_j(x)||\vec{\rho}(x)|^2 + |\vec{\kappa}_j(x)|^{-2}|\vec{\rho}(x)||\nabla\vec{\rho}(x)|\Big\}$$

$$\leq CA^4\left\{\eta_j^{(2)}(x)\eta_{j-1}^{(1)}(x) + \|\vec{g}\|_{\text{BMO}}\eta_{j-1}^{(1)}(x) + \eta_{j-1}^{(1)}(x)^2\right\}$$

by $|\vec{\kappa}_j(x)| \geq 0.1$, $(22.84)^*$, (22.85) and (22.93)

$\leq$ {the right-hand side of (22.97)} by (22.58).

$\square$

Proof of Claim 3. (Recall (22.95).) If $|\vec{\kappa}_j(x)| < 0.9$, then

$$\vec{\phi}_{j-1}(x) = \vec{0} \text{ by } (22.95).$$

If $|\vec{\kappa}_j(x)| \geq 0.9$, then (22.80) follows from

$$|\vec{\phi}_{j-1}(x)| \leq |v(\mu(x))| \leq |\mu(x)| \text{ by } (22.95)$$

and from (22.96).

$\square$

Proof of Claim 4. (Recall (22.95).) If $|\vec{\kappa}_j(x)| < 0.9$, then

$$\nabla\vec{\phi}_{j-1}(x) = \vec{0} \text{ by } (22.95).$$

If $|\vec{\kappa}_j(x)| \geq 0.9$, then

$$|\nabla\vec{\phi}_{j-1}(x)| \leq |\nabla\vec{\psi}_j(x)||\mu(x)| + |\nabla\vec{\kappa}_j(x)||\mu(x)| + \frac{dv}{dt}(\mu(x))|\nabla\mu(x)|$$

$$+|\mu(x)||\vec{\tau}(x)|^{-1}|\nabla\vec{\tau}(x)|$$

$$= \left\{|\nabla\vec{\psi}_j(x)| + |\nabla\vec{\kappa}_j(x)| + |\vec{\tau}(x)|^{-1}|\nabla\vec{\tau}(x)|\right\}|\mu(x)| + \frac{dv}{dt}(\mu(x))|\nabla\mu(x)|$$

$$\leq 2^{-j+1}|\mu(x)| + |\nabla\mu(x)| \text{ by } (22.73), (22.84)^* \text{ and } (22.87)\text{--}(22.88)$$

$\leq$ {the right-hand side of (22.81)} by (22.96) and (22.97).

$\square$

Proof of Claim 5.

$$\text{supp } \vec{\phi}_{j-1} \subset \text{supp } \vec{\psi}_j \text{ by } (22.77)$$

$$\subset \{x \in \mathbf{R}^n : |\vec{\kappa}_j(x)| \geq 0.9\}$$

$$\subset \left\{x \in \mathbf{R}^n : \sum_{I:\ell(I)\geq 2^j} \lambda_I|\vec{b}_I(x) + \vec{p}_I(x)| + \sum_{k\geq j}|\vec{\phi}_k(x)| \geq 0.9\right\}$$

$$= \left\{x \in \mathbf{R}^n : \sum_{I:\ell(I)\geq 2^j} \lambda_I|\vec{b}_I(x) + \vec{p}_I(x)| \geq 0.9\right\}$$

by the hypothesis of induction.

$\square$

Notes. A constructive proof of the Fefferman-Stein decomposition of $\text{BMO}(\mathbf{R}^1)$ was first given by P. W. Jones [80a] and [83].

The arguments in Sections 22, 23, 24, 26 and 27 are due to A. Uchiyama [82c] and [84]. (See also A. Uchiyama [82b].) But, we must emphasize the contributions from S. Janson [77], J.-A. Chao [74], J.-A. Chao–M. Taibleson [73] and M. Taibleson [75]. Our Lemma 22.1 is the dual of their geometrical investigation of certain vector-valued martingales. (Compare S. Janson [77] and A. Uchiyama [82b].)

As for the extensions of the agument in this section, see M. Christ–D. Geller [84], M. Frazier [85], D. Adams–M. Frazier [88] and A. Miyachi [pre2].

The idea of Lemma 22.6 is due to S.-Y. Chang–R. Fefferman [80]. The idea to use the curvature of the sphere was inspired by D. Sarason [75], S.-Y. Chang [76] and D. Marshall [76].

XXIII. Vector-valued unimodular BMO functions

By the same argument as in Section 22 we can show the following.

Theorem 23.1. *Let $\vec{g} \in \mathrm{BMO}(\mathbf{R}^n, \mathbf{R}^{n+1})$. Then there exist $\vec{h} \in S_{\vec{R}*}$ and $\vec{v} \in \mathrm{BMO}(\mathbf{R}^n, \mathbf{R}^{n+1})$ such that*

$$\|\vec{h}\|_{\mathrm{BMO}} \leq C\|\vec{g}\|_{\mathrm{BMO}}, \tag{23.1}$$

$$\|\vec{v}\|_{\mathrm{BMO}} \leq C\|\vec{g}\|_{\mathrm{BMO}}^2, \tag{23.2}$$

$$|\vec{g}(x) - \vec{h}(x) - \vec{v}(x)| \equiv 1. \tag{23.3}$$

Lemma 22.6′ . Let $\vec{g} \in \mathrm{BMO}(\mathbf{R}^n, \mathbf{R}^{n+1})$. and

$$\operatorname{supp} \vec{g} \subset B(0, 1). \tag{23.4}$$

Let $\varepsilon \in (0, 1)$. Then there exist $\{\lambda_I\}_I \subset [0, +\infty)$ and $\{\vec{b}_I\}_I \subset \mathcal{D}(\mathbf{R}^n, \mathbf{R}^{n+1})$, where I is taken over all dyadic cubes in $\mathbf{R}^n$, such that

$$\lambda_I = 0 \text{ and } \vec{b}_I(x) \equiv \vec{0} \quad \text{if } \ell(I) > 1/\varepsilon, \tag{22.38′}$$

(22.39)–(22.42) and

$$\left\| \vec{g} - \sum_I \lambda_I \vec{b}_I \right\|_{\mathrm{BMO}} \leq C(n)\varepsilon \|\vec{g}\|_{\mathrm{BMO}} \tag{22.44′}$$

hold.

Proof. Let $\phi \in \mathcal{D}(\mathbf{R}^n, \mathbf{R})$ be as in the proof of Lemma 22.6. Let

$$\lambda_I' = \begin{cases} \left\{ |I|^{-1} \displaystyle\iint_{T(I)} |\vec{g} * (\phi)_t(y)|^2 \, dy dt/t \right\}^{1/2} & \text{if } \ell(I) \leq 1/\varepsilon \\ 0 & \text{otherwise.} \end{cases}$$

Using this λ_I', define $\vec{b}_I'$, λ_I and $\vec{b}_I$ by the same procedure as in the proof of Lemma 22.6. Then $(22.38)'$ is clear. (22.39)–(22.42) follow from the same reason as in the proof of Lemma 22.6. $(22.44)'$ follows from

$$\left\| \vec{g} - \sum_I \lambda_I \vec{b}_I \right\|_{L^\infty} = \left\| \iint_{\mathbf{R}^n \times \left(2^{[\log_2(1/\varepsilon)]}, +\infty\right)} \vec{g} * (\phi)_t(y)(\phi)_t(\cdot - y) dy dt/t \right\|_{L^\infty}$$

$$\leq C(n)\varepsilon^n \|\vec{g}\|_{L^1} \leq C(n)\varepsilon^n \|\vec{g}\|_{\mathrm{BMO}} \quad \text{by } (23.4).$$

$\square$

Lemma 23.1. *Let* $m \geq 2$,

$$\vec{k} \in L^\infty(\mathbf{R}^n, \mathbf{R}^m), \quad |\vec{k}(x)| \equiv 1. \tag{23.5}$$

Let $B \subset \mathbf{R}^n$ *be any ball. Then*

$$1 - C(n,m)\|\vec{k}\|^2_{\mathrm{BMO}} \leq \left|\mathrm{av}(\vec{k}, B)\right| \leq 1.$$

Proof. Let $\vec{k}(x) = (k_1(x), \cdots, k_m(x))$. We may assume that $\|\vec{k}\|_{\mathrm{BMO}}$ is small enough and

$$\mathrm{av}(\vec{k}, B) = (a_1, 0, \cdots, 0), \quad a_1 \geq 0.$$

By Lemma 1.9

$$\left|\left\{x \in B : \left|\vec{k}(x) - \mathrm{av}(\vec{k}, B)\right| > \lambda\right\}\right|/|B| \leq C_0 e^{-\lambda/(C_0\|\vec{k}\|_{\mathrm{BMO}})}. \tag{23.6}$$

So, $\left|\left\{x \in B : \left|\vec{k}(x) - \mathrm{av}(\vec{k}, B)\right| \leq 0.1\right\}\right|/|B| \geq 1 - C_0 e^{-0.1/(C_0\|\vec{k}\|_{\mathrm{BMO}})} > 0$.
This, combined with (23.5), implies

$$0.9 \leq a_1 \leq 1.$$

Then

$$|\{x \in B : k_1(x) < 0\}|/|B| \leq C_0 e^{-0.9/(C_0\|\vec{k}\|_{\mathrm{BMO}})}.$$

Then

$$1 \geq a_1 = \mathrm{av}(k_1, B) = \int_{\{x \in B : k_1(x) \geq 0\}} \left\{1 - k_2(x)^2 - \cdots - k_m(x)^2\right\}^{1/2} dx/|B|$$

$$+ \int_{\{x \in B : k_1(x) < 0\}} k_1(x) dx/|B| \qquad \text{by (23.5)}$$

$$\geq 1 - \int_B \left\{k_2(x)^2 + \cdots + k_m(x)^2\right\} dx/|B| - C_0 e^{-0.9/(C_0\|\vec{k}\|_{\mathrm{BMO}})}$$

$$\geq 1 - \int_0^\infty C_0 e^{-\lambda/(C_0\|\vec{k}\|_{\mathrm{BMO}})} 2\lambda d\lambda - C_0 e^{-0.9/(C_0\|\vec{k}\|_{\mathrm{BMO}})} \qquad \text{by (23.6)}$$

$$\geq 1 - C\|\vec{k}\|^2_{\mathrm{BMO}}.$$

$\square$

Lemma 23.2. *Let* $\{\vec{k}_j\}_{j=1}^\infty \subset L^\infty(\mathbf{R}^n, \mathbf{R}^m)$, $\varepsilon > 0$,

$$|\vec{k}_j(x)| \equiv 1, \tag{23.7}$$

$$\|\vec{k}_j\|_{\mathrm{BMO}} \leq \varepsilon, \tag{23.8}$$

$$\vec{k}_j \to \vec{k} \in L^\infty(\mathbf{R}^n, \mathbf{R}^m) \ \textit{weak}^* \ \textit{in } L^\infty \ (j \to \infty). \tag{23.9}$$

Then

$$1 - C(n,m)\varepsilon^2 \leq |\vec{k}(x)| \leq 1. \tag{23.10}$$

Proof. Let

$$\vec{k}_j(x,t) = \mathrm{av}(\vec{k}_j, B(x,t)),$$
$$\vec{k}(x,t) = \mathrm{av}(\vec{k}, B(x,t)).$$

Then (23.7)–(23.8) and Lemma 23.1 imply

$$1 - C\varepsilon^2 \le |\vec{k}_j(x,t)| \le 1.$$

Since

$$\vec{k}_j(x,t) \to \vec{k}(x,t) \ (j \to \infty)$$

for all $(x,t) \in \mathbf{R}_+^{n+1}$, we have

$$1 - C\varepsilon^2 \le |\vec{k}(x,t)| \le 1.$$

Then letting $t \to +0$ implies (23.10). $\qquad\square$

Now, we begin **the proof of Theorem 23.1.** We may assume that

$$\|\vec{g}\|_{\mathrm{BMO}} \text{ is small enough,} \tag{23.11}$$

because otherwise we can take

$$\vec{v}(x) = \vec{g}(x), \quad \vec{h}(x) \equiv (1,0,\cdots,0).$$

The case when supp $\vec{g}$ is compact. We may assume (23.4). Let $A > 0$ be a sufficiently large number depending only on the dimension n. By (23.11) we may assume (22.56). The following procedure is essentially the same with that in Section 22. Instead of Lemma 22.6 we use Lemma 22.6′ with ε in (23.12). The construction of $\vec{h}$ is the same as in Section 22. The construction of $\vec{v}$ is almost the same as in Section 22, except that this time $\vec{g} - \sum \lambda_I \vec{b}_I$ is included in $\vec{v}$. The point is that we change the definition of $\vec{k}_j$ by adding the unit vector $(1,0,\cdots,0)$. (See (22.69)′ below.)
Let

$$\varepsilon = \|\vec{g}\|_{\mathrm{BMO}}. \tag{23.12}$$

Applying Lemma 22.6′ with this ε gives $\{\lambda_I\}_I$ and $\{\vec{b}_I\}_I$ that satisfy (22.38)′, (22.39)–(22.42) and (22.44)′. Let

$$\vec{p}_I(x) \equiv \vec{0} \text{ if } \ell(I) > 1/\varepsilon,$$
$$\vec{\phi}_j(x) \equiv \vec{0} \text{ if } j > [\log_2(1/\varepsilon)].$$

Next, following the procedure in Section 22, we construct

$$\{\vec{p}_I\}_{\ell(I)\le 1/\varepsilon}, \ \{\vec{\phi}_j\}_{j\le[\log_2(1/\varepsilon)]} \subset C^1(\mathbf{R}^n, \mathbf{R}^{n+1}) \tag{23.13}$$

so that (22.62)–(22.64) hold with

$$|\vec{\kappa}_j(x)| \equiv 1 \qquad (22.65)'$$

and (22.66)–(22.67) hold with

$$\vec{\kappa}_j = (1, 0, \cdots, 0) + \sum_{I:\ell(I)\geq 2^j} \lambda_I(\vec{b}_I + \vec{p}_I) - \sum_{k\geq j} \vec{\phi}_k \qquad (22.69)'$$

in place of (22.69).

The procedure of the construction of (23.13) is exactly the same as in Section 22. It is by induction on j. (This time, we have $\psi_j(x) \equiv 1$ because of (22.65)'.) The reason that the inductions of (22.62)–(22.64) and (22.66)–(22.67) can be done is the same as in Section 22. The induction of (22.65)' can be done because of (22.90).

Once these $\{\vec{p}_I\}$ and $\{\vec{\phi}_j\}$ are constructed, we define

$$\vec{h} = (-1, 0, \cdots, 0) - \sum_I \lambda_I \vec{p}_I,$$

$$\vec{v} = \sum_{k=-\infty}^{\infty} \vec{\phi}_k + (\vec{g} - \sum_I \lambda_I \vec{b}_I).$$

The proof of (23.1) is the same as that of (22.8). By the same argument as the proof of (22.9) we can show

$$\left\| \sum \vec{\phi}_k \right\|_{\mathrm{BMO}} \leq C \|\vec{g}\|_{\mathrm{BMO}}^2. \qquad (23.14)$$

(22.44)' with (23.12) implies

$$\left\| \vec{g} - \sum_I \lambda_I \vec{b}_I \right\|_{\mathrm{BMO}} \leq C\varepsilon \|\vec{g}\|_{\mathrm{BMO}} = C\|\vec{g}\|_{\mathrm{BMO}}^2. \qquad (23.15)$$

Thus, (23.14)–(23.15) imply (23.2). (23.3) follows from letting $j \to \infty$ in (22.65)'.

The case when supp $\vec{g}$ **is not compact.** Take $\{\vec{g}_j\}_{j=1}^{\infty} \subset \mathrm{BMO}(\mathbf{R}^n, \mathbf{R}^{n+1})$ so that

$$\text{supp } \vec{g}_j \text{ is compact,} \quad \|\vec{g}_j\|_{\mathrm{BMO}} \leq C(n)\|\vec{g}\|_{\mathrm{BMO}}, \qquad (23.16)$$

$$\vec{g}_j \to \vec{g} \text{ weak}^* \text{ in BMO } (j \to \infty). \qquad (23.17)$$

By the previous result there exist $\{\vec{h}_j\}_{j=1}^{\infty} \subset S_{\vec{R}*}$ and $\{\vec{v}_j\}_{j=1}^{\infty} \subset \mathrm{BMO}(\mathbf{R}^n, \mathbf{R}^{n+1})$ such that

$$\|\vec{h}_j\|_{\mathrm{BMO}} \leq C(n)\|\vec{g}\|_{\mathrm{BMO}}, \qquad (23.18)$$

$$\|\vec{v}_j\|_{\mathrm{BMO}} \leq C(n)\|\vec{g}\|_{\mathrm{BMO}}^2, \qquad (23.19)$$

$$|\vec{g}_j(x) - \vec{h}_j(x) - \vec{v}_j(x)| \equiv 1. \qquad (23.20)$$

We may assume

$$\vec{h}_j \to \vec{h} \in S_{\vec{R}_*} \qquad \text{weak}^* \text{ in BMO } (j \to \infty), \tag{23.21}$$

$$\vec{v}_j \to \vec{v}' \in \text{BMO}(\mathbf{R}^n, \mathbf{R}^{n+1}) \qquad \text{weak}^* \text{ in BMO } (j \to \infty), \tag{23.22}$$

$$\vec{g}_j - \vec{h}_j - \vec{v}_j \to \vec{k} \in L^\infty(\mathbf{R}^n, \mathbf{R}^{n+1}) \text{ weak}^* \text{ in } L^\infty \ (j \to \infty). \tag{23.23}$$

Then

$$\{\vec{g} - \vec{h} - \vec{v}'\} - \vec{k} \equiv \text{const.} \tag{23.24}$$

(23.18)–(23.19) and (23.21)–(23.22) imply (23.1) and

$$\|\vec{v}'\|_{\text{BMO}} \leq C\|\vec{g}\|_{\text{BMO}}^2. \tag{23.25}$$

Since (23.16), (23.18)–(23.19) and (23.11) imply

$$\|\vec{g}_j - \vec{h}_j - \vec{v}_j\|_{\text{BMO}} \leq C\|\vec{g}\|_{\text{BMO}} + C\|\vec{g}\|_{\text{BMO}}^2 \leq C\|\vec{g}\|_{\text{BMO}},$$

Lemma 23.2, (23.20) and (23.23) imply

$$\left| |\vec{k}(x)| - 1 \right| \leq C\|\vec{g}\|_{\text{BMO}}^2. \tag{23.26}$$

Put

$$\vec{v}(x) = \{\vec{g}(x) - \vec{h}(x) - \vec{v}'(x) - \vec{k}(x)\} + \vec{v}'(x) + \left(|\vec{k}(x)| - 1 \right) \vec{k}(x)/|\vec{k}(x)|.$$

Then, (23.2) follows from (23.24)–(23.26). (23.3) follows from

$$\vec{g} - \vec{h} - \vec{v} = \vec{k}/|\vec{k}|.$$

$\square$

XXIV. Extension of the Fefferman-Stein decomposition of BMO, 1

The advantage of the argument in Section 22 is that it can be applied to a large class of Fourier multipliers besides Riesz transforms. We have the following.

Theorem 24.1. *Let* $\{\theta_1, \cdots, \theta_m\} \subset C^\infty(S^{n-1}, \mathbf{C})$,

$$\bar{\theta}_j(\xi) \equiv \theta_j(-\xi), \tag{24.1}$$

$$\mathrm{rank} \begin{bmatrix} \mathrm{Re}\,\theta_1(\xi) & \mathrm{Re}\,\theta_2(\xi) & \cdots & \mathrm{Re}\,\theta_m(\xi) \\ \mathrm{Im}\,\theta_1(\xi) & \mathrm{Im}\,\theta_2(\xi) & \cdots & \mathrm{Im}\,\theta_m(\xi) \end{bmatrix} \equiv 2. \tag{24.2}$$

Let

$$\vec{g} \in \mathrm{BMO}(\mathbf{R}^n, \mathbf{R}^m), \quad \mathrm{supp}\,\vec{g} \subset B(0,1). \tag{24.3}$$

Then there exist $\vec{h} \in S_{\{\theta_1, \cdots, \theta_m\}^*_{(r)}}$ *such that*

$$\left| \vec{g}(x) - \vec{h}(x) \right| \leq C\left(\{\theta_1, \cdots, \theta_m\}\right) \|\vec{g}\|_{\mathrm{BMO}} \left(1 + |x|\right)^{-n-1}, \tag{24.4}$$

where

$$S_{\{\theta_1, \cdots, \theta_m\}^*_{(r)}} = \left\{ \vec{g} = (g_j)_{j=1, \cdots, m} \in \mathrm{BMO}(\mathbf{R}^n, \mathbf{R}^m) : \right.$$

$$\left. \sum_{j=1}^{m} \tilde{m}_{\theta_j} g_j = 0 \text{ in } \mathrm{BMO}(\mathbf{R}^n, \mathbf{R})/\mathbf{R} \right\}. \tag{24.5}$$

By the same argument as Theorem 22.1 was reduced to Lemma 22.2, Theorem 24.1 can be reduced to the following :

Lemma 24.1. *Let* $\{\theta_1, \cdots, \theta_m\} \subset C^\infty(S^{n-1}, \mathbf{C})$ *be as in Theorem 24.1. Let* $\vec{g} \in \mathrm{BMO}(\mathbf{R}^n, \mathbf{R}^m)$ *be as in Theorem 24.1. Then there exist* $\vec{h} \in S_{\{\theta_1, \cdots, \theta_m\}^*_{(r)}}$ *and* $\vec{v} \in \mathrm{BMO}(\mathbf{R}^n, \mathbf{R}^m)$ *such that*

$$\|\vec{h}\|_{\mathrm{BMO}} \leq C\left(\{\theta_1, \cdots, \theta_m\}\right) \|\vec{g}\|_{\mathrm{BMO}},$$

$$\|\vec{v}\|_{\mathrm{BMO}} \leq C\left(\{\theta_1, \cdots, \theta_m\}\right) \|\vec{g}\|_{\mathrm{BMO}}^2,$$

$$\mathrm{supp}\,\vec{v} \subset B(0,3),$$

$$\left| \vec{g}(x) - \vec{h}(x) - \vec{v}(x) \right| \leq \chi_{B(0,3)}(x) + |x|^{-n-1} \chi_{B(0,3)^c}(x).$$

This lemma can be proved by the same argument as the proof of Lemma 22.2

with $\mathbf{R}^{n+1}$ replaced by $\mathbf{R}^m$,

with $S_{\bar{R}_*}$ replaced by $S_{\{\theta_1,\cdots,\theta_m\}_{(r)}^*}$,

with Lemma 22.1 replaced by Lemma 24.3 below

and with $A > 0$ (in the proof of Lemma 22.2) dependent on $\{\theta_1, \cdots, \theta_m\}$. Since the rest is completely the same with the proof of Lemma 22.2 , we omit the details of the proof of Lemma 24.1.

In order to state Lemma 24.3, we need the following.

Lemma 24.2. *Let* $\{\theta_1, \cdots, \theta_m\} \subset C^\infty(S^{n-1}, \mathbf{C})$ *be as in Theorem 24.1. Let* $\bar{\nu} = (\nu_1, \cdots, \nu_m) \in S^{m-1}$. *Then there exists*

$$\{\Theta_{1,\bar{\nu}}, \cdots, \Theta_{m,\bar{\nu}}\} \subset C^\infty(S^{n-1}, \mathbf{C}) \tag{24.6}$$

such that

$$\bar{\Theta}_{j,\bar{\nu}}(\xi) \equiv \Theta_{j,\bar{\nu}}(-\xi), \tag{24.7}$$

$$\sum_{j=1}^m \theta_j(\xi)\Theta_{j,\bar{\nu}}(\xi) \equiv 1, \tag{24.8}$$

$$\sum_{j=1}^m \nu_j\Theta_{j,\bar{\nu}}(\xi) \equiv 0, \tag{24.9}$$

$$\left\| \nabla_\xi^k \{\Theta_{j,\bar{\nu}}(\xi/|\xi|)\} \right\|_{L^\infty(S^{n-1})} \leq C(\{\theta_1, \cdots, \theta_m\}, k) \quad \text{for any } k \in \mathbf{N}_0. \tag{24.10}$$

Remark 24.1. If (24.1)–(24.2) hold and if

$$\theta_1 \equiv 1, \quad \mathrm{Re}\,\theta_2 \equiv \mathrm{Re}\,\theta_3 \equiv \cdots \equiv \mathrm{Re}\,\theta_m \equiv 0,$$

then we can take

$$\begin{bmatrix} \Theta_{1,\bar{\nu}}(\xi) \\ \Theta_{2,\bar{\nu}}(\xi) \\ \vdots \\ \Theta_{m,\bar{\nu}}(\xi) \end{bmatrix} = \left\{ \sum_{j=2}^m (\nu_j^2 - \nu_1^2\theta_j(\xi)^2) \right\}^{-1} \begin{bmatrix} \sum_{j=2}^m \nu_j^2 + \sum_{j=2}^m \nu_1\nu_j\theta_j(\xi) \\ -\nu_1\nu_2 - \nu_1^2\theta_2(\xi) \\ \vdots \quad \vdots \\ -\nu_1\nu_m - \nu_1^2\theta_m(\xi) \end{bmatrix}.$$

The following trivial lemma is the substitute of Lemma 22.1.

Lemma 24.3. *Let* $\{\theta_1, \cdots, \theta_m\} \subset C^\infty(S^{n-1}, \mathbf{C})$ *be as in Theorem 24.1. Let* $\bar{\nu} = (\nu_1 \cdots \nu_m) \in S^{m-1}$. *Let* $\{\Theta_{j,\bar{\nu}}\}_{j=1,\cdots,m}$ *be as in Lemma 24.2. Let* $b \in L^2(\mathbf{R}^n, \mathbf{R}) \cap \mathrm{BMO}(\mathbf{R}^n, \mathbf{R})$. *Let*

$$\vec{p} = \begin{bmatrix} \nu_1 b - \mathsf{m}_{\Theta_{1,\vec{\nu}}}\left(\sum_{j=1}^{m} \mathsf{m}_{\theta_j}\nu_j b\right) \\ \vdots \qquad \vdots \\ \nu_m b - \mathsf{m}_{\Theta_{m,\vec{\nu}}}\left(\sum_{j=1}^{m} \mathsf{m}_{\theta_j}\nu_j b\right) \end{bmatrix} \in L^2(\mathbf{R}^n, \mathbf{R}^m) \cap \mathrm{BMO}(\mathbf{R}^n, \mathbf{R}^m).$$

(24.11)

Then

$$\vec{p} \in S_{\{\theta_1,\cdots,\theta_m\}^*_{(r)}}, \tag{24.12}$$

$$\vec{p}(x) \cdot \vec{\nu} \equiv b(x). \tag{24.13}$$

Proof of Lemma 24.2. By (24.1)–(24.2) there exists a partition of unity of S^{n-1}

$$\{\psi_{j,k}\}_{j,k \in \{1,\cdots,m\}} \subset C^\infty(S^{n-1}, \mathbf{R})$$

such that

$$\psi_{j,k}(\xi) \equiv \psi_{k,j}(\xi), \quad \psi_{j,j}(\xi) \equiv 0, \tag{24.14}$$

$$\psi_{j,k}(\xi) \equiv \psi_{j,k}(-\xi), \tag{24.15}$$

$$\sum_{j=1}^{m} \sum_{k=j+1}^{m} \psi_{j,k}(\xi) \equiv 1, \tag{24.16}$$

$$\mathrm{rank}\begin{bmatrix} \mathrm{Re}\,\theta_j(\xi) & \mathrm{Re}\,\theta_k(\xi) \\ \mathrm{Im}\,\theta_j(\xi) & \mathrm{Im}\,\theta_k(\xi) \end{bmatrix} = 2 \text{ for any } \xi \in \mathrm{supp}\,\psi_{j,k}. \tag{24.17}$$

For $j, k \in \{1, \cdots, m\}$, $\xi \in S^{n-1}$ and $(\mu_1\ \mu_2) \in \mathbf{R}^2$ let

$$\Theta_{j,k}(\xi, (\mu_1, \mu_2)) = \begin{cases} \dfrac{\mu_2\psi_{j,k}(\xi)}{\mu_2\theta_j(\xi) - \mu_1\theta_k(\xi)} & \text{if } \xi \in \mathrm{supp}\,\psi_{j,k},\ (\mu_1, \mu_2) \neq (0,0), \\ \dfrac{\psi_{j,k}(\xi)}{2\theta_j(\xi)} & \text{if } \xi \in \mathrm{supp}\,\psi_{j,k},\ \mu_1 = \mu_2 = 0, \\ 0 & \text{if } \xi \notin \mathrm{supp}\,\psi_{j,k}. \end{cases}$$

(Note that if $\xi \in \mathrm{supp}\,\psi_{j,k}$ and if $(\mu_1, \mu_2) \in \mathbf{R}^2 \setminus \{(0,0)\}$, then (24.17) implies

$$\mu_2\theta_j(\xi) - \mu_1\theta_k(\xi) \neq 0.)$$

Then

$$\overline{\Theta}_{j,k}\,(\xi, (\mu_1, \mu_2)) \equiv \Theta_{j,k}\,(-\xi, (\mu_1, \mu_2))\ \text{by (24.1) and (24.15)}, \tag{24.7*}$$

$$\theta_j(\xi)\Theta_{j,k}\,(\xi, (\mu_1, \mu_2)) + \theta_k(\xi)\Theta_{k,j}\,(\xi, (\mu_2, \mu_1)) \equiv \psi_{j,k}(\xi) \tag{24.8*}$$

$$\text{by (24.14)},$$

$$\mu_1\Theta_{j,k}\,(\xi, (\mu_1, \mu_2)) + \mu_2\Theta_{k,j}\,(\xi, (\mu_2, \mu_1)) \equiv 0\ \text{by (24.14)}, \tag{24.9*}$$

For $j \in \{1, \cdots, m\}$, $\xi \in S^{n-1}$ and

$$\vec{\nu} = (\nu_1, \cdots, \nu_m) \in S^{m-1}$$

let

$$\Theta_{j,\vec{\nu}}(\xi) = \sum_{k=1}^{m} \Theta_{j,k}\left(\xi, (\nu_j, \nu_k)\right).$$

Then, (24.7) is clear from (24.7)*. (24.8)–(24.9) follow from

$$\sum_{j=1}^{m} \theta_j(\xi)\Theta_{j,\vec{\nu}}(\xi) = \sum_{j=1}^{m}\sum_{k=1}^{m} \theta_j(\xi)\Theta_{j,k}\left(\xi, (\nu_j, \nu_k)\right)$$

$$= \sum_{j=1}^{m}\sum_{k=j+1}^{m} \left\{\theta_j(\xi)\Theta_{j,k}\left(\xi, (\nu_j, \nu_k)\right) + \theta_k(\xi)\Theta_{k,j}\left(\xi, (\nu_k, \nu_j)\right)\right\}$$

$$\text{by } \Theta_{j,j} \equiv 0$$

$$= \sum\sum \psi_{j,k}(\xi) \qquad \text{by (24.8)*}$$

$$\equiv 1 \text{ by (24.16)},$$

$$\sum_{j=1}^{m} \nu_j\Theta_{j,\vec{\nu}}(\xi) = \sum_{j=1}^{m}\sum_{k=j+1}^{m} \left\{\nu_j\Theta_{j,k}\left(\xi, (\nu_j, \nu_k)\right) + \nu_k\Theta_{k,j}\left(\xi, (\nu_k, \nu_j)\right)\right\}$$

$$= \sum\sum 0 \equiv 0 \text{ by (24.9)*}.$$

(24.10) follows from

$$\sup\left\{\left|\nabla^u_\xi \left\{\Theta_{j,k}\left(\xi/|\xi|, (\mu_1, \mu_2)\right)\right\}\right| : \xi \in S^{n-1}, \ (\mu_1, \mu_2) \in \mathbf{R}^2\right\}$$
$$= \sup\left\{\left|\nabla^u_\xi \left\{\Theta_{j,k}\left(\xi/|\xi|, (\mu_1, \mu_2)\right)\right\}\right| : \xi \in S^{n-1}, \ (\mu_1, \mu_2) \in S^1 \cup \{(0,0)\}\right\}$$
$$< \infty \text{ by compactness.}$$

$\square$

Proof of Lemma 24.3. (24.12) is clear from (24.8). (24.13) follows from the fact that

$$\sum_{j=1}^{m} \nu_j m_{\Theta_{j,\vec{\nu}}} h \equiv 0$$

for any $h \in L^2(\mathbf{R}^n, \mathbf{R})$ by (24.9). $\square$

Next, removing the restriction (24.1), we consider $\text{BMO}(\mathbf{R}^n, \mathbf{C}^m)$. The result corresponding to Theorem 24.1 is the following.

Theorem 24.2. Let $\{\theta_1, \cdots, \theta_m\} \subset C^\infty(S^{n-1}, \mathbf{C})$ and let

$$\text{rank}\begin{bmatrix} \theta_1(\xi) & \theta_2(\xi) & \cdots & \theta_m(\xi) \\ \theta_1(-\xi) & \theta_2(-\xi) & \cdots & \theta_m(-\xi) \end{bmatrix} \equiv 2. \qquad (24.18)$$

Let

$$\vec{g} \in \mathrm{BMO}(\mathbf{R}^n, \mathbf{C}^m), \quad \operatorname{supp}\vec{g} \subset B(0,1). \tag{24.19}$$

Then there exists $\vec{h} \in S_{\{\theta_1,\cdots,\theta_m\}^*}$ *such that*

$$\left|\vec{g}(x) - \vec{h}(x)\right| \leq C\left(\{\theta_1, \cdots, \theta_m\}\right) \|\vec{g}\|_{\mathrm{BMO}} (1 + |x|)^{-n-1}, \tag{24.20}$$

where

$$S_{\{\theta_1,\cdots,\theta_m\}^*} = \Big\{ \vec{g} = (g_j)_{j=1,\cdots,m} \in \mathrm{BMO}(\mathbf{R}^n, \mathbf{C}^m) :$$

$$\sum_{j=1}^{m} \tilde{m}_{\theta_j} g_j = 0 \text{ in } \mathrm{BMO}(\mathbf{R}^n, \mathbf{C})/\mathbf{C} \Big\}. \tag{24.21}$$

The argument of the proof of Theorem 24.2 is the same as the proofs of Theorems 22.1 and 24.1. First, by the same reasoning as Theorem 22.1 was reduced to Lemma 22.2, Theorem 24.2 can be reduced to the following.

Lemma 24.1′ . Let $\{\theta_1, \cdots, \theta_m\} \subset C^\infty(S^{n-1}, \mathbf{C})$ be as in Theorem 24.2. Let $\vec{g} \in \mathrm{BMO}(\mathbf{R}^n, \mathbf{C}^m)$ and let $\operatorname{supp}\vec{g} \subset B(0,1)$. Then there exist $\vec{h} \in S_{\{\theta_1,\cdots,\theta_m\}^*}$ and $\vec{v} \in \mathrm{BMO}(\mathbf{R}^n, \mathbf{C}^m)$ that satisfy the same conditions as in Lemma 24.1.

The above lemma can be proved by the same argument as the proof of Lemma 22.2
with $\mathbf{R}^{n+1}$ replaced by $\mathbf{R}^{2m}(= \mathbf{C}^m)$,
with $S_{\vec{R}^*}$ replaced by $S_{\{\theta_1,\cdots,\theta_m\}^*}$,
with Lemma 22.1 replaced by Lemma 24.3′ below
and with $A > 0$ (in the proof of Lemma 22.2) dependent on $\{\theta_1, \cdots, \theta_m\}$. We omit the details.

In order to state Lemma 24.3′ we need the following.

Lemma 24.2′ . Let $\{\theta_1, \cdots, \theta_m\} \subset C^\infty(S^{n-1}, \mathbf{C})$ be as in Theorem 24.2. Let $\vec{\nu} = (\nu_1, \cdots, \nu_{2m}) \in S^{2m-1}$. Then there exists $\{\Theta_{1,\vec{\nu}}, \cdots, \Theta_{m,\vec{\nu}}\} \subset C^\infty(S^{n-1}, \mathbf{C})$ such that

$$\sum_{j=1}^{m} \theta_j(\xi) \Theta_{j,\vec{\nu}}(\xi) \equiv 1, \tag{24.22}$$

$$\sum_{j=1}^{m} \{\nu_{2j-1} \operatorname{Re}\left(\Theta_{j,\vec{\nu}}(\xi) + \Theta_{j,\vec{\nu}}(-\xi)\right) + \nu_{2j} \operatorname{Im}\left(\Theta_{j,\vec{\nu}}(\xi) + \Theta_{j,\vec{\nu}}(-\xi)\right)\} \equiv 0,$$

$$\tag{24.23}$$

$$\sum_{j=1}^{m} \{\nu_{2j-1} \operatorname{Im}\left(\Theta_{j,\vec{\nu}}(\xi) - \Theta_{j,\vec{\nu}}(-\xi)\right) - \nu_{2j} \operatorname{Re}\left(\Theta_{j,\vec{\nu}}(\xi) - \Theta_{j,\vec{\nu}}(-\xi)\right)\} \equiv 0,$$

$$\tag{24.24}$$

$$\left\|\nabla_\xi^k \{\Theta_{j,\vec{\nu}}(\xi/|\xi|)\}\right\|_{L^\infty(S^{n-1})} \leq C\left(\{\theta_1, \cdots, \theta_m\}, k\right) \text{ for any } k \in \mathbf{N}_0. \tag{24.25}$$

Lemma 24.3′ . Let $\{\theta_1, \cdots, \theta_m\} \subset C^\infty(S^{n-1}, \mathbf{C})$ be as in Theorem 24.2. Let $\vec{\nu} = (\nu_1, \cdots, \nu_{2m}) \in S^{2m-1}$. Let $\{\Theta_{1,\vec{\nu}}, \cdots, \Theta_{m,\vec{\nu}}\} \subset C^\infty(S^{n-1}, \mathbf{C})$ be as in Lemma 24.2′. Let $b \in L^2(\mathbf{R}^n, \mathbf{R}) \cap \mathrm{BMO}(\mathbf{R}^n, \mathbf{R})$. Let

$$
\vec{p} = \begin{bmatrix} (\nu_1 + i\nu_2)b - \mathrm{m}_{\Theta_{1,\vec{\nu}}}(\mathrm{Re}\, b') - i\mathrm{m}_{\Theta_{1,\vec{\nu}}}(\mathrm{Im}\, b') \\ \cdots\cdots\cdots\cdots\cdots\cdots\cdots\cdots\cdots\cdots\cdots\cdots \\ \cdots\cdots\cdots\cdots\cdots\cdots\cdots\cdots\cdots\cdots\cdots\cdots \\ \cdots\cdots\cdots\cdots\cdots\cdots\cdots\cdots\cdots\cdots\cdots\cdots \\ (\nu_{2m-1} + i\nu_{2m})b - \mathrm{m}_{\Theta_{m,\vec{\nu}}}(\mathrm{Re}\, b') - i\mathrm{m}_{\Theta_{m,\vec{\nu}}}(\mathrm{Im}\, b') \end{bmatrix}
$$

$$
\in L^2(\mathbf{R}^n, \mathbf{C}^m) \cap \mathrm{BMO}(\mathbf{R}^n, \mathbf{C}^m), \tag{24.26}
$$

where

$$
b' = \sum_{j=1}^m \mathrm{m}_{\theta_j}\left((\nu_{2j-1} + i\nu_{2j})b\right) \in L^2(\mathbf{R}^n, \mathbf{C}) \cap \mathrm{BMO}(\mathbf{R}^n, \mathbf{C}), \tag{24.27}
$$

$$
\vec{\nu}' = (-\nu_2, \nu_1, -\nu_4, \nu_3, \cdots, -\nu_{2m}, \nu_{2m-1}) \in S^{2m-1}. \tag{24.28}
$$

Then

$$
\vec{p} \in S_{\{\theta_1, \cdots, \theta_m\}*}, \tag{24.29}
$$

$$
\vec{p}(x) \cdot \vec{\nu} \equiv b(x). \tag{24.30}
$$

(As for the definition of the inner product $\vec{p}(x) \cdot \vec{\nu}$ see Remark 21.1.)

Proof of Lemma 24.2′. By (24.18) there exist $\{\psi_{j,k}\}_{j,k \in \{1, \cdots, m\}} \subset C^\infty(S^{n-1}, \mathbf{R})$ that satisfy (24.14)–(24.16) and

$$
\mathrm{rank}\begin{bmatrix} \theta_j(\xi) & \theta_k(\xi) \\ \theta_j(-\xi) & \theta_k(-\xi) \end{bmatrix} = 2 \text{ for any } \xi \in \mathrm{supp}\,\psi_{j,k}. \tag{24.31}
$$

Let

$$
j, k \in \{1, \cdots, m\} \text{ and } j < k.
$$

We freeze these j and k temporarily. For

$$
(\xi, (\mu_1, \mu_2, \mu_3, \mu_4)) \in S^{n-1} \times S^3
$$

let

$$
B_{j,k}\left(\xi, (\mu_1, \mu_2, \mu_3, \mu_4)\right) =
$$

$$
\begin{bmatrix}
\mathrm{Re}\theta_j(\xi) & -\mathrm{Im}\theta_j(\xi) & \mathrm{Re}\theta_k(\xi) & -\mathrm{Im}\theta_k(\xi) & 0 & 0 & 0 & 0 \\
\mathrm{Im}\theta_j(\xi) & \mathrm{Re}\theta_j(\xi) & \mathrm{Im}\theta_k(\xi) & \mathrm{Re}\theta_k(\xi) & 0 & 0 & 0 & 0 \\
0 & 0 & 0 & 0 & \mathrm{Re}\theta_j(-\xi) & -\mathrm{Im}\theta_j(-\xi) & \mathrm{Re}\theta_k(-\xi) & -\mathrm{Im}\theta_k(-\xi) \\
0 & 0 & 0 & 0 & \mathrm{Im}\theta_j(-\xi) & \mathrm{Re}\theta_j(-\xi) & \mathrm{Im}\theta_k(-\xi) & \mathrm{Re}\theta_k(-\xi) \\
\mu_1 & \mu_2 & \mu_3 & \mu_4 & \mu_1 & \mu_2 & \mu_3 & \mu_4 \\
-\mu_2 & \mu_1 & -\mu_4 & \mu_3 & \mu_2 & -\mu_1 & \mu_4 & -\mu_3
\end{bmatrix}.
$$

Then (24.31) implies

$$\mathrm{rank}B_{j,k}\left(\xi,(\mu_1,\mu_2,\mu_3,\mu_4)\right)=6 \text{ for any } (\xi,(\mu_1,\mu_2,\mu_3,\mu_4))$$
$$\in \mathrm{supp}\,\psi_{j,k}\times S^3.$$

So, for any

$$\left(\xi^0,(\mu_1^0,\mu_2^0,\mu_3^0,\mu_4^0)\right)\in S^{n-1}\times S^3$$

there exist its small neighbourhood $U(\subset S^{n-1}\times S^3)$ and

$$\Theta_{j,k,U}\left(\xi,(\mu_1,\mu_2,\mu_3,\mu_4)\right),\ \Theta_{k,j,U}\left(\xi,(\mu_1,\mu_2,\mu_3,\mu_4)\right)\in C^\infty(U\cup U^*)$$

such that

$$U\cap U^*=\emptyset,$$

$$B_{j,k}\left(\xi,(\mu_1,\mu_2,\mu_3,\mu_4)\right)\begin{bmatrix}\mathrm{Re}\,\Theta_{j,k,U}\left(\xi,(\mu_1,\mu_2,\mu_3,\mu_4)\right)\\\mathrm{Im}\,\Theta_{j,k,U}\left(\xi,(\mu_1,\mu_2,\mu_3,\mu_4)\right)\\\mathrm{Re}\,\Theta_{k,j,U}\left(\xi,(\mu_1,\mu_2,\mu_3,\mu_4)\right)\\\mathrm{Im}\,\Theta_{k,j,U}\left(\xi,(\mu_1,\mu_2,\mu_3,\mu_4)\right)\\\mathrm{Re}\,\Theta_{j,k,U}\left(-\xi,(\mu_1,\mu_2,\mu_3,\mu_4)\right)\\\mathrm{Im}\,\Theta_{j,k,U}\left(-\xi,(\mu_1,\mu_2,\mu_3,\mu_4)\right)\\\mathrm{Re}\,\Theta_{k,j,U}\left(-\xi,(\mu_1,\mu_2,\mu_3,\mu_4)\right)\\\mathrm{Im}\,\Theta_{k,j,U}\left(-\xi,(\mu_1,\mu_2,\mu_3,\mu_4)\right)\end{bmatrix}=\psi_{j,k}(\xi)\begin{bmatrix}1\\0\\1\\0\\0\\0\end{bmatrix}$$

$$(24.32)$$

for any $(\xi,(\mu_1,\mu_2,\mu_3,\mu_4))\in U$, where

$$U^*=\{(-\xi,(\mu_1,\mu_2,\mu_3,\mu_4)):(\xi,(\mu_1,\mu_2,\mu_3,\mu_4))\in U\}.$$

(If $\xi^0\notin \mathrm{supp}\,\psi_{j,k}$, then it is enough to take $U,\Theta_{j,k,U}$ and $\Theta_{k,j,U}$ so that

$$U\cap(\mathrm{supp}\,\psi_{j,k}\times S^3)=\emptyset,\quad \Theta_{j,k,U}\equiv\Theta_{k,j,U}\equiv 0.)$$

Note that (24.32) implies that if

$$(\xi,(\mu_1,\mu_2,\mu_3,\mu_4))\in U\cup U^*,$$

then

$$\theta_j(\xi)\Theta_{j,k,U}\left(\xi,(\mu_1,\mu_2,\mu_3,\mu_4)\right)+\theta_k(\xi)\Theta_{k,j,U}\left(\xi,(\mu_1,\mu_2,\mu_3,\mu_4)\right)\equiv\psi_{j,k}(\xi),$$

$$(24.33)$$

$$\mu_1\,\mathrm{Re}\left(\Theta_{j,k,U}\left(\xi,(\mu_1,\mu_2,\mu_3,\mu_4)\right)+\Theta_{j,k,U}\left(-\xi,(\mu_1,\mu_2,\mu_3,\mu_4)\right)\right)$$
$$+\mu_2\,\mathrm{Im}\left(\Theta_{j,k,U}\left(\xi,(\mu_1,\mu_2,\mu_3,\mu_4)\right)+\Theta_{j,k,U}\left(-\xi,(\mu_1,\mu_2,\mu_3,\mu_4)\right)\right)$$
$$+\mu_3\,\mathrm{Re}\left(\Theta_{k,j,U}\left(\xi,(\mu_1,\mu_2,\mu_3,\mu_4)\right)+\Theta_{k,j,U}\left(-\xi,(\mu_1,\mu_2,\mu_3,\mu_4)\right)\right)$$
$$+\mu_4\,\mathrm{Im}\left(\Theta_{k,j,U}\left(\xi,(\mu_1,\mu_2,\mu_3,\mu_4)\right)+\Theta_{k,j,U}\left(-\xi,(\mu_1,\mu_2,\mu_3,\mu_4)\right)\right)\equiv 0,$$

$$(24.34)$$

$$\mu_1\,\mathrm{Im}\left(\Theta_{j,k,U}\left(\xi,(\mu_1,\mu_2,\mu_3,\mu_4)\right)-\Theta_{j,k,U}\left(-\xi,(\mu_1,\mu_2,\mu_3,\mu_4)\right)\right)$$

$$-\mu_2 \operatorname{Re}\left(\Theta_{j,k,U}\left(\xi, (\mu_1, \mu_2, \mu_3, \mu_4)\right) - \Theta_{j,k,U}\left(-\xi, (\mu_1, \mu_2, \mu_3, \mu_4)\right)\right)$$
$$+\mu_3 \operatorname{Im}\left(\Theta_{k,j,U}\left(\xi, (\mu_1, \mu_2, \mu_3, \mu_4)\right) - \Theta_{k,j,U}\left(-\xi, (\mu_1, \mu_2, \mu_3, \mu_4)\right)\right)$$
$$-\mu_4 \operatorname{Re}\left(\Theta_{k,j,U}\left(\xi, (\mu_1, \mu_2, \mu_3, \mu_4)\right) - \Theta_{k,j,U}\left(-\xi, (\mu_1, \mu_2, \mu_3, \mu_4)\right)\right) \equiv 0,$$
$$\tag{24.35}$$

By this procedure, for each point of $S^{n-1} \times S^3$, there corresponds its neighbourhood U. From these U's we pick up $\{U_1, \cdots, U_p\}$ which is a finite covering of $S^{n-1} \times S^3$. Let $\{\psi_1, \cdots, \psi_p\} \subset C^\infty(S^{n-1} \times S^3, \mathbf{R})$ be such that

$$\sum_{q=1}^{p} \psi_q\left(\xi, (\mu_1, \mu_2, \mu_3, \mu_4)\right) \equiv 1 \quad \text{on } S^{n-1} \times S^3, \tag{24.36}$$

$$\operatorname{supp} \psi_q \subset U_q \cup U_q^*, \quad \psi_q\left(\xi, (\mu_1, \mu_2, \mu_3, \mu_4)\right) \equiv \psi_q\left(-\xi, (\mu_1, \mu_2, \mu_3, \mu_4)\right)$$
$$\text{for } q = 1, \cdots, p. \tag{24.37}$$

For

$$\left(\xi, (\mu_1, \mu_2, \mu_3, \mu_4)\right) \in S^{n-1} \times S^3$$

let

$$\Theta_{j,k}\left(\xi, (\mu_1, \mu_2, \mu_3, \mu_4)\right)$$
$$= \sum_{q=1}^{p} \psi_q\left(\xi, (\mu_1, \mu_2, \mu_3, \mu_4)\right) \Theta_{j,k,U_q}\left(\xi, (\mu_1, \mu_2, \mu_3, \mu_4)\right),$$
$$\Theta_{k,j}\left(\xi, (\mu_1, \mu_2, \mu_3, \mu_4)\right)$$
$$= \sum_{q=1}^{p} \psi_q\left(\xi, (\mu_1, \mu_2, \mu_3, \mu_4)\right) \Theta_{k,j,U_q}\left(\xi, (\mu_1, \mu_2, \mu_3, \mu_4)\right),$$

where $\psi_q \Theta_{j,k,U_q}$ is defined to be 0 if $\left(\xi, (\mu_1, \mu_2, \mu_3, \mu_4)\right) \notin \operatorname{supp} \psi_q$. We call the formulae (24.33), (24.34) and (24.35) with

$$\Theta_{j,k} \text{ and } \Theta_{k,j} \text{ in place of } \Theta_{j,k,U} \text{ and } \Theta_{k,j,U}$$

by the names

$$(24.33)^*, \ (24.34)^* \text{ and } (24.35)^*,$$

respectively. Then, $(24.33)^*$–$(24.35)^*$ hold on $S^{n-1} \times S^3$ by (24.33)–(24.37). We extend $\Theta_{j,k}$ and $\Theta_{k,j}$ onto

$$\left(\xi, (\mu_1, \mu_2, \mu_3, \mu_4)\right) \in S^{n-1} \times \mathbf{R}^4$$

so that

$$\Theta_{j,k}\left(\xi, r(\mu_1, \mu_2, \mu_3, \mu_4)\right) = \Theta_{j,k}\left(\xi, (\mu_1, \mu_2, \mu_3, \mu_4)\right),$$
$$\Theta_{k,j}\left(\xi, r(\mu_1, \mu_2, \mu_3, \mu_4)\right) = \Theta_{k,j}\left(\xi, (\mu_1, \mu_2, \mu_3, \mu_4)\right)$$
$$\text{if } (\mu_1, \mu_2, \mu_3, \mu_4) \neq (0,0,0,0) \text{ and } r > 0,$$

$$\Theta_{j,k}\left(\xi, (0,0,0,0)\right) = \begin{cases} \dfrac{\overline{\theta_j(\xi)}\psi_{j,k}(\xi)}{|\theta_j(\xi)|^2 + |\theta_k(\xi)|^2} & \text{if } \xi \in \operatorname{supp} \psi_{j,k}, \\ 0 & \text{otherwise,} \end{cases}$$

$$\Theta_{k,j}\left(\xi, (0,0,0,0)\right) = \begin{cases} \dfrac{\overline{\theta_k(\xi)}\psi_{j,k}(\xi)}{|\theta_j(\xi)|^2 + |\theta_k(\xi)|^2} & \text{if } \xi \in \operatorname{supp} \psi_{j,k}, \\ 0 & \text{otherwise.} \end{cases}$$

Then $(24.33)^*$–$(24.35)^*$ still hold on $S^{n-1} \times \mathbf{R}^4$.

Using these $\{\Theta_{j,k}\}_{j,k\in\{1,\cdots,m\},j\neq k}$, we define the desired $\{\Theta_{j,\vec{\nu}}\}$. For $j \in \{1,\cdots,m\}$, $\xi \in S^{n-1}$ and for $\vec{\nu} \in S^{2m-1}$ let

$$\Theta_{j,\vec{\nu}}(\xi) = \sum_{k=1}^{j-1} \Theta_{j,k}\left(\xi, (\nu_{2k-1}, \nu_{2k}, \nu_{2j-1}, \nu_{2j})\right)$$
$$+ \sum_{k=j+1}^{m} \Theta_{j,k}\left(\xi, (\nu_{2j-1}, \nu_{2j}, \nu_{2k-1}, \nu_{2k})\right).$$

Then (24.22) follows from

$$\sum_{j=1}^{m} \theta_j(\xi)\Theta_{j,\vec{\nu}}(\xi) = \sum_{j=1}^{m}\sum_{k=1}^{j-1} \theta_j(\xi)\Theta_{j,k}\left(\xi, (\nu_{2k-1}, \nu_{2k}, \nu_{2j-1}, \nu_{2j})\right)$$
$$+ \sum_{j=1}^{m}\sum_{k=j+1}^{m} \theta_j(\xi)\Theta_{j,k}\left(\xi, (\nu_{2j-1}, \nu_{2j}, \nu_{2k-1}, \nu_{2k})\right)$$
$$= \sum_{j=1}^{m}\sum_{k=j+1}^{m} \Big\{ \theta_k(\xi)\Theta_{k,j}\left(\xi, (\nu_{2j-1}, \nu_{2j}, \nu_{2k-1}, \nu_{2k})\right)$$
$$+ \theta_j(\xi)\Theta_{j,k}\left(\xi, (\nu_{2j-1}, \nu_{2j}, \nu_{2k-1}, \nu_{2k})\right) \Big\}$$
$$= \sum\sum \psi_{j,k}(\xi) \text{ by } (24.33)^*$$
$$\equiv 1 \text{ by } (24.16).$$

Put

$$\nu_{j,k} = (\nu_{2j-1}, \nu_{2j}, \nu_{2k-1}, \nu_{2k}), \ j,k \in \{1,2,\cdots,m\}.$$

(24.23) follows from

$$\sum_{j=1}^{m} \{\nu_{2j-1} \operatorname{Re}\left(\Theta_{j,\bar{\nu}}(\xi) + \Theta_{j,\bar{\nu}}(-\xi)\right) + \nu_{2j} \operatorname{Im}\left(\Theta_{j,\bar{\nu}}(\xi) + \Theta_{j,\bar{\nu}}(-\xi)\right)\}$$

$$= \sum_{j=1}^{m}\sum_{k=1}^{j-1} \Big\{ \nu_{2j-1} \operatorname{Re}\left(\Theta_{j,k}\left(\xi, \nu_{k,j}\right) + \Theta_{j,k}\left(-\xi, \nu_{k,j}\right)\right)$$

$$+ \nu_{2j} \operatorname{Im}\left(\Theta_{j,k}(\xi, \nu_{k,j}) + \Theta_{j,k}(-\xi, \nu_{k,j})\right) \Big\}$$

$$+ \sum_{j=1}^{m}\sum_{k=j+1}^{m} \Big\{ \nu_{2j-1} \operatorname{Re}\left(\Theta_{j,k}(\xi, \nu_{j,k}) + \Theta_{j,k}(-\xi, \nu_{j,k})\right)$$

$$+ \nu_{2j} \operatorname{Im}\left(\Theta_{j,k}(\xi, \nu_{j,k}) + \Theta_{j,k}(-\xi, \nu_{j,k})\right) \Big\}$$

$$= \sum_{j=1}^{m}\sum_{k=j+1}^{m} \Big\{ \nu_{2k-1} \operatorname{Re}\left(\Theta_{k,j}\left(\xi, \nu_{j,k}\right) + \Theta_{k,j}\left(-\xi, \nu_{j,k}\right)\right)$$

$$+ \nu_{2k} \operatorname{Im}\left(\Theta_{k,j}(\xi, \nu_{j,k}) + \Theta_{k,j}(-\xi, \nu_{j,k})\right)$$

$$+ \nu_{2j-1} \operatorname{Re}\left(\Theta_{j,k}(\xi, \nu_{j,k}) + \Theta_{j,k}(-\xi, \nu_{j,k})\right)$$

$$+ \nu_{2j} \operatorname{Im}\left(\Theta_{j,k}(\xi, \nu_{j,k}) + \Theta_{j,k}(-\xi, \nu_{j,k})\right) \Big\}$$

$$= \sum\sum 0 = 0 \quad \text{by } (24.34)^*.$$

We omit (24.24). (24.25) follows from the facts that

$$\Theta_{j,k}\left(\xi, (\mu_1, \mu_2, \mu_3, \mu_4)\right) \in C^{\infty}\left(S^{n-1} \times (\mathbf{R}^4 \backslash \{(0,0,0,0)\})\right),$$
$$\Theta_{j,k}\left(\xi, (0,0,0,0)\right) \in C^{\infty}\left(S^{n-1}\right)$$

and that

$$\sup\left\{ \left|\nabla_\xi^u \left\{\Theta_{j,k}\left(\xi/|\xi|, (\mu_1, \mu_2, \mu_3, \mu_4)\right)\right\}\right| : \xi \in S^{n-1}, \ (\mu_1, \mu_2, \mu_3, \mu_4) \in \mathbf{R}^4 \right\}$$
$$= \sup\left\{ \left|\nabla_\xi^u \left\{\Theta_{j,k}\left(\xi/|\xi|, (\mu_1, \mu_2, \mu_3, \mu_4)\right)\right\}\right| : \xi \in S^{n-1}, \right.$$

$$\left. (\mu_1, \mu_2, \mu_3, \mu_4) \in S^3 \cup \{(0,0,0,0)\} \right\}$$

$$< \infty \qquad \text{by compactness.}$$

$\square$

Proof of Lemma 24.3′. (24.29) follows from

$$\sum_{j=1}^{m} m_{\Theta_j} \left\{ (\nu_{2j-1} + i\nu_{2j})b - m_{\Theta_{j,\bar{\nu}}}(\operatorname{Re} b') - i m_{\Theta_{j,\bar{\nu}}}(\operatorname{Im} b') \right\}$$

$$= b' - \operatorname{Re} b' - i \operatorname{Im} b' \quad \text{by } (24.22)$$

$$= 0.$$

(24.30) follows from the fact that if $h \in L^2(\mathbf{R}^n, \mathbf{R})$ then

$$\vec{\nu} \cdot (\mathrm{m}_{\Theta_{1,\vec{\nu}}} h, \cdots, \mathrm{m}_{\Theta_{m,\vec{\nu}}} h)$$

$$= \frac{1}{2} \mathcal{F}^{-1} \left[\sum_{j=1}^{m} \left\{ \nu_{2j-1} \left(\operatorname{Re} \left(\Theta_{j,\vec{\nu}}(\xi) + \Theta_{j,\vec{\nu}}(-\xi) \right) + i \operatorname{Im} \left(\Theta_{j,\vec{\nu}}(\xi) - \Theta_{j,\vec{\nu}}(-\xi) \right) \right) \right. \right.$$

$$\left. \left. + \nu_{2j} \left(\operatorname{Im} \left(\Theta_{j,\vec{\nu}}(\xi) + \Theta_{j,\vec{\nu}}(-\xi) \right) - i \operatorname{Re} \left(\Theta_{j,\vec{\nu}}(\xi) - \Theta_{j,\vec{\nu}}(-\xi) \right) \right) \right\} \mathcal{F} h(\xi) \right]$$

$$= 0 \qquad \text{by (24.23)--(24.24)},$$

$$\vec{\nu} \cdot \left(i \mathrm{m}_{\Theta_{1,\vec{\nu}'}} h, \cdots, i \mathrm{m}_{\Theta_{m,\vec{\nu}'}} h \right) = -\vec{\nu}' \cdot \left(\mathrm{m}_{\Theta_{1,\vec{\nu}'}} h, \cdots, \mathrm{m}_{\Theta_{m,\vec{\nu}'}} h \right)$$

$$= 0 \text{ by the same reason as above.}$$

$\square$

XXV. Characterization of H^1 in terms of Fourier multipliers

The purpose of this section is to show the following.

Theorem 25.1. Let $\{\theta_1, \cdots, \theta_m\} \subset C^\infty(S^{n-1}, \mathbf{C})$ satisfy (24.1). Then the following three conditions are equivalent.

$$\sup \left\{ \frac{\|f\|_{H^1}}{\displaystyle\sum_{j=1}^m \|m_{\theta_j} f\|_{L^1}} : f \in H^1(\mathbf{R}^n, \mathbf{R}) \backslash \{0\} \right\} < \infty, \qquad (25.1)$$

$$\mathrm{BMO}(\mathbf{R}^n, \mathbf{R})/\mathbf{R} = \sum_{j=1}^m \tilde{m}_{\theta_j} L^\infty(\mathbf{R}^n, \mathbf{R}), \qquad (25.2)$$

$$\mathrm{rank} \begin{bmatrix} \mathrm{Re}\,\theta_1(\xi) & \cdots & \mathrm{Re}\,\theta_m(\xi) \\ \mathrm{Im}\,\theta_1(\xi) & \cdots & \mathrm{Im}\,\theta_m(\xi) \end{bmatrix} \equiv 2. \qquad (25.3)$$

Corollary 25.1. Let $\{\theta_1, \cdots, \theta_m\} \subset C^\infty(S^{n-1}, \mathbf{C})$ Then the fllowing three conditions are equivalent.

$$\sup \left\{ \frac{\|f\|_{H^1}}{\displaystyle\sum_{j=1}^m \|m_{\theta_j} f\|_{L^1}} : f \in H^1(\mathbf{R}^n, \mathbf{R}) \backslash \{0\} \right\} < \infty, \qquad (25.1)^*$$

$$\mathrm{BMO}(\mathbf{R}^n, \mathbf{R})/\mathbf{R} = \mathrm{Re} \sum_{j=1}^m \tilde{m}_{\theta_j} L^\infty(\mathbf{R}^n, \mathbf{C}), \qquad (25.2)^*$$

$$\mathrm{rank} \begin{bmatrix} \mathrm{Re}\,(\theta_1(\xi) + \theta_1(-\xi)) & \mathrm{Im}\,(\theta_1(\xi) + \theta_1(-\xi)) & \cdots & \cdots \\ \mathrm{Im}\,(\theta_1(\xi) - \theta_1(-\xi)) & -\mathrm{Re}\,(\theta_1(\xi) - \theta_1(-\xi)) & \cdots & \cdots \end{bmatrix}$$

$$\begin{bmatrix} \cdots & \cdots & \mathrm{Re}\,(\theta_m(\xi) + \theta_m(-\xi)) & \mathrm{Im}\,(\theta_m(\xi) + \theta_m(-\xi)) \\ \cdots & \cdots & \mathrm{Im}\,(\theta_m(\xi) - \theta_m(-\xi)) & -\mathrm{Re}\,(\theta_m(\xi) - \theta_m(-\xi)) \end{bmatrix} \equiv 2.$$
$$(25.3)^*$$

What we have to show is the implication "(25.1)→(25.3)". The following lemma is very easy. We omit its proof.

Lemma 25.1. *Let $\varepsilon \in (0, 1/2)$. Let*

$$f_\copyright = \varepsilon^{-1}\chi_{[0,\varepsilon]} - \chi_{[1,2]} \in L^1(\mathbf{R}^1). \tag{25.4}$$

Then

$$\|f_\copyright\|_{H^1(\mathbf{R}^1)} \approx \log(1/\varepsilon).$$

Lemma 25.2. *Let $a \in \mathbf{C}$. Let*

$$\theta^*(t) = \begin{cases} a & \text{if } t = 1 \\ \bar{a} & \text{if } t = -1. \end{cases}$$

Then

$$\sup\left\{ \frac{\|f\|_{H^1(\mathbf{R}^1)}}{\|m_{\theta^*} f\|_{L^1(\mathbf{R}^1)}} : f \in H^1(\mathbf{R}^1, \mathbf{R}) \backslash \{0\} \right\} = \infty.$$

Proof. We may assume $a \neq 0$. Let ε and $f_\copyright$ be as in Lemma 25.1. Let

$$f = m_{1/\theta^*} f_\copyright.$$

Then $f \in H^1(\mathbf{R}^1, \mathbf{R})$ and

$$\frac{\|f\|_{H^1(\mathbf{R}^1)}}{\|m_{\theta^*} f\|_{L^1(\mathbf{R}^1)}} = \frac{\|m_{1/\theta^*} f_\copyright\|_{H^1(\mathbf{R}^1)}}{\|f_\copyright\|_{L^1(\mathbf{R}^1)}} \geq \frac{C|a|^{-1}\|f_\copyright\|_{H^1(\mathbf{R}^1)}}{\|f_\copyright\|_{L^1(\mathbf{R}^1)}} \geq C|a|^{-1}\log(1/\varepsilon).$$

$\square$

As a result of Lemma 25.2 we have the following.

Lemma 25.3. *Let $\{a_1, \cdots, a_m\} \subset \mathbf{C}$. Let*

$$\theta_j^*(t) = \begin{cases} a_j & \text{if } t = 1 \\ \bar{a}_j & \text{if } t = -1. \end{cases}$$

Let

$$\sup\left\{ \frac{\|f\|_{H^1(\mathbf{R}^1)}}{\sum_{j=1}^m \|m_{\theta_j^*} f\|_{L^1(\mathbf{R}^1)}} : f \in H^1(\mathbf{R}^1, \mathbf{R}) \backslash \{0\} \right\} < \infty.$$

Then

$$\operatorname{rank} \begin{bmatrix} \operatorname{Re} a_1 & \cdots & \operatorname{Re} a_m \\ \operatorname{Im} a_1 & \cdots & \operatorname{Im} a_m \end{bmatrix} = 2.$$

Lemma 25.4. *Let* $\theta \in C^\infty(S^{n-1}, \mathbf{C})$. *Let*

$$\theta^*(t) = \begin{cases} \theta(1, 0, \cdots, 0) & \text{if } t = 1 \\ \theta(-1, 0, \cdots, 0) & \text{if } t = -1. \end{cases}$$

Let $h \in \mathcal{S}_0(\mathbf{R}^1, \mathbf{C})$, $\phi \in \mathcal{S}(\mathbf{R}^{n-1}, \mathbf{R})$,

$$\mathcal{F}\phi(\xi_1, \cdots, \xi_{n-1}) = 0 \ \text{if} \ \xi_1^2 + \cdots + \xi_{n-1}^2 \geq 1. \tag{25.5}$$

Let $t > 0$ *and*

$$f_{\boxed{t}}(x_1, x_2, \cdots, x_n) = h(x_1) t^{n-1} \phi(tx_2, \cdots, tx_n). \tag{25.6}$$

Then

$$\|m_\theta f_{\boxed{t}}\|_{L^1(\mathbf{R}^n)} \to \|m_{\theta^*} h\|_{L^1(\mathbf{R}^1)} \|\phi\|_{L^1(\mathbf{R}^{n-1})} \ (t \to +0). \tag{25.7}$$

Proof. Take $\varepsilon > 0$ so that

$$\operatorname{supp} \mathcal{F}h \subset [-1/\varepsilon, \ -\varepsilon] \cup [\varepsilon, \ 1/\varepsilon]. \tag{25.8}$$

Take $\psi \in \mathcal{S}_0(\mathbf{R}^n)$ so that

$$\mathcal{F}\psi(\xi) = \begin{cases} 1 & \text{if } \varepsilon \leq |\xi_1| \leq 1/\varepsilon \text{ and if } \xi_2^2 + \cdots + \xi_n^2 \leq 1 \\ 0 & \text{if } |\xi_1| \leq \varepsilon/2. \end{cases} \tag{25.9}$$

Let

$$\eta = \mathcal{F}^{-1}\{\mathcal{F}\psi(\xi)\, (\theta(\xi/|\xi|) - \theta(\operatorname{sign}\xi_1, 0, \cdots, 0))\} \in \mathcal{S}_0(\mathbf{R}^n).$$

Then since $\mathcal{F}\eta(\xi_1, 0, \cdots, 0) \equiv 0$,

$$\int_{\mathbf{R}^{n-1}} \eta(x_1, x_2, \cdots, x_n) dx_2 \cdots dx_n \equiv 0. \tag{25.10}$$

So

$$|\{\text{the left-hand side of } (25.7)\} - \{\text{the right-hand side of } (25.7)\}|$$
$$= \left| \|m_\theta f_{\boxed{t}}\|_{L^1(\mathbf{R}^n)} - \|m_{\theta^*} h\|_{L^1(\mathbf{R}^1)} \left\| t^{n-1} \phi(tx_2, \cdots, tx_n) \right\|_{L^1(\mathbf{R}^{n-1})} \right|$$
$$\leq \left\| \mathcal{F}^{-1}\{(\theta(\xi/|\xi|) - \theta(\operatorname{sign}\xi_1, 0, \cdots, 0)) \mathcal{F}f_{\boxed{t}}(\xi)\} \right\|_{L^1}$$
$$= \|\eta * f_{\boxed{t}}\|_{L^1} \text{ if } t \in (0, 1) \text{ by } (25.5), (25.8) \text{ and } (25.9)$$
$$\to 0 \ (t \to +0) \text{ by } (25.10).$$

$\square$

Lemma 25.5. *Let* $\{\theta_1, \cdots, \theta_m\} \subset C^\infty(S^{n-1}, \mathbf{C})$. *Let*

$$\theta_j^*(t) = \left\{ \begin{array}{ll} \theta_j(1, 0, \cdots, 0) & \text{if } t = 1 \\ \theta_j(-1, 0, \cdots, 0) & \text{if } t = -1. \end{array} \right.$$

Then

$$C(n) \sup \left\{ \frac{\|f\|_{H^1(\mathbf{R}^n)}}{\displaystyle\sum_{j=1}^m \|m_{\theta_j} f\|_{L^1(\mathbf{R}^n)}} : f \in H^1(\mathbf{R}^n, \mathbf{R}) \backslash \{0\} \right\}$$

$$\geq \sup \left\{ \frac{\|h\|_{H^1(\mathbf{R}^1)}}{\displaystyle\sum_{j=1}^m \|m_{\theta_j^*} h\|_{L^1(\mathbf{R}^1)}} : h \in H^1(\mathbf{R}^1, \mathbf{R}) \backslash \{0\} \right\}. \qquad (25.11)$$

Proof. Let $h \in \mathcal{S}_0(\mathbf{R}^1, \mathbf{R})$. Let ϕ and $f_{\boxdot}$ be as in Lemma 25.4. Let $\phi \not\equiv 0$. Let $\varepsilon > 0$ be as in the proof of Lemma 25.4. Then if $t \in (0, \varepsilon)$, then

$$\operatorname{supp} \mathcal{F} f_{\boxdot} \subset \left\{ (\xi_1, \xi_2, \cdots, \xi_n) \in \mathbf{R}^n : (\xi_2^2 + \cdots + \xi_n^2)^{1/2} \leq |\xi_1| \right\}.$$

So, Theorem 18.1 implies

$$\|f_{\boxdot}\|_{H^1(\mathbf{R}^n)} \approx \|h\|_{H^1(\mathbf{R}^1)} \|\phi\|_{L^1(\mathbf{R}^{n-1})}. \qquad (25.12)$$

Thus (25.7) and (25.12) imply

$$C(n) \liminf_{t \to +0} \frac{\|f_{\boxdot}\|_{H^1(\mathbf{R}^n)}}{\displaystyle\sum_{j=1}^m \|m_{\theta_j} f_{\boxdot}\|_{L^1(\mathbf{R}^n)}} \geq \frac{\|h\|_{H^1(\mathbf{R}^1)}}{\displaystyle\sum_{j=1}^m \|m_{\theta_j^*} h\|_{L^1(\mathbf{R}^1)}}.$$

Since $\mathcal{S}_0(\mathbf{R}^1, \mathbf{R})$ is dense in $H^1(\mathbf{R}^1, \mathbf{R})$ we have obtained (25.11). $\square$

Proof of Theorem 25.1. The implication "(25.1)→ (25.3) with $\xi = (1, 0, \cdots, 0)$" follows from Lemma 25.5 and 25.3. Then the impication "(25.1)→(25.3)" follows from rotation. The implication "(25.1)↔(25.2)" follows from Corollary 21.1 and Remark 21.5. The implication "(25.3)→(25.2), (25.1)" follows from Theorems 24.1, 21.3, "(25.3)→(21.13)" and from the latter half of Corrolary 21.6. $\square$

If we replace $H^1(\mathbf{R}^n, \mathbf{R})$ and $\text{BMO}(\mathbf{R}^n, \mathbf{R})$ by $H^1(\mathbf{R}^n, \mathbf{C})$ and $\text{BMO}(\mathbf{R}^n, \mathbf{C})$ respectively, then we have the following.

Theorem 25.2. *Let* $\{\theta_1, \cdots, \theta_m\} \subset C^\infty(S^{n-1}, \mathbf{C})$. *Then the following three conditions are equivalent :*

$$\sup\left\{\frac{\|f\|_{H^1}}{\displaystyle\sum_{j=1}^{m}\|m_{\theta_j}f\|_{L^1}} : f \in H^1(\mathbf{R}^n, \mathbf{C})\backslash\{0\}\right\} < \infty, \qquad (25.13)$$

$$\mathrm{BMO}(\mathbf{R}^n, \mathbf{C})/\mathbf{C} = \sum_{j=1}^{m}\tilde{m}_{\theta_j}L^\infty(\mathbf{R}^n, \mathbf{C}), \qquad (25.14)$$

$$\mathrm{rank}\left[\begin{array}{ccc}\theta_1(\xi) & \cdots & \theta_m(\xi)\\ \theta_1(-\xi) & \cdots & \theta_m(-\xi)\end{array}\right] \equiv 2. \qquad (25.15)$$

Remark 25.1. If (24.1) holds, then the conditions (25.3) and (25.15) are the same.

Lemma 25.2′. Let $a, b \in \mathbf{C}$. Let

$$\theta^*(t) = \begin{cases} a & if\ t = 1\\ b & if\ t = -1. \end{cases}$$

Then

$$\sup\left\{\frac{\|f\|_{H^1(\mathbf{R}^1)}}{\displaystyle\sum_{j=1}^{m}\|m_{\theta^*}f\|_{L^1(\mathbf{R}^1)}} : f \in H^1(\mathbf{R}^1, \mathbf{C})\backslash\{0\}\right\} = \infty.$$

Proof. If $ab = 0$, then this is clear. So, we assume $ab \neq 0$. Let ε and $f_\copyright$ be as in Lemma 25.1. Let

$$f = m_{1/\theta^*}f_\copyright.$$

Then

$$\frac{\|f\|_{H^1(\mathbf{R}^1)}}{\|m_{\theta^*}f\|_{L^1(\mathbf{R}^1)}} = \frac{\|m_{1/\theta^*}f_\copyright\|_{L^1(\mathbf{R}^1)}}{\|f_\copyright\|_{H^1(\mathbf{R}^1)}} \geq \frac{C\min\{1/|a|,\ 1/|b|\}\|f_\copyright\|_{H^1(\mathbf{R}^1)}}{\|f_\copyright\|_{L^1(\mathbf{R}^1)}}$$
$$\geq C\min\{1/|a|,\ 1/|b|\}\log(1/\varepsilon)\ \text{by Lemma 25.1.}$$

$\square$

Lemma 25.3′. Let $\{a_1, b_1, \cdots, a_m, b_m\} \subset \mathbf{C}$. Let

$$\theta_j^*(t) = \begin{cases} a_j & if\ t = 1\\ b_j & if\ t = -1. \end{cases}$$

Let

$$\sup \left\{ \frac{\|f\|_{H^1(\mathbf{R}^1)}}{\displaystyle\sum_{j=1}^{m} \|\mathsf{m}_{\theta_j^*} f\|_{L^1(\mathbf{R}^1)}} : f \in H^1(\mathbf{R}^1, \mathbf{C}) \right\} < \infty.$$

Then

$$\operatorname{rank} \begin{bmatrix} a_1 & \cdots & a_m \\ b_1 & \cdots & b_m \end{bmatrix} = 2.$$

This follows from Lemma 25.2'.

Lemma 25.5' . Let $\{\theta_1, \cdots, \theta_m\}$ and $\{\theta_1^*, \cdots, \theta_m^*\}$ be as in Lemma 25.5. Then

$$C(n) \sup \left\{ \frac{\|f\|_{H^1(\mathbf{R}^n)}}{\displaystyle\sum_{j=1}^{m} \|\mathsf{m}_{\theta_j} f\|_{L^1(\mathbf{R}^n)}} : f \in H^1(\mathbf{R}^n, \mathbf{C}) \backslash \{0\} \right\}$$

$$\geq \sup \left\{ \frac{\|h\|_{H^1(\mathbf{R}^1)}}{\displaystyle\sum_{j=1}^{m} \|\mathsf{m}_{\theta_j^*} h\|_{L^1(\mathbf{R}^1)}} : h \in H^1(\mathbf{R}^1, \mathbf{C}) \backslash \{0\} \right\}.$$

This can be proved by the same reason as Lemma 25.5.

Proof of Theorem 25.2. The implication "(25.13)→(25.15)" follows from Lemmas 25.5' and 25.3'. The implication "(25.13)↔(25.14)" follows from Theorem 21.1 and Remark 21.5. The implication "(25.15)→(25.14), (25.13)" follows from Theorems 24.2 , 21.3, "(25.15)→(25.13)" and from the latter half of Corollary 21.5.   □

Definition 25.1. Let $S(\mathbf{R}^n, \mathbf{C})'_c$ be the set of all continuous complex-valued linear functionals on $S(\mathbf{R}^n, \mathbf{C})$, where we regard $S(\mathbf{R}^n, \mathbf{C})_c$ as a topological linear space over the complex-number field. Let $S_0(\mathbf{R}^n, \mathbf{C})'_c$ be the set of all continuous complex-valued linear functionals on $S_0(\mathbf{R}^n, \mathbf{C})$, where we regard $S_0(\mathbf{R}^n, \mathbf{C})$ as a subspace of a topological linear space $S(\mathbf{R}^n, \mathbf{C})$. Then

$$S_0(\mathbf{R}^n, \mathbf{C})'_c = S(\mathbf{R}^n, \mathbf{C})'_c / \{\text{polynomials}\}.$$

Let

$$\mathcal{J} : f \in S(\mathbf{R}^n, \mathbf{C})'_c \to f + \{\text{polynomials}\} \in S_0(\mathbf{R}^n, \mathbf{C})'_c.$$

Definition 25.2. For

$$f \in S(\mathbf{R}^n, \mathbf{C})'_c \text{ and } \theta \in C^\infty(S^{n-1}, \mathbf{C})$$

let $\tilde{\mathsf{m}}_\theta f$ be the complex-valued linear functional on $S_0(\mathbf{R}^n, \mathbf{C})$ such that

$$\langle \phi, \tilde{m}_\theta f \rangle_{S_0(\mathbf{R}^n, \mathbf{C})} = \langle m_{\check{\theta}}\phi, f \rangle_{S(\mathbf{R}^n, \mathbf{C})} \text{ for any } \phi \in S_0(\mathbf{R}^n, \mathbf{C}),$$

where $\check{\theta}(\xi) = \theta(-\xi)$. Since $\phi \in S_0(\mathbf{R}^n, \mathbf{C}) \to m_{\check{\theta}}\phi \in S_0(\mathbf{R}^n, \mathbf{C})$ is a bounded operator from $S_0(\mathbf{R}^n, \mathbf{C})$ into itself, it follows that

$$\tilde{m}_\theta f \in S_0(\mathbf{R}^n, \mathbf{C})'_c.$$

Definition 25.3. Let $M(\mathbf{R}^n, \mathbf{C})$ be the set of all finite complex measures on $\mathbf{R}^n$ and let $\| \cdot \|_M$ be the total variation. Let

$$\mathcal{I} = \mathcal{J}|_{M(\mathbf{R}^n,\mathbf{C})}(= \text{the restriction of } \mathcal{J} \text{ to } M(\mathbf{R}^n,\mathbf{C})). \tag{25.16}$$

Then, $\mathcal{I}$ is injective. For $f \in S_0(\mathbf{R}^n, \mathbf{C})'_c$ let

$$\|f\|_M = \begin{cases} \|\mathcal{I}^{-1}f\|_M & \text{if } f \in \mathcal{I}M(\mathbf{R}^n, \mathbf{C}) \\ +\infty & \text{otherwise,} \end{cases}$$

$$\|f\|_{H^1} = \begin{cases} \|\mathcal{I}^{-1}f\|_{H^1} & \text{if } f \in \mathcal{I}H^1(\mathbf{R}^n, \mathbf{C}) \\ +\infty & \text{otherwise.} \end{cases}$$

Theorem 25.3. *Let $\{\theta_1, \cdots, \theta_m\} \subset C^\infty(S^{n-1}, \mathbf{C})$ satisfy (25.15). Let $f \in S(\mathbf{R}^n, \mathbf{C})'_c$. Then*

$$\|\mathcal{J}f\|_{H^1} \le C_{25.1}(\{\theta_1, \cdots, \theta_m\}) \sum_{j=1}^m \|\tilde{m}_{\theta_j}f\|_M, \tag{25.17}$$

namely, there exists a polynomial $P(x)$ such that

$$\|f - P\|_{H^1} \le C_{25.1} \sum_{j=1}^m \|\tilde{m}_{\theta_j}f\|_M.$$

Lemma 25.6. *Let $\{\theta_1, \cdots, \theta_m\} \subset C^\infty(S^{n-1}, \mathbf{C})$ and*

$$\inf_{\xi \in S^{n-1}} \sum_{j=1}^m |\theta_j(\xi)| > 0. \tag{25.18}$$

Let $f \in S(\mathbf{R}^n, \mathbf{C})'_c$ and

$$\sum_{j=1}^m \|\tilde{m}_{\theta_j}f\|_M < \infty. \tag{25.19}$$

Then

$$\{\mathcal{F}f\}|_{\mathbf{R}^n\backslash\{0\}} \in C(\mathbf{R}^n\backslash\{0\}), \tag{25.20}$$

$$\|\{\mathcal{F}F\}|_{\mathbf{R}^n\backslash\{0\}}\|_{L^\infty(\mathbf{R}^n\backslash\{0\})} \le C(\{\theta_1, \cdots, \theta_m\}) \sum_{j=1}^m \|\tilde{m}_{\theta_j}f\|_M. \tag{25.21}$$

and there exists a sequence $\{f_k\}_{k \in \mathbf{N}} \subset \mathcal{S}_0(\mathbf{R}^n, \mathbf{C})$ such that

$$\|\mathcal{F}f_k\|_{L^\infty} \le C(n) \left\|\{\mathcal{F}f\}|_{\mathbf{R}^n\backslash\{0\}}\right\|_{L^\infty(\mathbf{R}^n\backslash\{0\})}, \qquad (25.22)$$

$$\|m_{\theta_j}f_k\|_{L^1} \le C(n)\|\tilde{m}_{\theta_j}f\|_M \quad (j = 1, \cdots, m), \qquad (25.23)$$

$$\mathcal{F}f_k(\xi) \to \mathcal{F}f(\xi) \ (k \to \infty) \ \text{for any } \xi \in \mathbf{R}^n\backslash\{0\}. \qquad (25.24)$$

Proof. Note that (25.19) implies

$$\theta_j\left(\xi/|\xi|\right)\{\mathcal{F}f\}|_{\mathbf{R}^n\backslash\{0\}} \in \{\mathcal{F}M(\mathbf{R}^n, \mathbf{C})\}|_{\mathbf{R}^n\backslash\{0\}} \subset C(\mathbf{R}^n, \mathbf{C})|_{\mathbf{R}^n\backslash\{0\}}.$$

So, (25.18) implies (25.20) and

$$\sup_{\xi \in \mathbf{R}^n\backslash\{0\}} |\mathcal{F}f(\xi)| \le \sup_{\xi \in \mathbf{R}^n\backslash\{0\}} \min_{1 \le j \le m} \left\{|\theta_j\left(\xi/|\xi|\right)|^{-1} \|\tilde{m}_{\theta_j}f\|_M\right\}$$

$$\le \left\|\max\{|\theta_1|, \cdots, |\theta_m|\}^{-1}\right\|_{L^\infty(S^{n-1})} \max_{1 \le j \le m} \|\tilde{m}_{\theta_j}f\|_M,$$

which means (25.21).

Take $\phi, \psi \in \mathcal{S}(\mathbf{R}^n, \mathbf{R})$, depending only on n, so that

$$\mathcal{F}\phi(\xi) = 1 \ \text{on } B(0, 1.1),$$
$$\mathcal{F}\phi(\xi) = 0 \ \text{on } B(0, 1.9)^c,$$
$$\psi(0) = 1 \ \text{and supp} \, \mathcal{F}\psi \subset B(0, 1).$$

For $k \in \mathbf{N}$ let

$$g_k = \mathcal{F}^{-1}\left\{\left(\mathcal{F}\phi(2^{-k+1}\xi) - \mathcal{F}\phi(2^k\xi)\right)\mathcal{F}f(\xi)\right\} \in L^2(\mathbf{R}^n, \mathbf{C}).$$

Then

$$\text{supp}\,\mathcal{F}g_k \subset B(0, 2^k)\backslash B(0, 2^{-k}), \qquad (25.25)$$

$$\|m_{\theta_j}g_k\|_{L^1} = \left\|\left((\phi)_{2^{-k+1}} - (\phi)_{2^k}\right) * \mathcal{I}^{-1}(\tilde{m}_{\theta_j}f)\right\|_{L^1} \le 2\|\phi\|_{L^1}\|\tilde{m}_{\theta_j}f\|_M,$$

$$\qquad (25.26)$$

$$g_k = \sum_{j=1}^m \mathcal{F}^{-1}\left\{\left(\mathcal{F}\phi(2^{-k+1}\xi) - \mathcal{F}\phi(2^k\xi)\right)\bar{\theta}_j(\xi/|\xi|)\left(\sum_{i=1}^m |\theta_i(\xi/|\xi|)|^2\right)^{-1}\right\}$$

$$* \mathcal{I}^{-1}(\tilde{m}_{\theta_j}f) \in \mathcal{S}_0 * M \subset L^1. \qquad (25.27)$$

Then, if $h \le -k - 3$ and if

$$\xi \in \text{supp}\,\mathcal{F}g_k \cup \text{supp}\left\{(\mathcal{F}g_k) * (\mathcal{F}\psi)_{2^h}\right\} = \text{supp}\,\mathcal{F}g_k \cup \text{supp}\,\mathcal{F}(\psi(2^h\cdot)g_k(\cdot)),$$

then

$$\mathcal{F}\phi(2^{-k-1}\xi) - \mathcal{F}\phi(2^{k+2}\xi) = 1.$$

So,

$$\left\| m_{\theta_j} \left(g_k(\cdot) \right) - m_{\theta_j} \left(\psi(2^h \cdot) g_k(\cdot) \right) \right\|_{L^1}$$

$$= \left\| \mathcal{F}^{-1} \{ (\mathcal{F}\phi(2^{-k-1}\xi) - \mathcal{F}\phi(2^{k+2}\xi))\theta_j(\xi/|\xi|) \} * (g_k(\cdot) - \psi(2^h \cdot)g_k(\cdot)) \right\|_{L^1}$$

$$\leq \left\| \mathcal{F}^{-1} \{ \cdots \} \right\|_{L^1} \left\| g_k(\cdot) - \psi(2^h \cdot) g_k(\cdot) \right\|_{L^1} \quad \text{(recall (25.27))}$$

$$\to 0 \ (h \to -\infty) \ \text{by} \ \psi(0) = 1. \tag{25.28}$$

Thus, (25.26) and (25.28) imply that

$$\left\| m_{\theta_j} \left(\psi(2^h \cdot) g_k(\cdot) \right) \right\|_{L^1} \leq 3 \|\phi\|_{L^1} \|\tilde{m}_{\theta_j} f\|_M \quad (j = 1, \cdots, m)$$

$$\text{if } h \leq \mathring{k}(k, \{\theta_1, \cdots, \theta_m\}, f) \ (\leq -k - 3). \tag{25.29}$$

Let

$$f_k(x) = \psi(2^{\mathring{k}} x) g_k(x).$$

Then (25.29) implies (25.23). The formula

$$\mathcal{F} f_k(\xi) = (\mathcal{F}\psi)_{2^{\mathring{k}}} * \mathcal{F} g_k(\xi)$$

$$(= (\mathcal{F}\psi)_{2^{\mathring{k}}} * \{ \left(\mathcal{F}\phi(2^{-k+1}\cdot) - \mathcal{F}\phi(2^k \cdot) \right) \mathcal{F} f(\cdot) \} (\xi)), \tag{25.30}$$

(25.25), supp $\mathcal{F}\psi \subset B(0,1)$ and $\mathring{k} \leq -k - 3$ imply

$$f_k \in \mathcal{S}_0(\mathbf{R}^n, \mathbf{C}).$$

(25.22) is easy from (25.30). (25.24) follows from (25.30), (25.20), $\int \mathcal{F}\psi d\xi = 1$ and from $\mathring{k} \to -\infty \ (k \to \infty)$. □

Lemma 25.7. *Let* $\{f_k\} \subset H^1(\mathbf{R}^n, \mathbf{C})$, $\liminf\limits_{k \to \infty} \|f_k\|_{H^1} < \infty$ *and*

$$f_k \to f \in \mathcal{D}(\mathbf{R}^n, \mathbf{C})'_c \ (k \to \infty) \ in \ \mathcal{D}(\mathbf{R}^n, \mathbf{C})'_c.$$

Then $f \in H^1(\mathbf{R}^n, \mathbf{C})$ *and*

$$\|f\|_{H^1} \leq C(n) \liminf\limits_{k \to \infty} \|f_k\|_{H^1}.$$

Proof. Let $a > 0$. Then

$$G_a f(x) = \sup \{ |f * (\phi)_t(x)| : \phi \in \mathcal{D} \cap \mathcal{B}_a, \ t > 0 \}$$

$$= \sup \left\{ \lim\limits_{k \to \infty} |f_k * (\phi)_t(x)| : \phi \in \mathcal{D} \cap \mathcal{B}_a, \ t > 0 \right\}$$

$$\leq \liminf\limits_{k \to \infty} G_a f_k(x).$$

Thus

$$\int G_a f(x) dx \leq \int \liminf\limits_{k \to \infty} G_a f_k(x) dx \leq \liminf\limits_{k \to \infty} \int G_a f_k(x) dx.$$

□

Proof of Theorem 25.3 We may assume (25.19). Take a sequence $\{f_k\} \subset S_0(\mathbf{R}^n, \mathbf{C})$ as in Lemma 25.6. Let $g \in S(\mathbf{R}^n, \mathbf{C})'_c$ be such that

$$\mathcal{F}g \in L^\infty(\mathbf{R}^n, \mathbf{C}),$$
$$\{\mathcal{F}g\}|_{\mathbf{R}^n \setminus \{0\}} = \{\mathcal{F}f\}|_{\mathbf{R}^n \setminus \{0\}}. \tag{25.31}$$

Then,

$$f_k \to g \text{ in } S(\mathbf{R}^n, \mathbf{C})'_c \ (k \to \infty). \tag{25.32}$$

On the other hand, since

$$\|f_k\|_{H^1} \le C(\{\theta_1, \cdots, \theta_m\}) \sum_{j=1}^m \|\mathsf{m}_{\theta_j} f_k\|_{L^1} \text{ by Theorem 25.2}$$

$$\le C(\{\theta_1, \cdots, \theta_m\}) \sum_{j=1}^m \|\tilde{\mathsf{m}}_{\theta_j} f\|_M \text{ by (25.23)},$$

we have

$$\|g\|_{H^1} \le C \liminf \|f_k\|_{H^1} \text{ by Lemma 25.7 and (25.32)}$$

$$\le C \sum_{j=1}^m \|\tilde{\mathsf{m}}_{\theta_j} f\|_M. \tag{25.33}$$

(25.31) and (25.33) imply (25.17).   □

Remark 25.2. If $f \in S(\mathbf{R}^n, \mathbf{C})'_c$ and if $\mathcal{F}f$ is locally integrable near the origin, then $\mathsf{m}_\theta f = \mathcal{F}^{-1}\{\theta(\xi/|\xi|)\mathcal{F}f\}$ can be defined in the sense of distributions. In this case, instead of (25.17) we have

$$\|f\|_{H^1} \le C_{25.1} \sum_{j=1}^m \|\mathsf{m}_{\theta_j} f\|_M,$$

where

$$\|\mathsf{m}_{\theta_j} f\|_M = \infty \quad \text{if } \mathsf{m}_{\theta_j} f \in S(\mathbf{R}^n, \mathbf{C})'_c \setminus M(\mathbf{R}^n, \mathbf{C}).$$

Notes. The implication "(25.13)→(25.15)" (or "(25.1)→(25.3)") is due to S. Janson [77]. He treated the case $\theta_1 \equiv 1$ and showed that the condition

$$\inf_{\xi \in S^{n-1}} \sum_{j=2}^m |\theta_j(\xi) - \theta_j(-\xi)| > 0$$

is necessary in order for (25.13) to hold. See also A. Gandulfo–J. García-Cuerva–M. Taibleson [76]. (See also M. Christ [85].)

The argument of the proof of Lemma 25.6 is due to E. Stein [70] p. 231. The S_0-distribution is in Reimann–Rychener [75] p. 111.

In A. Uchiyama [82] the statement of Corollary 1 is ambiguous. We must add the hypothesis that h in that Corollary is a tempered distribution such that $\mathcal{F}h$ is integrable around the origin. (Or we must interpret the statement there in the sense of $\mathcal{S}_0(\mathbf{R}^n, \mathbf{C})'_c$.) The author apologizes this.

XXVI. Extension of the Fefferman-Stein decomposition of BMO, 2

In this section we extend the argument in Sections 22 and 24 to certain weighted BMO functions.

Definition 26.1. Let $w \in L^1_{\text{loc}}(\mathbf{R}^n, \mathbf{R})$ and let $w(x) > 0$ a.e. x. For a measurable set $E \subset \mathbf{R}^n$, for $\varepsilon > 0$ and for $\vec{g} \in L^1_{\text{loc}}(\mathbf{R}^n, \mathbf{R}^m)$ let

$$w(E) = \operatorname*{ess.\,sup}_{y \in E} w(y),$$

$$\|\vec{g}\|_{\mathrm{BMO}w,\varepsilon} = \sup_{B:\ell(B) \le \varepsilon} \inf_{\vec{c} \in \mathbf{R}^m} \int |\vec{g}(x) - \vec{c}|\, dx / (w(B)|B|),$$

where B is taken over all balls in $\mathbf{R}^n$ with its radius $\le \varepsilon$, and let

$$\|\vec{g}\|_{\mathrm{BMO}w,\varepsilon} = \|\vec{g}\|_{\mathrm{BMO}w,\varepsilon} + \sup_{x \in \mathbf{R}^n} \frac{|\vec{g}| * (\chi)_\varepsilon(x)}{w(B(x,\varepsilon))}, \quad \text{where } \chi = \chi_{B(0,1)}.$$

Theorem 26.1. *Let* $\{\theta_1, \cdots, \theta_m\} \subset C^\infty(S^{n-1}, \mathbf{C})$ *be as in Theorem 24.1. Then there exist* $\varepsilon_{26.1}(\{\theta_1, \cdots, \theta_m\}) > 0$ *and* $C_{26.1}(\{\theta_1, \cdots, \theta_m\}) < \infty$ *such that if*

$$\varepsilon \in (0, \varepsilon_{26.1}], \tag{26.1}$$

$$f \in L^1_{\text{loc}}(\mathbf{R}^n, \mathbf{R}) \backslash \{0\}, \quad Mf(x) \not\equiv +\infty, \tag{26.2}$$

$$w(x) = Mf(x)^{-\varepsilon} (1 + |x|)^{-n-1/10}, \tag{26.3}$$

$$\vec{g} \in L^2(\mathbf{R}^n, \mathbf{R}^m), \quad \|\vec{g}\|_{\mathrm{BMO}w,\varepsilon} < \infty, \tag{26.4}$$

then there exists $\vec{h} \in S_{\{\theta_1, \cdots, \theta_m\}*(r)}$ *such that*

$$\left| \vec{g}(x) - \vec{h}(x) \right| \le C_{26.1} \|\vec{g}\|_{\mathrm{BMO}w,\varepsilon} w(x), \tag{26.5}$$

where Mf *is the Hardy-Littlewood maximal function of* f *and where* $S_{\{\theta_1, \cdots, \theta_m\}*(r)}$ *is as in* (24.5).

We will apply this theorem in the next section. For the proof of this theorem we need a lot of preparation.

Lemma 26.1. *Assume*

$$\varepsilon \in (0, 10^{-100}n^{-1}]. \tag{26.6}$$

Assume (26.2)–(26.4). Then

$$\|\vec{g}\|_{L^1} \leq C(n) \left\{ \sup_{y \in B(0,1)} w(y) \right\} \|\vec{g}\|_{\mathrm{BMO}w,\varepsilon}, \tag{26.7}$$

$$\|\vec{g}\|_{\mathrm{BMO}} \leq C(n) \left\{ \sup_{y \in B(0,1)} w(y) \right\} \|\vec{g}\|_{\mathrm{BMO}w,\varepsilon}. \tag{26.8}$$

Proof. It is easy to see that

$$Mf(x) \geq C \left\{ \inf_{y \in B(0,1)} Mf(y) \right\} (1 + |x|)^{-n}.$$

So,

$$w(x) \leq C \left\{ \inf_{y \in B(0,1)} Mf(y) \right\}^{-\varepsilon} (1 + |x|)^{\varepsilon n - n - 1/10}$$

$$\leq C \left\{ \sup_{y \in B(0,1)} w(y) \right\} (1 + |x|)^{-n-1/11}.$$

So,

$$|\vec{g}| * (\chi)_\varepsilon(x) \leq \|\vec{g}\|_{\mathrm{BMO}w,\varepsilon} w\left(B(x,\varepsilon)\right)$$

$$\leq \|\vec{g}\|_{\mathrm{BMO}w,\varepsilon} C \left\{ \sup_{y \in B(0,1)} w(y) \right\} (1 + |x|)^{-n-1/11}.$$

So, (26.7) is clear. (26.8) follows from

$$\|\vec{g}\|_{\mathrm{BMO}} \leq \sup_{B:\ell(B)\leq\varepsilon} \inf_{\vec{c}\in\mathbf{R}^m} \mathrm{av}(|\vec{g}-\vec{c}|, B) + C \, \|\,|\vec{g}| * (\chi)_\varepsilon\|_{L^\infty}$$

$$\leq \|w\|_{L^\infty} \|\vec{g}\|_{\mathrm{BMO}w,\varepsilon} + C\|w\|_{L^\infty} \left\| \frac{|\vec{g}| * (\chi)_\varepsilon(\,\cdot\,)}{w(B(\,\cdot\,,\varepsilon))} \right\|_{L^\infty}$$

and from the above estimate. □

By the same reasonning as the proof of Theorem 22.1, Theorem 26.1 can be reduced to the following.

Lemma 26.2. *Let $\{\theta_1, \cdots, \theta_m\}$ be as in Theorem 24.1. Then there exists $C_{26.2}(\{\theta_1, \cdots, \theta_m\}) < \infty$ such that if (26.6) and (26.2)–(26.4) hold, then there exist*

$$\vec{h} \in S_{\{\theta_1, \cdots, \theta_m\}*(r)} \quad and \quad \vec{v} \in L^1(\mathbf{R}^n, \mathbf{R}^m)$$

that satisfy

$$\|\vec{h}\|_{\text{BMO}w,\varepsilon} \leq C_{26.2}\left\{\|\vec{g}\|_{\text{BMO}w,\varepsilon} + \varepsilon\right\}, \tag{26.9}$$

$$\|\vec{v}\|_{\text{BMO}w,\varepsilon} \leq C_{26.2}\left\{\|\vec{g}\|^2_{\text{BMO}w,\varepsilon} + \sqrt{\varepsilon}\right\}, \tag{26.10}$$

$$\left|\vec{g}(x) - \vec{h}(x) - \vec{v}(x)\right| \leq w(x). \tag{26.11}$$

Proof of "Lemma 26.2 $\to$ Theorem 26.1." We may assume

$$\|\vec{g}\|_{\text{BMO}w,\varepsilon} = 1/(4C_{26.2}) \ (< 1). \tag{26.12}$$

By taking $\varepsilon_{26.1}$ small enough we may assume

$$\varepsilon \leq 1/(4C_{26.2})^4. \tag{26.13}$$

First, applying Lemma 26.2 to $\vec{g}$ gives

$$\vec{h}_1 \in S_{\{\theta_1,\cdots,\theta_m\}*_{(r)}} \quad \text{and} \quad \vec{v}_1 \in L^1(\mathbf{R}^n, \mathbf{R}^m)$$

such that

$$\|\vec{h}_1\|_{\text{BMO}w,\varepsilon} \leq C_{26.2}\left\{\|\vec{g}\|_{\text{BMO}w,\varepsilon} + \varepsilon\right\} \leq 1/2 \tag{26.9}_1$$

$$\text{by } (26.12)\text{–}(26.13),$$

$$\|\vec{v}_1\|_{\text{BMO}w,\varepsilon} \leq C_{26.2}\left\{\|\vec{g}\|^2_{\text{BMO}w,\varepsilon} + \sqrt{\varepsilon}\right\} \leq 1/(8C_{26.2}) \tag{26.10}_1$$

$$\text{by } (26.12)\text{–}(26.13),$$

$$\left|\vec{g}(x) - \vec{h}_1(x) - \vec{v}_1(x)\right| \leq w(x). \tag{26.11}_1$$

Next, we apply Lemma 26.2 to $(4C_{26.2})^{-1}\|\vec{v}_1\|^{-1}_{\text{BMO}w,\varepsilon}\vec{v}_1$ and multiply $4C_{26.2}\|\vec{v}_1\|_{\text{BMO}w,\varepsilon}$ to the obtained $\vec{h}$ and $\vec{v}$. Then we get

$$\vec{h}_2 \in S_{\{\theta_1,\cdots,\theta_m\}*_{(r)}} \quad \text{and} \quad \vec{v}_2 \in L^1(\mathbf{R}^n, \mathbf{R}^m)$$

such that

$$\|\vec{h}_2\|_{\text{BMO}w,\varepsilon} \leq 4C_{26.2}\|\vec{v}_1\|_{\text{BMO}w,\varepsilon}/2 \leq 1/4, \tag{26.9}_2$$

$$\|\vec{v}_2\|_{\text{BMO}w,\varepsilon} \leq 4C_{26.2}\|\vec{v}_1\|_{\text{BMO}w,\varepsilon}/(8C_{26.2}) \leq 1/(16C_{26.2}), \tag{26.10}_2$$

$$\left|\vec{v}_1(x) - \vec{h}_2(x) - \vec{v}_2(x)\right| \leq 4C_{26.2}\|\vec{v}_1\|_{\text{BMO}w,\varepsilon}w(x) \leq w(x)/2. \tag{26.11}_2$$

Substituting $(26.11)_2$ into $(26.11)_1$ gives

$$\left|\vec{g}(x) - \vec{h}_1(x) - \vec{h}_2(x) - \vec{v}_2(x)\right| \leq w(x)\{1 + 1/2\}. \tag{26.14}_2$$

Repeating this process gives

$$\{\vec{h}_j\} \subset S_{\{\theta_1,\cdots,\theta_m\}*_{(r)}} \quad \text{and} \quad \{\vec{v}_j\} \subset L^1(\mathbf{R}^n, \mathbf{R}^m)$$

such that

$$\|\vec{h}_j\|_{\text{BMO}w,\varepsilon} \leq 1/2^j, \qquad (26.9)_j$$

$$\|\vec{v}_j\|_{\text{BMO}w,\varepsilon} \leq 1/(2^{j+2}C_{26.2}), \qquad (26.10)_j$$

$$\left|\vec{g}(x) - \sum_{k=1}^{j} \vec{h}_k(x) - \vec{v}_j(x)\right| \leq w(x)\{1 + 2^{-1} + \cdots + 2^{-j+1}\}. \qquad (26.14)_j$$

Let

$$\vec{h} = \sum_{k=1}^{\infty} \vec{h}_k \in L^1 \cap \text{BMO}.$$

(By $(26.9)_j$ and (26.7)–(26.8) $\sum \vec{h}_k$ converges in L^1 and in BMO.)
Then $\vec{h} \in S_{\{\theta_1,\cdots,\theta_m\}*(r)}$ by $\vec{h}_j \in S_{\{\theta_1,\cdots,\theta_m\}*(r)}$. Letting $j \to \infty$ in $(26.14)_j$
gives

$$\left|\vec{g}(x) - \vec{h}(x)\right| \leq 2w(x).$$

$\square$

For the proof of Lemma 26.2 we need preparations.

Lemma 26.3. *Assume (26.6) and (26.2)–(26.3). Then*

(i) $\displaystyle\int_B (w(B) - w(x)) \, dx/(w(B)|B|) \leq C(n)\varepsilon$ *if* $\ell(B) \leq \varepsilon$, $\qquad (26.15)$

   so $\|w\|_{\text{BMO}w,\varepsilon} \leq C(n)\varepsilon$,

(ii) *if I is a cube, B is a ball, $I \cap B \neq \emptyset$, $\ell(I)/\ell(B) \in [1/2, 2]$
   and if $\ell(B) \leq 1$, then*
   $$|1 - w(I)/w(B)| \leq C(n), \qquad (26.16)$$

(iii) *if $B(x,t) \subset B(y,s)$ and $s \leq 1$, then*
   $$w(B(y,s)) \leq C(n)(s/t)^{n\varepsilon} w(B(x,t)), \qquad (26.17)$$

(iv) *if $|x - y| \leq 1$ and $t \leq 1$, then*
   $$w(B(y,t)) \leq C(n)(1 + |x - y|/t)^{n\varepsilon} w(B(x,t)), \qquad (26.18)$$

(v) *if $t \leq 1$, then*
   $$\int_{\mathbf{R}^n} w(B(y,t))t^{-n}(1 + |x - y|/t)^{-n-1/9} \, dy \leq C(n)w(B(x,t)). \qquad (26.19)$$

Proof of (26.15). Let $B \subset \mathbf{R}^n$ be any ball. Then it is easy to see that

$$\left|\left\{x \in B : Mf(x) > \lambda \inf_{y \in B} Mf(y)\right\}\right| \leq C(n)|B|/\lambda \qquad (26.20)$$

for any $\lambda > 0$. Let

$$\ddot{w}(x) = Mf(x)^{-\varepsilon}.$$

Then (26.20) implies

$$|\{x \in B : \ddot{w}(x) < \lambda \ddot{w}(B)\}| \le C\lambda^{1/\varepsilon}|B| \tag{26.21}$$

for any $\lambda > 0$. So

$$\int_B (\ddot{w}(B) - \ddot{w}(x))\, dx/(\ddot{w}(B)|B|) \le C\varepsilon. \tag{26.22}$$

Let $B = B(x_B, r_B)$ and $r_B \le \varepsilon$. Since

$$\left|1 - (1 + |x|)^{n+1/10}/(1 + |x_B|)^{n+1/10}\right| \le C\varepsilon, \quad x \in B,$$

we have

$$w(x)/w(B) \ge \ddot{w}(x)/\ddot{w}(B) - C\varepsilon, \quad x \in B. \tag{26.23}$$

Thus

$$\int_B (w(B) - w(x))\, dx/(w(B)|B|)$$
$$\le \int_B (1 - \ddot{w}(x)/\ddot{w}(B) + C\varepsilon)\, dx/|B| \le C\varepsilon \qquad \text{by (26.22).}$$

$\square$

(26.16)–(26.17) are easy from (26.21). (26.18) follows from

$$w(B(y,t)) \le w(B(x, |x - y| + t)) \le C((|x - y| + t)/t)^{n\varepsilon} w(B(x,t))$$
$$\text{by (26.17).}$$

Proof of (26.19). Since

$$w(B(y,t)) \le C\ddot{w}(B(y,t))(1 + |y|)^{-n-1/10} \quad \text{by } t \le 1$$
$$\le C\ddot{w}(B(x, t + |x - y|))(1 + |y|)^{-n-1/10}$$
$$\le C((t + |x - y|)/t)^{n\varepsilon}\ddot{w}(B(x,t))(1 + |y|)^{-n-1/10} \quad \text{by (26.21)}$$
$$\le C(1 + |x - y|/t)^{n\varepsilon} w(B(x,t))(1 + |x|)^{n+1/10}(1 + |y|)^{-n-1/10},$$

{the left-hand side of (26.19)}

$$\le Cw(B(x,t))(1 + |x|)^{n+1/10} \int \frac{(1 + |x - y|/t)^{n\varepsilon - n - 1/9}}{(1 + |y|)^{n+1/10} t^n}\, dy$$
$$\le Cw(B(x,t)).$$

$\square$

Definition 26.2. Let $w \in L^1_{\text{loc}}(\mathbf{R}^n, \mathbf{R})$ and let $w(x) > 0$ a.e. x. For a signed measure μ on $\mathbf{R}^{n+1}_+$ and $\varepsilon > 0$ let

$$\|\mu\|_{Cw,\varepsilon} = \sup \{|\mu|(Q(B))/(w(B)|B|) : B \text{ is taken over all balls in}$$
$$\mathbf{R}^n \text{ with } \ell(B) \le \varepsilon\}.$$

Lemma 26.4. *Assume (26.6) and (26.2)–(26.3). Assume*

$$\log_2 \varepsilon \in \mathbf{Z}. \tag{26.24}$$

Let μ be a signed measure on $\mathbf{R}_+^{n+1}$. Then

$$\|\mu\|_{Cw^j,\varepsilon} \approx \sup \{|\mu|(I \times (0,\ell(I)])/(w(I)^j|I|) : I \text{ is taken over all} $$
$$\text{dyadic cubes in } \mathbf{R}^n \text{ with } \ell(I) \le \varepsilon\} \ (j = 1, 2). \tag{26.25}$$

Proof. For any ball B with $\ell(B) \le \varepsilon$, there exist dyadic cubes $\{I_k\}_{k=1,2,\cdots,3^n}$ such that

$$B \subset \bigcup_k I_k \subset 3\sqrt{n}B, \quad \ell(B) \le \ell(I_k) < 2\ell(B), \quad \ell(I_k) \le \varepsilon.$$

Then, {the left-hand side of (26.25)} $\le C$ {the right-hand side of (26.25)} follows from (26.16). The opposite inequality follows from a similar argument.☐

Lemma 26.5. *Assume (26.6) and (26.2)–(26.3). Let μ be a signed measure on $\mathbf{R}_+^{n+1}$. Then*

$$\|w(B(x,t))^{-1}\mu\|_{Cw,\varepsilon} \le C(n)\|\mu\|_{Cw^2,\varepsilon}.$$

Proof. We show only the case (26.24). The general case can be shown by changing the scale of the "dyadic" mesh slightly. Let I be any dyadic cube in $\mathbf{R}^n$ with $\ell(I) \le \varepsilon$. For $i \in \mathbf{N}_0$ let $\{I_{i,j}\}_{j=1,2,\cdots}$ be the maximal dyadic subcubes of I such that

$$w(I_{i,j}) \le 2^{-i}w(I). \tag{26.26}$$

Then if

$$(x,t) \in (I_{i,j} \times (0,\ell(I_{i,j})]) \setminus \bigcup_k (I_{i+1,k} \times (0,\ell(I_{i+1,k})]),$$

then

$$w(B(x,t)) \ge 2^{-i}w(I)/C \tag{26.27}$$

by the maximality of $\{I_{i+1,k}\}_k$ and by (26.16). Since

$$\sup_j |I_{i,j}| \le (C2^{-i})^{1/\varepsilon}|I| \to 0 \ (i \to \infty)$$

by (26.17), we have

$$I \times (0,\ell(I)] = \bigcup_i \bigcup_j \left\{ (I_{i,j} \times (0,\ell(I_{i,j})]) \setminus \bigcup_k (I_{i+1,k} \times (0,\ell(I_{i+1,k})]) \right\}. \tag{26.28}$$

Thus,

$$\iint_{I\times(0,\ell(I)]} w(B(x,t))^{-1}d|\mu|(x,t)/(w(I)|I|)$$

$$\leq C\sum_i\sum_j \iint_{I_{i,j}\times(0,\ell(I_{i,j})]\setminus\bigcup_k(I_{i+1,k}\times(0,\ell(I_{i+1,k})])} (2^{-i}w(I))^{-1}d|\mu|/(w(I)|I|)$$

$$\text{by (26.27)--(26.28)}$$

$$\leq C\sum_i\sum_j 2^i w(I)^{-2}|I|^{-1}\iint_{I_{i,j}\times(0,\ell(I_{i,j})]}d|\mu|$$

$$\leq C\sum_i\sum_j 2^i w(I)^{-2}|I|^{-1}\|\mu\|_{Cw^2,\varepsilon}w(I_{i,j})^2|I_{i,j}| \quad\text{by (26.25)}$$

$$\leq C\|\mu\|_{Cw^2,\varepsilon}|I|^{-1}\sum\sum 2^{-i}|I_{i,j}| \quad\text{by (26.26)}$$

$$\leq C\|\mu\|_{Cw^2,\varepsilon}. \tag{26.29}$$

So, Lemma 26.5 follows from (26.29) and Lemma 26.4. $\qquad\square$

Lemma 26.6. *Assume (26.6) and (26.2)--(26.3). Let $\{\lambda_I\}_I \subset [0,+\infty)$ and $\{p_I\}_I \subset C^1(\mathbf{R}^n,\mathbf{R})$, where I is taken over all dyadic cubes in $\mathbf{R}^n$. Let*

$$|p_I(x)| + \ell(I)|\nabla p_I(x)| \leq (1 + |x - x_I|/\ell(I))^{-n-1}, \quad \int p(x)dx = 0,$$

$$\lambda_I = 0, \ p_I(x) \equiv 0 \ \text{if} \ \ell(I) > \varepsilon.$$

Assume the right-hand side of the following (26.30) is finite. Then

$$\left\|\sum_I \lambda_I p_I\right\|_{\mathrm{BMO}w,\varepsilon}^2 \leq C(n)\left\|\sum_I \lambda_I^2|I|\delta_{(x_I,\ell(I))}\right\|_{Cw^2,\varepsilon}. \tag{26.30}$$

Proof. We may assume

$$\left\|\sum_I \lambda_I^2|I|\delta_{(x_I,\ell(I))}\right\|_{Cw^2,\varepsilon} = 1. \tag{26.31}$$

Then, $\sum_I \lambda_I^2|I| \ (= \sum_{\ell(I)\leq\varepsilon} \lambda_I^2|I|) < \infty$. So, $\sum \lambda_I p_I$ converges in L^2 by Lemma 22.5. Note that

$$\lambda_I \leq C\left\|\sum_I \lambda_I^2|I|\delta_{(x_I,\ell(I))}\right\|_{Cw^2,\varepsilon}^{1/2} w(I). \tag{26.32}$$

Let $B_0 = B(x_0,r_0)$ and $r_0 \leq \varepsilon$. Let

$$\sum \lambda_I p_I = \sum_{I:x_I\in 2B_0,\ell(I)\leq r_0} \lambda_I p_I + \sum_{I:x_I\notin 2B_0,\ell(I)\leq r_0} + \sum_{I:\ell(I)>r_0}$$

$$= q_1 + q_2 + q_3, \text{ say,}$$

where x_I is the center of a dyadic cube I. Then

$$\|q_2\|_{L^1(B_0)} \leq C \sum_{k=-\infty}^{[\log_2 r_0]} \sum_{I:x_I \notin 2B_0, \ell(I)=2^k} \int_{B_0} w(I)\,|p_I(x)|\,dx \quad \text{by (26.32)}$$

$$\leq C \sum_k |B_0| 2^{-kn} \int_{(2B_0)^c} w(B(y,2^k))(1+|x_0-y|/2^k)^{-n-1}dy$$

$$\leq C|B_0| \sum_k (2^k/r_0)^{1/2} 2^{-kn} \int_{(2B_0)^c} w(B(y,2^k))(1+|x_0-y|/2^k)^{-n-1/2}dy$$

$$\leq C|B_0| \sum_{k=-\infty}^{[\log_2 r_0]} (2^k/r_0)^{1/2} w(B(x_0,2^k)) \quad \text{by (26.19)}$$

$$\leq C|B_0|w(B_0). \tag{26.33}$$

$$\|\nabla q_3\|_{L^\infty(B_0)} \leq C \sum_{k=[\log_2 r_0]+1}^{[\log_2 \varepsilon]} \sum_{I:\ell(I)=2^k} w(I)\|\nabla p_I\|_{L^\infty(B_0)} \quad \text{by (26.32)}$$

$$\leq C \sum \sum w(I) 2^{-k}(1+|x_0-x_I|/2^k)^{-n-1}$$

$$\leq C \sum_k 2^{-k} 2^{-kn} \int_{\mathbf{R}^n} w(B(y,2^k))(1+|x_0-y|/2^k)^{-n-1}dy$$

$$\leq C \sum_k 2^{-k} w(B(x_0,2^k)) \quad \text{by (26.19)}$$

$$\leq C \sum_{k=[\log_2 r_0]+1}^{[\log_2 \varepsilon]} 2^{-k} C(2^k/r_0)^{n\varepsilon} w(B_0) \quad \text{by (26.17)}$$

$$\leq Cw(B_0)/r_0. \tag{26.34}$$

Therefore,

$$\int_{B_0} \left|\sum \lambda_I p_I(x) - q_3(x_0)\right| dx/|B_0|$$

$$\leq \|q_1\|_{L^1(B_0)}/|B_0| + \|q_2\|_{L^1(B_0)}/|B_0| + Cr_0\|\nabla q_3\|_{L^\infty(B_0)}$$

$$\leq \|q_1\|_{L^2(\mathbf{R}^n)}/|B_0|^{1/2} + Cw(B_0) + Cw(B_0) \quad \text{by (26.33)–(26.34)}$$

$$\leq C\left\{\sum_{\substack{I:x_I \in 2B_0, \\ \ell(I) \leq r_0}} \lambda_I^2 |I|\right\}^{1/2}/|B_0|^{1/2} + Cw(B_0) \quad \text{by (22.31)}$$

$$\leq Cw(B_0) \quad \text{by (26.31).} \tag{26.35}$$

If $r_0 = \varepsilon$, then $q_3 \equiv 0$. So,

$$\int_{B(x_0,\varepsilon)} \left|\sum \lambda_I p_I\right| dx/|B(x_0,\varepsilon)| \leq Cw(B(x_0,\varepsilon)). \tag{26.36}$$

Then (26.35)–(26.36) imply (26.30). □

Definition 26.3. Let

$$\phi_0 \in \mathcal{D}(\mathbf{R}^n, \mathbf{R}), \ \operatorname{supp} \phi_0 \subset B(0,1), \ \int \phi_0 dx = 0,$$

$$\int_0^{+\infty} \mathcal{F}\phi_0(t\xi)^2 dt/t = 1 \text{ if } \xi \neq 0.$$

We freeze this ϕ_0. Let

$$\Phi_0 = \mathcal{F}^{-1} \int_{1/2}^{+\infty} \mathcal{F}\phi_0(t\xi)^2 dt/t = \delta - \lim_{\varepsilon \to +0 \text{ in } \mathcal{D}'} \int_\varepsilon^{1/2} (\phi_0)_t * (\phi_0)_t dt/t,$$

where δ is the Dirac measure concentrated at the origin. For $g \in L^1_{\text{loc}}(\mathbf{R}^n)$ and for $j \in \mathbf{Z}$ let

$$g_{j\dagger}(x) = g * (\Phi_0)_{2^j}(x).$$

We omit the proofs of

$$\Phi_0 \in \mathcal{D}(\mathbf{R}^n, \mathbf{R}), \ \operatorname{supp} \Phi_0 \subset B(0,1) \ \text{and} \ \int \Phi_0 dx = 1. \qquad (26.37)$$

Lemma 26.7. *Assume* (26.6) *and* (26.2)–(26.4). *Let* $j \in \mathbf{Z}$ *and* $j \le \log_2 \varepsilon$. *Then*

$$w_{j\dagger}(x) \to w(x) \text{ a.e. } x \ (j \to -\infty), \qquad (26.38)$$
$$|1 - w_{j\dagger}(x)/w(B(x, 2^j))| \le C(n)\varepsilon, \qquad (26.39)$$
$$|\vec{g}_{[\log_2 \varepsilon]\dagger}(x)| \le C(n)\|\vec{g}\|_{\text{BMO}w,\varepsilon} w_{[\log_2 \varepsilon]\dagger}(x), \qquad (26.40)$$
$$2^j |\nabla \vec{g}_{j\dagger}(x)| \le C(n)\|\vec{g}\|_{\text{BMO}w,\varepsilon} w_{j\dagger}(x), \qquad (26.41)$$
$$2^j |\nabla w_{j\dagger}(x)| \le C(n)\varepsilon w_{j\dagger}(x). \qquad (26.42)$$

(26.39) is clear from (26.37) and (26.15). (26.40) is clear from

$$|\vec{g}_{[\log_2 \varepsilon]\dagger}(x)| \le C|\vec{g}| * (\chi)_\varepsilon(x) \le C\|\vec{g}\|_{\text{BMO}w,\varepsilon} w(B(x, \varepsilon))$$

and (26.39). (26.41) follows from

$$2^j \nabla \vec{g}_{j\dagger}(x) = \int \vec{g}(x-y)(\nabla \Phi_0)_{2^j}(y) dy = \int (\vec{g}(x-y) - \vec{c}) (\nabla \Phi_0)_{2^j}(y) dy.$$

(26.42) follows from (26.41) with $\vec{g} = w$ and from (26.15).

Lemma 26.8. *Assume* (26.6) *and* (26.2)–(26.4). *Then there exist* $\{\lambda_{\vec{g},I}\}_I \subset [0, +\infty)$ *and* $\{\vec{b}_I\}_I \subset \mathcal{D}(\mathbf{R}^n, \mathbf{R}^m)$, *where* I *is taken over all dyadic cubes in* $\mathbf{R}^n$, *such that*

$$\lambda_{\vec{g},I} = 0 \ \ and \ \ \vec{b}_I \equiv \vec{0} \ \ if \ \ \ell(I) > \varepsilon/2, \tag{26.43}$$

$$\mathrm{supp}\,\vec{b}_I \subset 3I, \ \ \int \vec{b}_I dx = \vec{0}, \ \ \|\nabla^2 \vec{b}_I\|_{L^\infty} \leq \ell(I)^{-2}, \tag{26.44}$$

$$\left\| \sum_I \lambda_{\vec{g},I}^2 |I| \delta_{(x_I,\ell(I))} \right\|_{Cw^2,\varepsilon} \leq C(n) \|\vec{g}\|_{\mathrm{BMO}w,\varepsilon}^2, \tag{26.45}$$

$$\vec{g} = \vec{g}_{[\log_2 \varepsilon]\dagger} + \sum_I \lambda_{\vec{g},I} \vec{b}_I \ \ in \ \ L^2. \tag{26.46}$$

Proof. We follow the argument of the proof of Lemma 22.6. Recall Definition 22.1. Let ϕ_0 be as in Definition 26.3. For a dyadic cube I in $\mathbf{R}^n$ let

$$\lambda'_{\vec{g},I} = \begin{cases} \left\{ |I|^{-1} \displaystyle\iint_{T(I)} |\vec{g} * (\phi_0)_t|^2 \, dydt/t \right\}^{1/2} & if \ \ \ell(I) \leq \varepsilon/2, \\[2em] 0 & otherwise, \end{cases}$$

$$\vec{b}'_I(x) = \begin{cases} \displaystyle\iint_{T(I)} \vec{g} * (\phi_0)_t(y)(\phi_0)_t(x-y)dydt/t/\lambda'_{\vec{g},I} & if \ \ \lambda'_{\vec{g},I} \neq 0, \\[2em] \vec{0} & otherwise. \end{cases}$$

As in the proof of Lemma 22.6 let

$$\lambda_{\vec{g},I} = C_{22.4}(\phi_0)\lambda'_{\vec{g},I}, \ \ \vec{b}_I = \vec{b}'_I/C_{22.4}(\phi_0).$$

Then (26.43)–(26.44) are easy. Take any ball B with $\ell(B) \leq \varepsilon$. Then

$$\iint_{Q(B)} |\vec{g} * (\phi_0)_t|^2 \, dydt/t$$

$$= \iint_{Q(B)} |(\vec{g} - \mathrm{av}(\vec{g}, B)\chi_{2B}) * (\phi_0)_t(y)|^2 \, dydt/t$$

$$\leq C \|\vec{g} - \mathrm{av}(\vec{g}, B)\|_{L^2(2B)}^2$$

$$\leq C \|\vec{g}\|_{\mathrm{BMO}(4B)}^2 |B| \ \ \text{by Lemma 1.9}$$

$$\leq C \|\vec{g}\|_{\mathrm{BMO}w,\varepsilon}^2 w(4B)^2 |B|$$

$$\leq C \|\vec{g}\|_{\mathrm{BMO}w,\varepsilon}^2 w(B)^2 |B|,$$

from which (26.45) follows. (26.46) is easy from the definitions of ϕ_0 and Φ_0. $\square$

Lemma 26.9. *Assume* (26.6) *and* (26.2)–(26.3). *Then there exist* $\{\lambda_{w,I}\}_I \subset [0, +\infty)$ *and* $\{b_I\}_I \subset \mathcal{D}(\mathbf{R}^n, \mathbf{R})$, *where I is taken over all dyadic cubes in* $\mathbf{R}^n$, *such that*

$$\lambda_{w,I} = 0, \quad b_I \equiv 0 \quad \text{if} \quad \ell(I) > \varepsilon/2, \tag{26.47}$$

$$\operatorname{supp} b_I \subset 3I, \quad \int b_I dx = 0, \quad \|\nabla^2 b_I\|_{L^\infty} \le \ell(I)^{-2}, \tag{26.48}$$

$$\left\| \sum_I \lambda_{w,I}^2 |I| \delta_{(x_I, \ell(I))} \right\|_{Cw^2, \varepsilon} \le C(n)\varepsilon^2, \tag{26.49}$$

$$w_{j\dagger} = w_{[\log_2 \varepsilon]\dagger} + \sum_{I:\ell(I) \ge 2^j} \lambda_{w,I} b_I \quad \text{if} \quad j \le \log_2(\varepsilon/2). \tag{26.50}$$

This follows from Lemma 26.8 with $\vec{g} = w$ and from (26.15).

In the following Definition 26.4 and Lemma 26.10 for the sake of simplicity we write

$$\| \cdots \|_{Cw^2, \varepsilon} \quad \text{for} \quad \left\| \sum_I \lambda_I^2 |I| \delta_{(x_I, \ell(I))} \right\|_{Cw^2, \varepsilon}.$$

Definition 26.4. Assume (26.6) and (26.2)–(26.3). Let $\{\lambda_I\}_I \subset [0, +\infty)$, where I is taken over all dyadic cubes in $\mathbf{R}^n$. Let

$$\lambda_I = 0 \quad \text{if} \quad \ell(I) > \varepsilon \quad \text{and let} \quad \left\| \sum_I \lambda_I^2 |I| \delta_{(x_I, \ell(I))} \right\|_{Cw^2, \varepsilon} \le 1. \tag{26.51}$$

For $j \in \mathbf{Z}$ let $\eta_j^{(1)}(x)$, $\eta_j^{(2)}(x)$ and $\eta_j^{(0)}(x,y)$ be as in Definition 22.2. Let

$$\eta_j^{(3)'}(x) = \left\{ \sum_{I:\ell(I)=2^j, x_I \in B(x,1)} \lambda_I^2 \left(1 + |x - x_I|/2^j\right)^{-n-1/2} \right\}^{1/2},$$

$$\eta_j^{(3)''}(x) = \left\{ \sum_{I:\ell(I)=2^j, x_I \in B(x,2)} \lambda_I^2 \left(1 + |x - x_I|/2^j\right)^{-n-1/2} \right\}^{1/2},$$

$$\eta_j^{(4)'}(x) = \left\{ \sum_{k \ge j} 0.99^{k-j} \eta_k^{(3)'}(x)^2 \right\}^{1/2},$$

$$\eta_j^{(4)''}(x) = \left\{ \sum_{k \ge j} 0.99^{k-j} \eta_k^{(3)''}(x)^2 \right\}^{1/2},$$

$$\eta_j^{(5)'}(x) = \begin{cases} \eta_j^{(4)'}(x) + 0.99^{[\log_2 \varepsilon]-j} \varepsilon^{1/4} \| \cdots \|_{Cw^2, \varepsilon}^{1/4} w(B(x,1)) & \text{if } j \le \log_2 \varepsilon, \\ 0 & \text{if } j > \log_2 \varepsilon, \end{cases}$$

$$\eta_j^{(5)''}(x) = \begin{cases} \eta_j^{(4)''}(x) + 0.99^{[\log_2 \varepsilon]-j} \varepsilon^{1/4} \| \cdots \|_{Cw^2, \varepsilon}^{1/4} w(B(x,2)) & \text{if } j \le \log_2 \varepsilon, \\ 0 & \text{if } j > \log_2 \varepsilon. \end{cases}$$

Lemma 26.10. *Assume (26.6) and (26.2)–(26.3). Let $\{\lambda_I\}_I$ be as in Definition 26.4. Assume (26.51). Let $j \leq \log_2 \varepsilon$. Then*

$$\sum_{k \geq j} 0.9^{k-j} \eta_k^{(5)''}(x)^2 \leq 100 \eta_j^{(5)''}(x)^2, \tag{26.52}$$

$$2^{-jn} \int_{B(x,1)} \eta_j^{(5)'}(y)^2 \left(1 + |x-y|/2^j\right)^{-n-1/2} dy \leq C(n) \eta_j^{(5)''}(x)^2, \tag{26.53}$$

$$\eta_j^{(2)}(x) \leq C(n) \eta_j^{(5)'}(x), \tag{26.54}$$

$$\eta_j^{(2)}(x) \leq C(n) \| \cdots \|_{Cw^2,\varepsilon}^{1/2} w(B(x, 2^j)), \tag{26.55}$$

$$\eta_j^{(3)''}(x) \leq C(n) \| \cdots \|_{Cw^2,\varepsilon}^{1/2} w(B(x, 2^j)), \tag{26.56}$$

$$\eta_j^{(0)}(x,y) \leq C(n) \| \cdots \|_{Cw^2,\varepsilon}^{1/2} w\left(B(x, 2^j)\right) \left(1 + |x-y|/2^j\right)^{n\varepsilon}$$
$$\times \log_2\left(2 + |x-y|/2^j\right) \quad \text{if } |x-y| \leq 1, \tag{26.57}$$

$$\sum_{I:\ell(I)=2^j} \lambda_I \left(1 + |x-x_I|/2^j\right)^{-n-1} \min\left\{\eta_{j+1}^{(0)}(x, x_I), w\left(B(x, 2^{j+1})\right)\right\}$$
$$\leq C(n) \min\left\{\eta_j^{(5)''}(x)^2, (\| \cdots \|_{Cw^2,\varepsilon}^{1/2} + \varepsilon^{1/2}) w\left(B(x, 2^j)\right) \eta_j^{(1)}(x)\right\}, \tag{26.58}$$

$$\left\| \sum_{j=-\infty}^{\infty} \eta_j^{(5)''}(x)^2 \delta_{t=2^j} \right\|_{Cw^2,\varepsilon} \leq C(n) \left\{\| \cdots \|_{Cw^2,\varepsilon} + \| \cdots \|_{Cw^2,\varepsilon}^{1/2} \varepsilon^{1/3}\right\}$$
$$\left(\leq C(n) \left\{\| \cdots \|_{Cw^2,\varepsilon} + \varepsilon^{2/3}\right\}.\right) \tag{26.59}$$

We follow the argument of the proof of Lemma 22.7. (26.52) is easy.
Proof of (26.53).

{the left-hand side of (26.53)}/2

$$\leq \int_{B(x,1)} \left\{\eta_j^{(4)'}(y)^2 + \left(0.99^{[\log_2 \varepsilon]-j} \varepsilon^{1/4} \| \cdots \|_{Cw^2,\varepsilon}^{1/4}\right)^2 w(B(y,1))^2\right\}$$
$$\times 2^{-jn} \left(1 + |x-y|/2^j\right)^{-n-1/2} dy$$

$$\leq \sum_{k \geq j} 0.99^{k-j} \int_{B(x,1)} \sum_{\substack{I:\ell(I)=2^k, \\ x_I \in B(y,1)}} \lambda_I^2 \left(1 + |y-x_I|/2^k\right)^{-n-1/2}$$
$$\times 2^{-jn} \left(1 + |x-y|/2^j\right)^{-n-1/2} dy$$

$$+ \left(0.99^{[\log_2 \varepsilon]-j} \varepsilon^{1/4} \| \cdots \|_{Cw^2,\varepsilon}^{1/4}\right)^2 \int_{B(x,1)} w(B(x,2))^2$$
$$\times 2^{-jn} \left(1 + |x-y|/2^j\right)^{-n-1/2} dy$$

$$\leq C \sum_{k \geq j} 0.99^{k-j} \eta_k^{(3)''}(x)^2 + C \left\{0.99^{[\log_2 \varepsilon]-j} \varepsilon^{1/4} \| \cdots \|_{Cw^2,\varepsilon}^{1/4} w(B(x,2))\right\}^2$$

$$\text{by (22.54)}$$

$$\leq C\eta_j^{(5)''}(x)^2.$$

$\square$

Proof of (26.54).

$$\eta_k^{(1)}(x) = \sum_{\substack{I:\ell(I)=2^k,\\ x_I \in B(x,1)}} \lambda_I \left(1 + |x - x_I|/2^k\right)^{-n-1/2} + \sum_{\substack{I:\ell(I)=2^k,\\ x_I \notin B(x,1)}} \cdots\cdots$$

$$\leq C\eta_k^{(3)'}(x) + \sum_{\substack{I:\ell(I)=2^k,\\ x_I \notin B(x,1)}} C\|\cdots\|_{Cw^2,\varepsilon}^{1/2} w(I)\left(1 + |x - x_I|/2^k\right)^{-n-1/2}$$

$$\text{by the Schwarz inequality and (26.32)}$$

$$\leq \cdots + C\|\cdots\|_{Cw^2,\varepsilon}^{1/2} \int_{y \notin B(x,1)} \frac{w(B(y,2^k))}{2^{kn}\left(1 + |x - y|/2^k\right)^{n+1/2}} dy$$

$$\leq \cdots + C\|\cdots\|_{Cw^2,\varepsilon}^{1/2} 2^{k(1/2-1/9)} \int \frac{w(B(y,2^k))}{2^{kn}\left(1 + |x - y|/2^k\right)^{n+1/9}} dy$$

$$\leq \cdots + C\|\cdots\|_{Cw^2,\varepsilon}^{1/2} 2^{k/4} w(B(x,1)) \quad \text{by (26.19).} \tag{26.60}$$

Summing up $(26.60)\times0.99^{k-j}$ with respect to $k = j, j+1, \cdots, [\log_2 \varepsilon]$ gives (26.54), because $\|\cdots\|_{Cw^2,\varepsilon} \leq 1$. $\square$

Proof of (26.55). If $j \leq k \leq \log_2 \varepsilon$, then

$$\eta_k^{(1)}(x) \leq \sum_{I:\ell(I)=2^k} C\|\cdots\|_{Cw^2,\varepsilon}^{1/2} w(I) \left(1 + |x - x_I|/2^k\right)^{-n-1/2} \quad \text{by (26.32)}$$

$$\leq \int C\|\cdots\|_{Cw^2,\varepsilon}^{1/2} w(B(y,2^k)) 2^{-kn}\left(1 + |x - y|/2^k\right)^{-n-1/2} dy$$

$$\leq C\|\cdots\|_{Cw^2,\varepsilon}^{1/2} w(B(x,2^k)) \quad \text{by (26.19)}$$

$$\leq C\|\cdots\|_{Cw^2,\varepsilon}^{1/2} 2^{(k-j)n\varepsilon} w(B(x,2^j)) \quad \text{by (26.17).} \tag{26.61}$$

Summing up $(26.61)\times0.99^{k-j}$ with respect to k (recall (26.6)) gives (26.55). $\square$

Proof of (26.56).

$$\eta_j^{(3)''}(x)^2 \leq \int_{B(x,2)} C\|\cdots\|_{Cw^2,\varepsilon} w(B(y,2^j))^2 2^{-jn}\left(1 + |x - y|/2^j\right)^{-n-1/2} dy$$

$$\text{by (26.32)}$$

$$\leq C\|\cdots\|_{Cw^2,\varepsilon} \int_{B(x,2)} w(B(x,2^j))^2 2^{-jn}\left(1 + |x - y|/2^j\right)^{-n-1/2+2n\varepsilon} dy$$

$$\text{by (26.18)}$$

$$\leq C\|\cdots\|_{Cw^2,\varepsilon} w(B(x,2^j))^2.$$

□

Proof of (26.57). This follows from the fact that if

$$2^j \leq 2^k \leq |x - y| \leq 1,$$

then

$$\eta_k^{(2)}(x) + \eta_k^{(2)}(y) \leq C\| \cdots \|_{Cw^2,\varepsilon}^{1/2} \{w(B(x, 2^k)) + w(B(y, 2^k))\} \quad \text{by (26.55)}$$
$$\leq C\| \cdots \|_{Cw^2,\varepsilon}^{1/2} w(B(x, |x - y| + 2^k))$$
$$\leq C\| \cdots \|_{Cw^2,\varepsilon}^{1/2} (1 + |x - y|/2^j)^{n\varepsilon} w(B(x, 2^j)) \quad \text{by (26.17).}$$

□

Proof of (26.58).

the left-hand side of (26.58)

$$\leq \sum_{\substack{I:\ell(I)=2^j, \\ x_I \in B(x,1)}} \lambda_I \left(1 + |x - x_I|/2^j\right)^{-n-1} \eta_{j+1}^{(0)}(x, x_I)$$

$$+ \sum_{\substack{I:\ell(I)=2^j, \\ x_I \notin B(x,1)}} \lambda_I \left(1 + |x - x_I|/2^j\right)^{-n-1} w(B(x, 2^{j+1}))$$

$$= (26.62) + (26.63), \quad \text{say.}$$

Then

$$(26.63) \quad \leq \quad C2^{j/2} w\left(B(x, 2^{j+1})\right) \sum \lambda_I \left(1 + |x - x_I|/2^j\right)^{-n-1/2}$$
$$\leq \quad C2^{j/2} w(B(x, 2^j)) \eta_j^{(1)}(x)$$
$$\leq \quad C \min \left\{ 0.99^{2(\lceil \log_2 \varepsilon \rceil - j)} \varepsilon^{1/2} w(B(x, 2^j)) \| \cdots \|_{Cw^2,\varepsilon}^{1/2} w(B(x, 2^j)), \right.$$
$$\left. \varepsilon^{1/2} w(B(x, 2^j)) \eta_j^{(1)}(x) \right\} \qquad (26.64)$$

by $2^{j/2} \leq 0.99^{2(\lceil \log_2 \varepsilon \rceil - j)} \varepsilon^{1/2}$ and (26.55) (and $\eta_j^{(1)} \leq \eta_j^{(2)}$).

$$(26.62) \quad \leq \quad \sum_{\substack{I:\ell(I)=2^j, \\ x_I \in B(x,1)}} \lambda_I \left(1 + |x - x_I|/2^j\right)^{-n-1} C\| \cdots \|_{Cw^2,\varepsilon}^{1/2} w(B(x, 2^{j+1}))$$

$$\times \left(1 + |x - x_I|/2^{j+1}\right)^{n\varepsilon} \log_2(2 + |x - x_I|/2^{j+1}) \quad \text{by (26.57)}$$

$$\leq \quad C\| \cdots \|_{Cw^2,\varepsilon}^{1/2} \eta_j^{(1)}(x) w(B(x, 2^j)). \qquad (26.65)$$

$$(26.62) \leq \sum_{I:\ell(I)=2^j,\, x_I \in B(x,1)} \lambda_I \left(1 + |x - x_I|/2^j\right)^{-n-1}$$

$$\times \sum_{k=j+1}^{[\log_2(2^{j+1}+|x-x_I|)]} \left(\eta_k^{(2)}(x) + \eta_k^{(2)}(x_I)\right)$$

$$\leq \sum_{k>j} \sum_{\substack{I:\ell(I)=2^j,\\ x_I \in B(x,1)\setminus B(x,2^k - 2^{j+1})}} \lambda_I \left(1 + |x - x_I|/2^j\right)^{-n-1}\left(\eta_k^{(2)}(x) + \eta_k^{(2)}(x_I)\right)$$

$$\leq \left\{\sum\sum \lambda_I^2 \left(1 + |x - x_I|/2^j\right)^{-n-1}\right\}^{1/2}$$

$$\times \left\{\sum\sum \left(\eta_k^{(2)}(x)^2 + \eta_k^{(2)}(x_I)^2\right)\left(1 + |x - x_I|/2^j\right)^{-n-1}\right\}^{1/2}$$

$$\leq C \left\{\sum\sum 2^{(j-k)/2}\lambda_I^2 \left(1 + |x - x_I|/2^j\right)^{-n-1/2}\right\}^{1/2}$$

$$\times \left\{\sum\sum \left(\eta_k^{(2)}(x)^2 + \eta_k^{(2)}(x_I)^2\right)\frac{2^{(j-k)(n+1)}}{\left(1 + |x - x_I|/2^k\right)^{n+1}}\right\}^{1/2}$$

$$\leq C\eta_j^{(3)'}(x)\left\{\sum_{k>j} 2^{j-k}2^{-kn} \int_{B(x,1)} \frac{\eta_k^{(2)}(x)^2 + \eta_k^{(2)}(y)^2}{(1 + |x - y|/2^k)^{n+1}}\, dy\right\}^{1/2}$$

$$\leq C\eta_j^{(3)'}(x)\left\{\sum_{k>j} 2^{j-k}2^{-kn} \int_{B(x,1)} \frac{\eta_k^{(5)'}(x)^2 + \eta_k^{(5)'}(y)^2}{(1 + |x - y|/2^k)^{n+1}}\, dy\right\}^{1/2} \quad \text{by (26.54)}$$

$$\leq C\eta_j^{(3)'}(x)\left\{\sum_{k>j} 2^{j-k} \left(\eta_k^{(5)'}(x)^2 + \eta_k^{(5)''}(x)^2\right)\right\}^{1/2} \quad \text{by (26.53)}$$

$$\leq C\eta_j^{(5)''}(x)^2 \tag{26.66}$$

by (26.52). Then (26.58) follows from (26.64)–(26.66). $\qquad\square$

Proof of (26.59). Take any ball $B_0 = B(x_0, r_0)$ with $r_0 \leq \varepsilon$. Then

$$\sum_{j=-\infty}^{[\log_2 r_0]} \int_{B_0} \eta_j^{(3)''}(x)^2 dx$$

$$\leq \sum_{j=-\infty}^{[\log_2 r_0]} \sum_{\substack{I:\ell(I)=2^j,\\ x_I \in B(x_0,3)}} \lambda_I^2 \int_{B_0} \left(1 + |x_I - x|/2^j\right)^{-n-1/2} dx$$

$$\leq C \sum\sum \lambda_I^2 \min\left\{|I|,\, |B_0|\, (\text{dist}(x_I, B_0)/\ell(I))^{-n-1/2}\right\}$$

$$\leq C \sum\sum \lambda_I^2 |I|(1 + |x_I - x_0|/r_0)^{-n-1/2}$$

$$= C \iint_{\mathbf{R}^n \times (0,r_0]} (1 + |x - x_0|/r_0)^{-n-1/2} \sum_{I:x_I \in B(x_0,3)} \lambda_I^2 |I|\delta_{(x_I,\ell(I))}$$

$$\leq C \int_{y \in B(x_0,3)} r_0^{-n}(1+|y-x_0|/r_0)^{-n-1/2}dy \iint_{Q(B(y,r_0))} \sum \lambda_I^2 |I| \delta_{(x_I,\ell(I))}$$

$$\leq C \int (1+|y-x_0|/r_0)^{-n-1/2} \| \cdots \|_{Cw^2,\varepsilon} w(B(y,r_0))^2 dy$$

$$\leq C\| \cdots \|_{Cw^2,\varepsilon} w(B_0)^2 \int (1+|y-x_0|/r_0)^{-n-1/2+2n\varepsilon} dy \quad \text{by (26.18)}$$

$$\leq C\| \cdots \|_{Cw^2,\varepsilon} w(B_0)^2 |B_0|. \tag{26.67}$$

If $x \in B_0$ and $k \in \mathbf{N}$, then

$$\eta^{(3)''}_{[\log_2 r_0]+k}(x)^2 \leq C\| \cdots \|_{Cw^2,\varepsilon} w\big(B(x, 2^{[\log_2 r_0]+k})\big)^2 \quad \text{by (26.56)}$$

$$\leq C\| \cdots \|_{Cw^2,\varepsilon} 2^{2kn\varepsilon} w(B_0)^2 \quad \text{by (26.17)}. \tag{26.68}$$

so, if $m \in \mathbf{N}_0$, then

$$\iint_{Q(B_0)} \sum_{j=-\infty}^{\infty} \eta^{(3)''}_{j+m}(x)^2 \delta_{t=2^j}$$

$$\leq \sum_{j=-\infty}^{[\log_2 r_0]+m} \int_{B_0} \eta^{(3)''}_{j}(x)^2 dx$$

$$\leq \sum_{j=-\infty}^{[\log_2 r_0]} \cdots\cdots\cdots + \sum_{j=[\log_2 r_0]+1}^{[\log_2 r_0]+m} \cdots\cdots$$

$$\leq C\| \cdots \|_{Cw^2,\varepsilon} w(B_0)^2 |B_0| + mC\| \cdots \|_{Cw^2,\varepsilon} 2^{2mn\varepsilon} w(B_0)^2 |B_0|$$

$$\text{by (26.67)–(26.68)}$$

$$\leq C(m+1)2^{2mn\varepsilon}\| \cdots \|_{Cw^2,\varepsilon} w(B_0)^2 |B_0|.$$

Thus

$$\left\| \sum_{j=-\infty}^{\infty} \eta^{(3)''}_{j+m}(x)^2 \delta_{t=2^j} \right\|_{Cw^2,\varepsilon} \leq C(m+1)2^{2mn\varepsilon}\| \cdots \|_{Cw^2,\varepsilon}. \tag{26.69}$$

Thus

$$\left\| \sum_{j=-\infty}^{\infty} \eta^{(4)''}_{j}(x)^2 \delta_{t=2^j} \right\|_{Cw^2,\varepsilon}$$

$$\leq \sum_{m \geq 0} 0.99^m \left\| \sum_{j=-\infty}^{\infty} \eta^{(3)''}_{j+m}(x)^2 \delta_{t=2^j} \right\|_{Cw^2,\varepsilon}$$

$$\leq \sum_{m \geq 0} 0.99^m C(m+1)2^{2mn\varepsilon}\| \cdots \|_{Cw^2,\varepsilon} \quad \text{by (26.69)}$$

$$\leq C\| \cdots \|_{Cw^2,\varepsilon} \quad \text{by (26.6)}. \tag{26.70}$$

On the other hand

$$\iint_{Q(B_0)} \sum_{j=-\infty}^{[\log_2 \varepsilon]} \left\{ 0.99^{[\log_2 \varepsilon]-j} \varepsilon^{1/4} w(B(x,2)) \right\}^2 \delta_{t=2^j}$$

$$= \varepsilon^{1/2} w(B(x_0,3))^2 |B_0| \sum_{j=-\infty}^{[\log_2 r_0]} 0.99^{2[\log_2 \varepsilon]-2j}$$

$$\leq C\varepsilon^{1/2} r_0^{-2n\varepsilon} w(B_0)^2 |B_0| 0.99^{2\log_2(\varepsilon/r_0)} \quad \text{by (26.17)}$$

$$\leq C\varepsilon^{1/3} w(B_0)^2 |B_0|.$$

Thus,

$$\left\| \sum_{j=-\infty}^{[\log_2 \varepsilon]} \left\{ 0.99^{[\log_2 \varepsilon]-j} \varepsilon^{1/4} w(B(x,2)) \right\}^2 \delta_{t=2^j} \right\|_{Cw^2,\varepsilon} \leq C\varepsilon^{1/3}. \quad (26.71)$$

Combining (26.70) and (26.71) gives (26.59). $\qquad\square$

Proof of Lemma 26.2. We follow the argument of the proof of Lemma 22.2. The letter I denotes dyadic cubes in $\mathbf{R}^n$ and the letter j denotes integers. Let A be a sufficiently large number depending only on $\{\theta_1, \cdots, \theta_m\}$. We may assume

$$\varepsilon \leq A^{-100} \quad (26.72)$$

and

$$\|\vec{g}\|_{\mathrm{BMO}w,\varepsilon} \leq A^{-100} \quad (26.73)$$

because otherwise we can take

$$\vec{v} = \vec{g} \quad \text{and} \quad \vec{h} \equiv \vec{0}.$$

Applying Lemmas 26.8–9 gives $\{\lambda_{\vec{g},I},\ \vec{b}_I,\ \lambda_{w,I},\ b_I\}_I$ that satisfy (26.43)–(26.50). Let

$$\lambda_I = \begin{cases} \lambda_{\vec{g},I} + \lambda_{w,I} & \text{if } \ell(I) \leq \varepsilon/2 \\ \|\vec{g}\|_{\mathrm{BMO}w,\varepsilon} w(I) & \text{if } \log_2 \ell(I) = [\log_2 \varepsilon] \\ 0 & \text{otherwise.} \end{cases} \quad (26.74)$$

Then

$$\left\| \sum_{I:\ell(I)\leq\varepsilon/2} \lambda_I^2 |I| \delta_{(x_I,\ell(I))} \right\|_{Cw^2,\varepsilon} \leq C \left\{ \|\vec{g}\|_{\mathrm{BMO}w,\varepsilon}^2 + \varepsilon^2 \right\}, \quad (26.75)$$

$$\left\| \sum_{\text{all } I} \lambda_I^2 |I| \delta_{(x_I,\ell(I))} \right\|_{Cw^2,\varepsilon} \leq C \left\{ \|\vec{g}\|_{\mathrm{BMO}w,\varepsilon}^2 + \varepsilon^2 \right\}, \quad (26.76)$$

by (26.45) and (26.49). Let $\eta_j^{(1)}$, $\eta_j^{(2)}$, $\eta_j^{(5)''}$ and $\eta_j^{(0)}$ be as in Definitions 22.2 and 26.4. Then (26.55), (26.76) and (26.59) imply

$$\eta_j^{(2)}(x) \leq C\left\{\|\vec{g}\|_{\mathrm{BMO}w,\varepsilon} + \varepsilon\right\} w\left(B(x,2^j)\right), \tag{26.77}$$

$$\left\|\sum_{j=-\infty}^{\infty} \eta_j^{(5)''}(x)^2 \delta_{t=2^j}\right\|_{Cw^2,\varepsilon} \leq C\left\{\|\vec{g}\|_{\mathrm{BMO}w,\varepsilon}^2 + \varepsilon^{2/3}\right\}. \tag{26.78}$$

Note that

$$\begin{aligned}
\left|\vec{g}_{[\log_2 \varepsilon]\dagger}(x)\right| &\leq C\|\vec{g}\|_{\mathrm{BMO}w,\varepsilon} w_{[\log_2 \varepsilon]\dagger}(x) \text{ by (26.40)} \\
&\leq w_{[\log_2 \varepsilon]\dagger}(x) \text{ by (26.73).}
\end{aligned} \tag{26.79}$$

(For the definitions of $\vec{g}_{j\dagger}$ and $w_{j\dagger}$ recall Definition 26.3.)
    Let

$$\vec{p}_I(x) \equiv \vec{0} \text{ if } \ell(I) > \varepsilon/2, \tag{26.80}$$
$$\vec{\phi}_j(x) \equiv \vec{0} \text{ if } 2^j > \varepsilon/2. \tag{26.81}$$

We will construct

$$\{\vec{p}_I\}_{\ell(I)\leq\varepsilon/2}, \ \{\vec{\phi}_j\}_{j\leq\log_2(\varepsilon/2)} \subset C^1(\mathbf{R}^1, \mathbf{R}^m)$$

so that the following hold:

$$\left|\vec{p}_I(x)\right| + \ell(I)\left|\nabla\vec{p}_I(x)\right| \leq A\left(1 + |x - x_I|/\ell(I)\right)^{-n-1}, \tag{26.82}$$

$$\int \vec{p}_I(x)dx = \vec{0}, \tag{26.83}$$

$$\vec{p}_I \in S_{\{\theta_1,\cdots,\theta_m\}*(r)}, \tag{26.84}$$

$$\left|\vec{\kappa}_j(x)\right| \leq w_{j\dagger}(x), \tag{26.85}$$

$$\left|\vec{\phi}_j(x)\right| \leq A^{10} \min\left\{\eta_j^{(5)''}(x)^2/w\left(B(x,2^j)\right), \left(\|\vec{g}\|_{\mathrm{BMO}w,\varepsilon} + \varepsilon^{1/2}\right)\eta_j^{(1)}(x)\right\}, \tag{26.86}$$

$$\left|\nabla\vec{\phi}_j(x)\right| \leq 2^{-j}A^{10}\left(\|\vec{g}\|_{\mathrm{BMO}w,\varepsilon} + \varepsilon^{1/2}\right)\eta_j^{(1)}(x), \tag{26.87}$$

where

$$\vec{\kappa}_j = \begin{cases} \vec{g}_{[\log_2 \varepsilon]\dagger} + \displaystyle\sum_{I:\ell(I)\geq 2^j}\lambda_I(\vec{b}_I + \vec{p}_I) - \sum_{k\geq j}\vec{\phi}_k & \text{if } j \leq \log_2 \varepsilon, \\[4mm] \vec{0} & \text{if } j > \log_2 \varepsilon. \end{cases} \tag{26.88}$$

We grant this construction temporarily and finish the proof of Lemma 26.2.
    Put

$$\vec{h} = -\sum_I \lambda_I \vec{p}_I \quad \text{and} \quad \vec{v} = \sum_{k=-\infty}^{\infty} \vec{\phi}_k,$$

where I is taken over all dyadic cubes. (The convergences of these $\sum$'s follow from the same reasons as in Section 22.) Then, (26.84) implies

$$\vec{h} \in S_{\{\theta_1,\cdots,\theta_m\}*(r)}.$$

(26.9) follows from (26.30) and (26.75). Letting $j \to -\infty$ in (26.85) gives (26.11). (Recall (26.38).) So, we show only (26.10).

Proof of (26.10). First note that

$$\left\| \sum \left| \vec{\phi}_j(x) \right| \delta_{t=2^j} \right\|_{Cw,\varepsilon} \leq A^{10} \left\| \sum \eta_j^{(5)''}(x)^2 w\big(B(x,2^j)\big)^{-1} \delta_{t=2^j} \right\|_{Cw,\varepsilon}$$
$$\text{by (26.86)}$$

$$\leq A^{11} \left\| \sum \eta_j^{(5)''}(x)^2 \delta_{t=2^j} \right\|_{Cw^2,\varepsilon} \quad \text{by Lemma 26.5}$$

$$\leq A^{12} \left(\|\vec{g}\|_{\text{BMO}w,\varepsilon}^2 + \varepsilon^{2/3} \right) \quad \text{by (26.78),} \quad (26.89)$$

$$\left| \nabla \vec{\phi}_k(x) \right| \leq 2^{-k} A^{10} \left(\|\vec{g}\|_{\text{BMO}w,\varepsilon} + \varepsilon^{1/2} \right) \eta_k^{(1)}(x) \quad \text{by (26.87)}$$

$$\leq 2^{-k} A^{11} \left(\|\vec{g}\|_{\text{BMO}w,\varepsilon}^2 + \varepsilon \right) w\big(B(x,2^k)\big) \quad \text{by (26.77).} \quad (26.90)$$

Take any ball $B_0 = B(x_0, r_0)$ with $r_0 \leq \varepsilon$. Then

$$\int_{B_0} \left| \vec{v}(x) - \sum_{k \geq [\log_2 r_0]+1} \vec{\phi}_k(x_0) \right| dx/|B_0|$$

$$\leq \int_{B_0} \sum_{k \leq [\log_2 r_0]} |\vec{\phi}_k(x)| dx/|B_0| + \int_{B_0} \sum_{k \geq [\log_2 r_0]+1} |\vec{\phi}_k(x) - \vec{\phi}_k(x_0)| dx/|B_0|$$

$$\leq \left\| \sum |\vec{\phi}_k| \delta_{t=2^k} \right\|_{Cw,\varepsilon} w(B_0)$$
$$+ A^{11} \left(\|\vec{g}\|_{\text{BMO}w,\varepsilon}^2 + \varepsilon \right) \sum_{k \geq [\log_2 r_0]+1} 2^{-k} w\big(B(x_0,2^k)\big) r_0 \quad \text{by (26.90)}$$

$$\leq A^{12} \left(\|\vec{g}\|_{\text{BMO}w,\varepsilon}^2 + \varepsilon^{2/3} \right) w(B_0)$$
$$+ A^{11} \left(\|\vec{g}\|_{\text{BMO}w,\varepsilon}^2 + \varepsilon \right) \sum_{k \geq [\cdots]+1} 2^{-k} C(2^k/r_0)^{n\varepsilon} w(B_0) r_0$$
$$\text{by (26.89) and (26.17)}$$

$$\leq A^{13} \left(\|\vec{g}\|_{\text{BMO}w,\varepsilon}^2 + \varepsilon^{2/3} \right) w(B_0).$$

(Note that if $r_0 = \varepsilon$, $\displaystyle\sum_{k \geq [\log_2 r_0]+1} \vec{\phi}_k(x_0) = \vec{0}$.) $\qquad\qquad\square$

The construction of $\{\vec{p}_I\}$ and $\{\vec{\phi}_j\}$. We follow the argument of Section 22. Note that (26.85) for the case $j = [\log_2 \varepsilon]$ follows from (26.79). Let $j \leq [\log_2 \varepsilon]$ and suppose that $\{\vec{p}_I\}_{\ell(I) \geq 2^j}$ and $\{\vec{\phi}_k\}_{k \geq j}$ have been constructed so that (26.82)–(26.87) hold. We will give the construction of

$$\{\vec{p}_I\}_{\ell(I)=2^{j-1}}, \quad \vec{\phi}_{j-1} \subset C^1(\mathbf{R}^n, \mathbf{R}^m). \tag{26.91}$$

Claim 1. There exists $\psi_j \in C^1(\mathbf{R}^n)$ such that

$$\psi_j(x) \begin{cases} = 1 & \text{if} \quad |\vec{\kappa}_j(x)| \, / w_{j\dagger}(x) \geq 0.99 \\ = 0 & \text{if} \quad |\vec{\kappa}_j(x)| \, / w_{j\dagger}(x) \leq 0.9 \\ \in [0,1] & \text{otherwise}, \end{cases} \tag{26.92}$$

$$\|\nabla \psi_j\|_{L^\infty} \leq 2^{-j}. \tag{26.93}$$

Let I be a dyadic cube with $\ell(I) = 2^{j-1}$. Applying Lemma 24.3 with

$$b(x) = \left\{ -\lambda_{\vec{g},I} \vec{b}_I(x) \cdot U(\vec{\kappa}_j(x_I)) + \lambda_{w,I} b_I(x) \right\} / \lambda_I,$$
$$\vec{v} = U(\vec{\kappa}_j(x_I))$$

gives

$$\vec{p}_I \in S_{\{\theta_1, \cdots, \theta_m\}*(r)}$$

so that

$$\left(\lambda_I \vec{p}_I(x) + \lambda_{\vec{g},I} \vec{b}_I(x) \right) \cdot U(\vec{\kappa}_j(x_I)) = \lambda_{w,I} b_I(x). \tag{26.94}$$

The formula (24.11) and Lemma 22.3 imply (26.82)–(26.83).

Put

$$\vec{\rho}(x) = \sum_{I:\ell(I)=2^{j-1}} \left(\lambda_{\vec{g},I} \vec{b}_I(x) + \lambda_I \vec{p}_I(x) \right), \tag{26.95}$$

$$\vec{\tau}(x) = \vec{\kappa}_j(x) + \vec{\rho}(x), \tag{26.96}$$

$$\nu(x) = \sum_{I:\ell(I)=2^{j-1}} \lambda_{w,I} b_I(x), \tag{26.97}$$

$$\vec{\phi}_{j-1}(x) = \psi_j(x) \{|\vec{\tau}(x)| - |\vec{\kappa}_j(x)| - \nu(x)\} U(\vec{\tau}(x)). \tag{26.98}$$

(See Fig. 26.1.) Then

$$w_{j-1\dagger}(x) = w_{j\dagger}(x) + \nu(x), \tag{26.99}$$

$$\vec{\kappa}_{j-1}(x) = \vec{\tau}(x) - \vec{\phi}_{j-1}(x). \tag{26.100}$$

Claim 2.

$$|\vec{\kappa}_{j-1}(x)| \leq w_{j-1\dagger}(x). \tag{26.101}$$

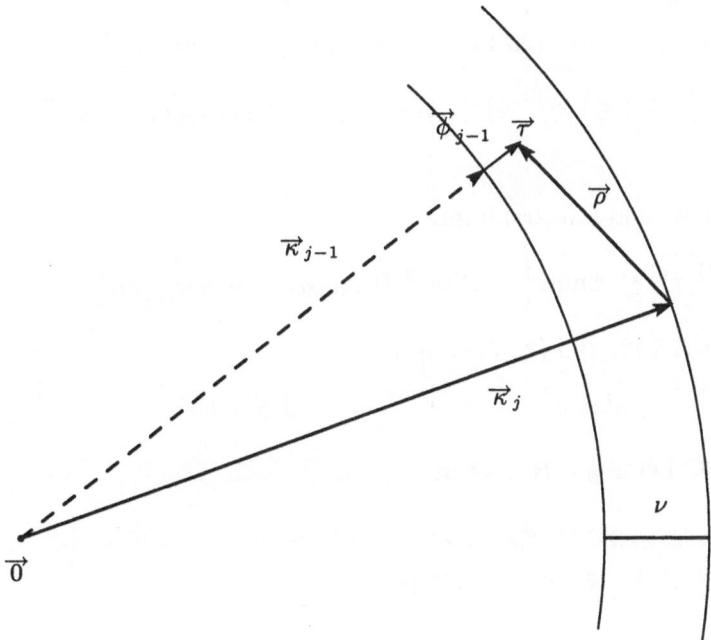

Fig. 26.1: The construction of $\vec{\phi}_{j-1}$

Claim 3.

$$\left|\vec{\phi}_{j-1}(x)\right| \leq A^{10} \min\left\{ \frac{\eta_{j-1}^{(5)''}(x)^2}{w\big(B(x, 2^{j-1})\big)}, \big(\|\vec{g}\|_{\text{BMO}w,\varepsilon} + \varepsilon^{1/2}\big)\eta_{j-1}^{(1)}(x)\right\}.$$

$$(26.102)$$

Claim 4.

$$\left|\nabla\vec{\phi}_{j-1}(x)\right| \leq 2^{-j+1}A^{10}\big(\|\vec{g}\|_{\text{BMO}w,\varepsilon} + \varepsilon^{1/2}\big)\eta_{j-1}^{(1)}(x).$$

$$(26.103)$$

This concludes the construction of (26.91). $\qquad\qquad\qquad\square$

Finally, we prove Claims 1–4.

First, note that (26.44), (26.48) and (26.82) imply

$$\left|\vec{b}_I(x)\right| + |\vec{p}_I(x)| + |b_I(x)| + \ell(I)\Big(\left|\nabla\vec{b}_I(x)\right| + |\nabla\vec{p}_I(x)| + |\nabla b_I(x)|\Big)$$

$$\leq 2A\left(1 + |x - x_I|/\ell(I)\right)^{-n-1},$$

$$(26.104)$$

that (26.39) and (26.72) imply

$$0.999w\big(B(x, 2^j)\big) \leq w_{j\dagger}(x) \leq 1.001w\big(B(x, 2^j)\big) \text{ if } j \leq [\log_2 \varepsilon],$$

$$(26.105)$$

that (26.55), (26.76), (26.105), (26.72) and (26.73) imply

$$\left(\eta_j^{(1)}(x) \le\right) \eta_j^{(2)}(x) \le A\left(\|\vec{g}\|_{\mathrm{BMO}w,\varepsilon} + \varepsilon\right) w_{j\dagger}(x) \left(\le A^{-98} w_{j\dagger}(x)\right),$$
(26.106)

that (26.54) and (26.106) imply

$$\eta_j^{(1)}(x)^2 \le A \min\left\{\eta_j(5)'(x)^2, \left(\|\vec{g}\|_{\mathrm{BMO}w,\varepsilon} + \varepsilon\right) w_{j\dagger}(x)\eta_j^{(1)}(x)\right\} \quad (26.107)$$

and that (26.42) and (26.72) imply

$$2^j |\nabla w_{j\dagger}(x)| \le A^{-99} w_{j\dagger}(x) \quad \text{if } j \le [\log_2 \varepsilon].$$
(26.108)

Claim 5. Let $x, y \in \mathbf{R}^n$. Then

$$\left|\vec{g}_{[\log_2 \varepsilon]\dagger}(x) - \vec{g}_{[\log_2 \varepsilon]\dagger}(y)\right| \le A\eta_j^{(0)}(x,y) \quad \text{if } j \le [\log_2 \varepsilon], \quad (26.109)$$

$$|\vec{\kappa}_j(x) - \vec{\kappa}_j(y)| \le A^2 \eta_j^{(0)}(x,y), \quad (26.110)$$

$$|\nabla \vec{\kappa}_j(x)| \le 2^{-j} A^2 \eta_j^{(2)}(x). \quad (26.110)^*$$

Proof. Note that (26.74) implies

$$\|\vec{g}\|_{\mathrm{BMO}w,\varepsilon} w_{[\log_2 \varepsilon]\dagger}(x) \le C\eta_{[\log_2 \varepsilon]}^{(1)}(x). \quad (26.111)$$

Thus, if $|x - y| < 2^j$, then

{the left-hand side of (26.109)}

$$\le C\varepsilon^{-1}\|\vec{g}\|_{\mathrm{BMO}w,\varepsilon} \sup\left\{w_{[\log_2 \varepsilon]\dagger}(z) : z \in B(x, 2^j)\right\} |x - y| \text{ by (26.41)}$$

$$\le C\varepsilon^{-1}\|\vec{g}\|_{\mathrm{BMO}w,\varepsilon} w_{[\log_2 \varepsilon]\dagger}(x)|x - y| \text{ by (26.108)}$$

$$\le C\varepsilon^{-1}\eta_{[\log_2 \varepsilon]}^{(1)}(x)|x - y| \text{ by (26.111)}$$

$$\le C2^{-j}\eta_j^{(2)}(x)|x - y|$$

$$= C\eta_j^{(0)}(x,y).$$

If $2^j \le |x - y| \le \varepsilon$, then

{the left-hand side of (26.109)}

$$\le C\varepsilon^{-1}\eta_{[\log_2 \varepsilon]}^{(1)}(x)|x - y| \text{ by the same argument as above}$$

$$\le C\eta_{[\log_2 |x-y|]}^{(2)}(x)$$

$$\le C\eta_j^{(0)}(x,y).$$

If $\varepsilon < |x - y|$, then

{the left-hand side of (26.109)}

$$\leq \left| \vec{g}_{[\log_2 \epsilon]\dagger}(x) \right| + \left| \vec{g}_{[\log_2 \epsilon]\dagger}(y) \right|$$

$$\leq C \|\vec{g}\|_{\mathrm{BMO}w,\epsilon} \left\{ w_{[\log_2 \epsilon]\dagger}(x) + w_{[\log_2 \epsilon]\dagger}(y) \right\} \quad \text{by (26.40)}$$

$$\leq C \left(\eta^{(1)}_{[\log_2 \epsilon]}(x) + \eta^{(1)}_{[\log_2 \epsilon]}(y) \right) \quad \text{by (26.111)}$$

$$\leq C \eta^{(0)}_j(x,y).$$

Thus, we get (26.109).

Recall (26.88). Since

$$\sum_{I:\ell(I) \geq 2^j} \lambda_I (\vec{b}_I + \vec{p}_I) - \sum_{k \geq j} \vec{\phi}_k$$

can be treated by the same way as in the proof of (22.84), (26.110) follows from (26.109) and the argument of the proof of (22.84). (26.110)* follows from (26.110). □

Proof of Claim 1. (26.108) and (26.110)* (+ refeq(26.106)) imply

$$\mathrm{dist} \left(\{ x \in \mathbf{R}^n : |\vec{\kappa}_j(x)| / w_{j\dagger}(x) > 0.99 \}, \right.$$
$$\left. \{ x \in \mathbf{R}^n : |\vec{\kappa}_j(x)| / w_{j\dagger}(x) < 0.9 \} \right) > A2^j.$$

This implies the existence of the desired ψ_j. □

Claim 6.

$$|\vec{\rho}(x)| + |\nu(x)| + 2^{j-1}(|\nabla\vec{\rho}(x)| + |\nabla\nu(x)|) \leq A^2 \eta^{(1)}_{j-1}(x)$$
$$(\leq A^{-96} w_{j-1\dagger}(x)). \quad (26.112)$$
$$0.999 w_{j\dagger}(x) \leq w_{j-1\dagger}(x) \leq 1.001 w_{j\dagger}(x). \quad (26.113)$$

Claim 7.

$$2^{j-1} |\nabla\vec{\tau}(x)| \leq A^2 \eta^{(2)}_{j-1}(x). \quad (26.114)$$

If $|\vec{\kappa}_j(x)| \geq 0.9 w_{j\dagger}(x)$, then $|\vec{\tau}(x)| \geq 0.89 w_{j\dagger}(x)$. $\quad (26.115)$

Claim 8.

$$\vec{\kappa}_{j-1}(x) = \{ \psi_j(x) (|\vec{\kappa}_j(x)| + \nu(x)) + (1 - \psi_j(x)) |\vec{\tau}(x)| \} U(\vec{\tau}(x)). (26.116)$$

If $|\vec{\kappa}_j(x)| \geq 0.99 w_{j\dagger}(x)$, then $|\vec{\kappa}_{j-1}(x)| = |\vec{\kappa}_j(x)| + \nu(x)$. $\quad (26.117)$

(26.112) is clear from (26.104). (26.113) is clear from (26.112) and (26.99). (26.114) is clear from (26.110)* and (26.112). (26.115) is clear from (26.112). (26.116) is clear from (26.100) and (26.98). (26.117) is clear from (26.116) and (26.92).

Proof of Claim 2. If $|\vec{\kappa}_j(x)| \geq 0.99 w_{j\dagger}(x)$, then

$$
\begin{aligned}
|\vec{\kappa}_{j-1}(x)| &= |\vec{\kappa}_j(x)| + \nu(x) \text{ by } (26.117) \\
&\le w_{j\dagger}(x) + \nu(x) = w_{j-1\dagger}(x) \text{ by } (26.85) \text{ and } (26.99).
\end{aligned}
$$

If $|\vec{\kappa}_j(x)| < 0.99 w_{j\dagger}(x)$, then

$$
\begin{aligned}
|\vec{\kappa}_{j-1}(x)| &= \psi_j(x)\,(|\vec{\kappa}_j(x)| + \nu(x)) + (1 - \psi_j(x))\,|\vec{\tau}(x)| \text{ by } (26.116) \\
&\le \max\{|\vec{\kappa}_j(x)| + \nu(x), |\vec{\tau}(x)|\} \\
&\le \max\{w_{j\dagger}(x) + \nu(x), |\vec{\kappa}_j(x)| + |\vec{\rho}(x)|\} \\
&\le \max\{w_{j-1\dagger}(x), 0.99 w_{j\dagger}(x) + A^{-96} w_{j-1\dagger}(x)\} \\
&\qquad\qquad\qquad\qquad\qquad\text{by } (26.99) \text{ and } (26.112) \\
&\le w_{j-1\dagger}(x) \text{ by } (26.113).
\end{aligned}
$$

$\square$

Claim 9. If

$$|\vec{\kappa}_j(x)| \ge 0.1 w_{j\dagger}(x), \tag{26.118}$$

then

$$
\begin{aligned}
&|\vec{\tau}(x)| - |\vec{\kappa}_j(x)| \\
&\quad = |\vec{\kappa}_j(x)|\, v\big(2\,|\vec{\kappa}_j(x)|^{-1}\,\vec{\rho}(x) \cdot U(\vec{\kappa}_j(x)) + |\vec{\kappa}_j(x)|^{-2}\,|\vec{\rho}(x)|^2\big). \tag{26.119}
\end{aligned}
$$

(Recall Definition 22.4.)

The proof is the same as the proof of (22.91).

Claim 10. If (26.118) holds, then

$$
\begin{aligned}
&|\vec{\rho}(x) \cdot U(\vec{\kappa}_j(x)) - \nu(x)| \\
&\quad \le A^4 \min\Big\{\eta_{j-1}^{(5)''}(x)^2 / w_{j-1\dagger}(x), \big(\|\vec{g}\|_{\mathrm{BMO}w,\varepsilon} + \varepsilon^{1/2}\big)\,\eta_{j-1}^{(1)}(x)\Big\}, \tag{26.120}
\end{aligned}
$$

$$|(\nabla\vec{\rho}(x)) \cdot U(\vec{\kappa}_j(x)) - \nabla\nu(x)| \le 2^{-j+1} A^4 \big(\|\vec{g}\|_{\mathrm{BMO}w,\varepsilon} + \varepsilon^{1/2}\big)\,\eta_{j-1}^{(1)}(x). \tag{26.121}$$

Proof.

{the left-hand side of (26.120)}

$$
\begin{aligned}
&= \left| \sum_{I:\ell(I)=2^{j-1}} \left\{ (\lambda_{\vec{g},I}\vec{b}_I(x) + \lambda_I\vec{p}_I(x)) \cdot U(\vec{\kappa}_j(x)) - \lambda_{w,I} b_I(x) \right\} \right| \\
&= \left| \sum (\lambda_{\vec{g},I}\vec{b}_I(x) + \lambda_I\vec{p}_I(x)) \cdot (U(\vec{\kappa}_j(x)) - U(\vec{\kappa}_j(x_I))) \right| \text{ by } (26.94) \\
&\le \sum \lambda_I \left(|\vec{b}_I(x)| + |\vec{p}_I(x)| \right) \min\{2\,|\vec{\kappa}_j(x) - \vec{\kappa}_j(x_I)| / |\vec{\kappa}_j(x)|, 2\} \\
&\le 20 w_{j\dagger}(x)^{-1} \sum \frac{2 A \lambda_I}{(1 + |x - x_I|/2^{j-1})^{n+1}} \min\left\{ A^2 \eta_j^{(0)}(x, x_I), w_{j\dagger}(x) \right\}
\end{aligned}
$$

$$\text{by (26.118), (26.104) and (26.110)}$$

$$\leq CA^3 w_{j\dagger}(x)^{-1}$$

$$\times \min\left\{\eta^{(5)''}_{j-1}(x)^2, \left(\left\|\sum \lambda_I^2 |I| \delta_{(x_I, \ell(I))}\right\|^{1/2}_{Cw^2,\varepsilon} + \varepsilon^{1/2}\right) w_{j-1\dagger}(x)\eta^{(1)}_{j-1}(x)\right\}$$

$$\text{by (26.58) and (26.105)}$$

$$\leq \{\text{the right-hand side of (26.120)}\} \quad \text{by (26.76) and (26.113).}$$

(26.121) can be proved similarly. $\qquad\square$

If $\vec{\kappa}_j(x) \neq \vec{0}$, then put

$$\mu(x) = 2|\vec{\kappa}_j(x)|^{-1}\,\vec{\rho}(x) \cdot U(\vec{\kappa}_j(x)) + |\vec{\kappa}_j(x)|^{-2}\,|\vec{\rho}(x)|^2. \qquad (26.122)$$

Then (26.98), (26.119) and (26.92) imply

$$\vec{\phi}_{j-1}(x) = \begin{cases} \psi_j(x)\{|\vec{\kappa}_j(x)|\,\upsilon(\mu(x)) - \nu(x)\}U(\vec{\tau}(x)) & \text{if } |\vec{\kappa}_j(x)| \geq 0.9w_{j\dagger}(x) \\ \vec{0} & \text{if } |\vec{\kappa}_j(x)| < 0.9w_{j\dagger}(x). \end{cases}$$
$$(26.123)$$

Claim 11. If (26.118) holds, then

$$|\mu(x)| \leq A^3\eta^{(1)}_{j-1}(x)w_{j\dagger}(x)^{-1} \ (< 0.5), \qquad (26.124)$$

$$|\nabla\mu(x)| \leq 2^{-j+1}A^3\eta^{(1)}_{j-1}(x)w_{j\dagger}(x)^{-1}, \qquad (26.125)$$

$$|2^{-1}|\vec{\kappa}_j(x)|\,\mu(x) - \nu(x)|$$
$$\leq A^6 \min\left\{\eta^{(5)''}_{j-1}(x)^2/w_{j-1\dagger}(x), \left(\|\vec{g}\|_{\text{BMOw},\varepsilon} + \varepsilon^{1/2}\right)\eta^{(1)}_{j-1}(x)\right\}, \qquad (26.126)$$

$$|2^{-1}|\vec{\kappa}_j(x)|\,\nabla\mu(x) - \nabla\nu(x)| \leq 2^{-j+1}A^7\left(\|\vec{g}\|_{\text{BMOw},\varepsilon} + \varepsilon^{1/2}\right)\eta^{(1)}_{j-1}(x),$$
$$(26.127)$$

$$\left||\vec{\kappa}_j(x)|\,\upsilon(\mu(x)) - \nu(x)\right| \leq A^2\{\text{the right-hand side of (26.126)}\}, \qquad (26.128)$$

$$\left||\vec{\kappa}_j(x)|\frac{d\upsilon}{dt}(\mu(x))\nabla\mu(x) - \nabla\nu(x)\right| \leq A\{\text{the right-hand side of (26.127)}\}.$$
$$(26.129)$$

Proof. (26.124) is clear from (26.112) and (26.118).

$$\{\text{the left-hand side of (26.125)}\}$$

$$\leq 4\left\{|\vec{\kappa}_j(x)|^{-2}|\nabla\vec{\kappa}_j(x)||\vec{\rho}(x)| + |\vec{\kappa}_j(x)|^{-1}|\nabla\vec{\rho}(x)|\right.$$

$$\left. + |\vec{\kappa}_j(x)|^{-3}|\nabla\vec{\kappa}_j(x)||\vec{\rho}(x)|^2 + |\vec{\kappa}_j(x)|^{-2}|\vec{\rho}(x)||\nabla\vec{\rho}(x)|\right\}$$

$$\leq 4\left(2^{-j+1}|\vec{\rho}(x)| + |\nabla\vec{\rho}(x)|\right)|\vec{\kappa}_j(x)|^{-1}$$

$$\times \left(|\vec{\kappa}_j(x)|^{-1}2^{j-1}|\nabla\vec{\kappa}_j(x)| + 1 + |\vec{\kappa}_j(x)|^{-2}2^{j-1}|\nabla\vec{\kappa}_j(x)||\vec{\rho}(x)|\right.$$

$$+ |\vec{\kappa}_j(x)|^{-1} |\vec{\rho}(x)|\big)$$

$$\leq 4 \left(2^{-j+1} A^2 \eta_{j-1}^{(1)}(x)\right) (0.1 w_{j\dagger}(x))^{-1} \cdot 2$$

by (26.112), (26.118) and (26.110)*

$\leq$ {the right-hand side of (26.125)}.

{the left-hand side of (26.126)}

$$\leq |\vec{\rho}(x) \cdot U(\vec{\kappa}_j(x)) - \nu(x)| + 2^{-1} |\vec{\kappa}_j(x)|^{-1} |\vec{\rho}(x)|^2$$

$\leq$ {the right-hand side of (26.126)}

by (26.120), (26.112) and (26.107).

{the left-hand side of (26.127)}

$$\leq |\nabla\vec{\rho}(x) \cdot U(\vec{\kappa}_j(x)) - \nabla\nu(x)| + 2 \left\{ |\vec{\kappa}_j(x)|^{-1} |\nabla\vec{\kappa}_j(x)| |\vec{\rho}(x)| \right.$$

$$\left. + |\vec{\kappa}_j(x)|^{-2} |\nabla\vec{\kappa}_j(x)| |\vec{\rho}(x)|^2 + |\vec{\kappa}_j(x)|^{-1} |\vec{\rho}(x)| |\nabla\vec{\rho}(x)| \right\}$$

$$\leq |\cdots| + 2 |\vec{\rho}(x)| \left\{ |\vec{\kappa}_j(x)|^{-1} |\nabla\vec{\kappa}_j(x)| + |\vec{\kappa}_j(x)|^{-2} |\nabla\vec{\kappa}_j(x)| |\vec{\rho}(x)| \right.$$

$$\left. + |\vec{\kappa}_j(x)|^{-1} |\nabla\vec{\rho}(x)| \right\}$$

$$\leq |\cdots| + 2 A^2 \eta_{j-1}^{(1)}(x) \left\{ 2^{-j} A^4 \left(\|\vec{g}\|_{\mathrm{BMOw},\varepsilon} + \varepsilon \right) \right\}$$

by (26.112), (26.110)*, and (26.106)

$\leq$ {the right-hand side of (26.127)} by (26.121).

(26.128) and (26.129) are easy from (26.126)–(26.127),

$$v(\mu(x)) = 2^{-1}\mu(x) + O(\mu(x)^2),$$

$$\frac{dv}{dt}(\mu(x)) = 2^{-1} + O(\mu(x)),$$

(26.124)–(26.125) and from (26.107).   □

Proof of Claim 3. Recall (26.123). If $|\vec{\kappa}_j(x)| < 0.9 w_{j\dagger}(x)$, then $\vec{\phi}_{j-1}(x) = \vec{0}$ by (26.123). If $|\vec{\kappa}_j(x)| \geq 0.9 w_{j\dagger}(x)$, then the desired result follows from (26.128).   □

Proof of Claim 4. Recall (26.123). If $|\vec{\kappa}_j(x)| < 0.9 w_{j\dagger}(x)$, then $\nabla\vec{\phi}_j(x) = \vec{0}$ by (26.123). If $|\vec{\kappa}_j(x)| \geq 0.9 w_{j\dagger}(x)$, then (26.103) follows from

$$\left|\nabla\vec{\phi}_{j-1}(x)\right| \leq \left(|\nabla\psi_j(x)| + |\vec{\tau}(x)|^{-1} |\nabla\vec{\tau}(x)|\right) \left||\vec{\kappa}_j(x)|v(\mu(x)) - \nu(x)\right|$$

$$+ |\nabla\vec{\kappa}_j(x)| |v(\mu(x))| + \left||\vec{\kappa}_j(x)|\frac{dv}{dt}(\mu(x))\nabla\mu(x) - \nabla\nu(x)\right|$$

from (26.128)–(26.129), from

$$|\nabla\psi_j(x)| + |\tau(x)|^{-1} |\nabla\tau(x)| \leq 2 \cdot 2^{-j} \quad \text{(by (26.92) and (26.114)–(26.115))}$$

and from

$$|\nabla \vec{\kappa}_j(x)| \, |v(\mu(x))| \;\leq\; 2^{-j} A^2 \eta_j^{(2)}(x) A^3 \eta_{j-1}^{(1)}(x) w_{j\dagger}(x)^{-1}$$
$$\text{(by (26.110)}^* \text{ and (26.124))}$$
$$\leq\; 2^{-j} A^6 \left(\|\vec{g}\|_{\mathrm{BMO}w,\varepsilon} + \varepsilon \right) \eta_{j-1}^{(1)}(x).$$

$\square$

For $\vec{g} \in L^1_{\mathrm{loc}}(\mathbf{R}^n, \mathbf{C}^m)$ $\left(= L^1_{\mathrm{loc}}(\mathbf{R}^n, \mathbf{R}^{2m}) \right)$ we define $\|\vec{g}\|_{\mathrm{BMO}w,\varepsilon}$ and $\|\vec{g}\|_{\mathrm{BMO}w,\varepsilon}$ by Definition 26.1 with $\mathbf{R}^m$ replaced by $\mathbf{R}^{2m}$ $(= \mathbf{C}^m)$. In the case when $\{\theta_1, \cdots, \theta_m\}$ does not satisfy (24.1) we have the following.

Theorem 26.2. *Let* $\{\theta_1, \cdots, \theta_m\} \subset C^\infty(S^{n-1}, \mathbf{C})$ *be as in Theorem 24.2. Then there exist* $\varepsilon'_{26.1}(\{\theta_1, \cdots, \theta_m\}) > 0$ *and* $C'_{26.1}(\{\theta_1, \cdots, \theta_m\}) < \infty$ *such that if*

$$\varepsilon \in (0, \varepsilon'_{26.1}],$$

(26.2)–(26.3) *and*

$$\vec{g} \in L^2(\mathbf{R}^n, \mathbf{C}^m), \quad \|\vec{g}\|_{\mathrm{BMO}w,\varepsilon} < \infty \tag{26.130}$$

hold, then there exists $\vec{h} \in S_{\{\theta_1, \cdots, \theta_m\}*}$ *such that*

$$\left| \vec{g}(x) - \vec{h}(x) \right| \leq C'_{26.1} \|\vec{g}\|_{\mathrm{BMO}w,\varepsilon} w(x),$$

where $S_{\{\theta_1, \cdots, \theta_m\}*}$ *is defined in (24.21).*

By the same reasoning as the proof of Theorem 26.1, Theorem 26.2 can be reduced to the following.

Lemma 26.2′. . let $\{\theta_1, \cdots, \theta_m\}$ be as in Theorem 24.2. Then there exists $C'_{26.2}(\{\theta_1, \cdots, \theta_m\}) < \infty$ such that if (26.6), (26.2)–(26.3) and (26.130) hold, then there exist

$$\vec{h} \in S_{\{\theta_1, \cdots, \theta_m\}*} \text{ and } \vec{v} \in L^1(\mathbf{R}^n, \mathbf{C}^m)$$

that satisfy (26.9)–(26.11) (with $C_{26.2}$ replaced by $C'_{26.2}$).

The proof of Lemma 26.2′ is completely the same with that of Lemma 26.2. All we have to do is to replace $\mathbf{R}^m$ by $\mathbf{C}^m (= \mathbf{R}^{2m})$, to replace $S_{\{\theta_1, \cdots, \theta_m\}*(r)}$ $(\subset \mathrm{BMO}(\mathbf{R}^n, \mathbf{R}^m))$ by $S_{\{\theta_1, \cdots, \theta_m\}*}$ $(\subset \mathrm{BMO}(\mathbf{R}^n, \mathbf{C}^m))$ and to use Lemma 24.3′ instead of Lemma 24.3 in the construction of $\vec{p}_I \in S_{\{\theta_1, \cdots, \theta_m\}*}$ that satisfies (26.94).

XXVII. Characterization of H^p in terms of Fourier multipliers

The purpose of this section is to show the following.

Theorem 27.1. *Let $\{\theta_1, \cdots, \theta_m\} \subset C^\infty(S^{n-1}, \mathbf{C})$ satisfy (24.18). Then there exists $p_{27.1}(\{\theta_1, \cdots, \theta_m\}) < 1$ such that if*

$$p \in (p_{27.1}, 1],$$

then

$$\sup\left\{ \frac{\|f\|_{H^p}}{\sum\limits_{j=1}^{m} \|\mathsf{m}_{\theta_j} f\|_{L^p}} : f \in H^p(\mathbf{R}^n, \mathbf{C}) \cap L^2(\mathbf{R}^n, \mathbf{C}) \cap \{0\}^c \right\} < \infty. \quad (27.1)$$

Corollary 27.1. *Let $\{\theta_1, \cdots, \theta_m\} \subset C^\infty(S^{n-1}, \mathbf{C})$ satisfy (25.3)*. Then there exists $p_{27.2}(\{\theta_1, \cdots, \theta_m\}) < 1$ such that if*

$$p \in (p_{27.2}, 1],$$

then

$$\sup\left\{ \frac{\|f\|_{H^p}}{\sum\limits_{j=1}^{m} \|\mathsf{m}_{\theta_j} f\|_{L^p}} : f \in H^p(\mathbf{R}^n, \mathbf{R}) \cap L^2(\mathbf{R}^n, \mathbf{R}) \cap \{0\}^c \right\} < \infty. \quad (27.2)$$

Lemma 27.1. *Assume (26.6) and (26.2)–(26.4). Assume*

$$\operatorname{supp} \bar{g} \subset B(0, 1).$$

Then

$$\|\bar{g}\|_{\mathrm{BMO}w,\varepsilon} \le C_{27.1}(n) w(B(0, 1))^{-1} \|\bar{g}\|_{\Lambda_{1/2}}.$$

Proof. Take any ball $B_0 = B(x_0, r_0)$ with $r_0 \leq \varepsilon$. Then

$$\int_{B_0} |\vec{g}(x) - \vec{g}(x_0)| \, dx \leq C(n) \|\vec{g}\|_{\Lambda_{1/2}} r_0^{1/2} |B_0|$$

$$\leq C(n) \|\vec{g}\|_{\Lambda_{1/2}} w(B(0,1))^{-1} w(B_0) |B_0| \text{ by (26.17)}$$

and

$$|\vec{g}| * (\chi)_\varepsilon (x_0) \leq C(n) \|\vec{g}\|_{L^\infty} \chi_{B(0,2)}(x_0)$$

$$\leq C(n) \varepsilon^{-n\varepsilon} \frac{w(B(x_0, \varepsilon))}{w(B(0,1))} \|\vec{g}\|_{\Lambda_{1/2}} \text{ by (26.17)}$$

$$\leq C(n) w(B(0,1))^{-1} \|\vec{g}\|_{\Lambda_{1/2}} w(B(x_0, \varepsilon)).$$

$\square$

Lemma 27.2. Let $\{\theta_1, \cdots, \theta_m\} \subset C^\infty(S^{n-1}, \mathbf{C})$ satisfy (24.18). Then there exist $p_{27.3}(\{\theta_1, \cdots, \theta_m\}) \in (1/2, 1)$ and $C_{27.2}(\{\theta_1, \cdots, \theta_m\}) < \infty$ such that if

$$f \in L^2(\mathbf{R}^n, \mathbf{C}), \tag{27.3}$$

then

$$G_{1/2}(m_{\theta_k} f)(x) \leq C_{27.2} M_{p_{27.3}}\left(M_{1/2}\left(\sum_{j=1}^m |m_{\theta_j} f| \right) \right)(x) \ (k = 1, \cdots, m) \tag{27.4}$$

where G_a and M_p are defined in Sections 3 and 0, respectively.

Proof. Assume (27.3) and $f \not\equiv 0$. Let

$$\phi \in \mathcal{D}(\mathbf{R}^n, \mathbf{C}), \ \text{supp} \, \phi \subset B(0,1).$$

Then, by the argument of dilation and translation, for the proof of (27.4) it is enough to show

$$\left| \int \overline{m_{\theta_k} f(x)} \phi(x) dx \right| \leq C \|\phi\|_{\Lambda_{1/2}} M_{p_{27.3}}\left(M_{1/2}\left(\sum_{j=1}^m |m_{\theta_j} f| \right) \right)(0)$$

$$(k = 1, \cdots, m). \tag{27.5}$$

Since $\{\bar{\theta}_1, \cdots, \bar{\theta}_m\}$ satisfies (24.18), applying Theorem 26.2 to $\{\bar{\theta}_1, \cdots, \bar{\theta}_m\}$ gives $\varepsilon'_{26.1} = \varepsilon'_{26.1}(\{\bar{\theta}_1, \cdots, \bar{\theta}_m\})$ and $C'_{26.1} = C'_{26.1}(\{\bar{\theta}_1, \cdots, \bar{\theta}_m\})$. Let

$$w(x) = M\left(\left(\sum_{j=1}^m |m_{\theta_j} f| \right)^{1/2} \right)(x)^{-\varepsilon'_{26.1}} (1 + |x|)^{-n-1/10},$$

$$\vec{g}(x) = (0, \cdots, 0, \phi(x), 0, \cdots, 0) \in \mathcal{D}(\mathbf{R}^n, \mathbf{C}^m).$$

the k-th element

Theorem 26.2 implies the existence of

$$\vec{h} = (h_1, \cdots, h_m) \in S_{\{\bar{\theta}_1, \cdots, \bar{\theta}_m\}*}$$

such that

$$\left| \vec{g}(x) - \vec{h}(x) \right| \leq C'_{26.1} \|\vec{g}\|_{\mathrm{BMO}w, \varepsilon'_{26.1}} w(x).$$

So,

$$
\begin{aligned}
\left| \vec{g}(x) - \vec{h}(x) \right| &\leq C'_{26.1} C_{27.1} w(B(0,1))^{-1} \|\vec{g}\|_{\Lambda_{1/2}} w(x) \\
&\qquad \text{by Lemma 27.1} \\
&\leq C \left\{ \inf_{y \in B(0,1)} M \left(\left(\sum_{j=1}^m |\mathsf{m}_{\theta_j} f| \right)^{1/2} \right)(y) \right\}^{\varepsilon'_{26.1}} \\
&\qquad \times \|\phi\|_{\Lambda_{1/2}} M \left(\left(\sum_{j=1}^m |\mathsf{m}_{\theta_j} f| \right)^{1/2} \right)(x)^{-\varepsilon'_{26.1}} (1+|x|)^{-n-1/10} \\
&\leq C M_{1/2} \left(M_{1/2} \left(\sum |\mathsf{m}_{\theta_j} f| \right) \right)(0)^{\varepsilon'_{26.1}/2} \|\phi\|_{\Lambda_{1/2}} \\
&\qquad \times M_{1/2} \left(\sum |\mathsf{m}_{\theta_j} f| \right)(x)^{-\varepsilon'_{26.1}/2} (1+|x|)^{-n-1/10} \qquad (27.6) \\
&\quad (\in L^2(\mathbf{R}^n, \mathbf{C}^m) \text{ because } \varepsilon'_{26.1}(>0) \text{ is very small.})
\end{aligned}
$$

Since

$$\vec{h} \in S_{\{\bar{\theta}_1, \cdots, \bar{\theta}_m\}*}, \quad \vec{g} \in L^2(\mathbf{R}^n, \mathbf{C}^m), \quad \vec{g} - \vec{h} \in L^2(\mathbf{R}^n, \mathbf{C}^m),$$

it follows that

$$\vec{h} \in L^2(\mathbf{R}^n, \mathbf{C}^m)$$

and that

$$\sum_{j=1}^m \mathsf{m}_{\bar{\theta}_j} h_j = 0 \text{ in } L^2(\mathbf{R}^n, \mathbf{C}). \qquad (27.7)$$

Thus,

$$
\begin{aligned}
\left| \int \phi(x) \overline{\mathsf{m}_{\theta_k} f(x)} \, dx \right| &= \left| \int \sum_{j=1}^m g_j(x) \overline{\mathsf{m}_{\theta_j} f(x)} \, dx \right| \\
&= \left| \int \sum_{j=1}^m (g_j(x) - h_j(x)) \overline{\mathsf{m}_{\theta_j} f(x)} \, dx \right| \text{ by (27.7)} \\
&\leq \int \left\{ \left| \vec{g}(x) - \vec{h}(x) \right| \right\} \left\{ \sum_{j=1}^m |\mathsf{m}_{\theta_j} f(x)| \right\} dx \\
&\leq \int \left\{ C M_{\frac{1}{2}} \left(M_{\frac{1}{2}} \left(\sum |\mathsf{m}_{\theta_j} f| \right) \right)(0)^{\frac{\varepsilon'_{26.1}}{2}} \|\phi\|_{\Lambda_{1/2}} M_{\frac{1}{2}} \left(\sum |\mathsf{m}_{\theta_j} f| \right)(x)^{-\frac{\varepsilon'_{26.1}}{2}} \right.
\end{aligned}
$$

$$(1 + |x|)^{-n-1/10}\Big\} \times \Big\{M_{\frac{1}{2}}\Big(\sum|\mathsf{m}_{\theta_j}f|\Big)(x)\Big\}dx \text{ by (27.6)}$$

$$\leq CM_{\frac{1}{2}}\Big(M_{\frac{1}{2}}\Big(\sum|\mathsf{m}_{\theta_j}f|\Big)\Big)(0)^{\varepsilon'_{26.1}/2}\|\phi\|_{\Lambda_{1/2}}$$

$$\times M_{1-\varepsilon'_{26.1}/2}\Big(M_{\frac{1}{2}}\Big(\sum|\mathsf{m}_{\theta_j}f|\Big)\Big)(0)^{1-\varepsilon'_{26.1}/2}.$$

Thus, we get (27.5) with $p_{27.3} = 1 - \varepsilon'_{26.1}/2$.   □

Proof of Theorem 27.1. The condition (24.18) implies (16.2). So, Theorems 16.1 and 16.2 imply

$$\|f\|_{H^p} \approx \sum_{j=1}^{m}\|\mathsf{m}_{\theta_j}f\|_{H^p} \text{ for any } f \in L^2(\mathbf{R}^n, \mathbf{C}). \tag{27.8}$$

On the other hand, if

$$p \in (p_{27.3}, 1],$$

then

$$\sum_{j=1}^{m}\|\mathsf{m}_{\theta_j}f\|_{H^p} \leq C\sum_{j=1}^{m}\Big\|G_{\frac{1}{2}}(\mathsf{m}_{\theta_j}f)\Big\|_{L^p} \text{ by Theorem 4.1}$$

$$\leq C\Big\|M_{p_{27.3}}\Big(M_{\frac{1}{2}}\Big(\sum_{j=1}^{m}|\mathsf{m}_{\theta_j}f|\Big)\Big)\Big\|_{L^p} \text{ by Lemma 27.2}$$

$$\leq C\sum_{j=1}^{m}\|\mathsf{m}_{\theta_j}f\|_{L^p} \text{ by } 1/2 < p_{27.3} < p$$

and by the Hardy-Littlewood maximal theorem. (27.9)

Thus, combining (27.8)–(27.9) gives Theorem 27.1 with $p_{27.1} = p_{27.3}$.   □

Proof of Corollary 27.1. If f is real-valued, then Remark 16.7 implies

$$\|\mathsf{m}_{\theta_j}f\|_{L^p} \approx \|\mathsf{m}_{\theta_{j\$}}f\|_{L^p} + \|\mathsf{m}_{\theta_{j\notin}}f\|_{L^p}.$$

Thus, the desired result follows from applying Theorem 27.1 to the set of Fourier multipliers

$$\{\theta_{1\$}, \theta_{1\notin}, \theta_{2\$}, \theta_{2\notin}, \cdots, \theta_{m\$}, \theta_{m\notin}\}.$$

□

Notes. The argument in this section is due to A. Uchiyama [84]. The idea of this argument was inspired by P. W. Jones [83].

XXVIII. The one-dimensional case

Let

$$S_{H^*_{(r)}} = \left\{ h = h_0 + ih_1 \in \text{BMO}(\mathbf{R}^1, \mathbf{C}) = \text{BMO}(\mathbf{R}^1, \mathbf{R}^2) : \right.$$
$$\left. h_0, h_1 \in \text{BMO}(\mathbf{R}^1, \mathbf{R}), \ h_0 + \tilde{H}h_1 = 0 \text{ in } \text{BMO}(\mathbf{R}^1, \mathbf{R})/\mathbf{R} \right\},$$
$$(28.1)$$

where $\tilde{H}$ is defined in Definition 21.2 and Remark 21.4. Let

$$g \in \text{BMO}(\mathbf{R}^1, \mathbf{C}) \text{ and let supp } g \text{ be compact.} \qquad (28.2)$$

In Section 22 we gave a construction of

$$h \in S_{H^*_{(r)}}$$

such that

$$\|g - h\|_{L^\infty} \le C\|g\|_{\text{BMO}}. \qquad (28.3)$$

In this section we explain P. W. Jones's construction of $h \in S_{H^*_{(r)}}$ and apply his idea to $H^p(\mathbf{R}^1)$.

Let

$$P(x,t) = \frac{1}{\pi} \frac{t}{x^2 + t^2}, \quad Q(x,t) = \frac{1}{\pi} \frac{x}{x^2 + t^2}.$$

We leave the proof of the following lemma to readers.

Lemma 28.1. Let $b \in L^1_{loc}(\mathbf{R}^1, \mathbf{R})$,

$$\sup \left\{ b(x) : x \in \mathbf{R}^1 \right\} < \infty,$$

$$\int_{-\infty}^{+\infty} b(x)(1 + x^2)^{-1} dx > -\infty.$$

For $x, y \in \mathbf{R}$ and $s, t > 0$ let

$$B_b(x + is) = \int_{-\infty}^{+\infty} b(u) \left\{ P(x - u, s) \right.$$
$$\left. + i \left(Q(x - u, s) + \chi_{(-\infty, -1] \cup [1, +\infty)}(u)(\pi u)^{-1} \right) \right\} du, \qquad (28.4)$$

$$p_{b,y,t}(x + is) = P(x + is - y, t) \left\{ e^{B_b(x+is) - B_b(y+it)} - 1 \right\}. \qquad (28.5)$$

Then

$$p_{b,y,t}(x+is) \text{ is bounded and analytic as a function of } x+is \in \mathbf{R}_+^2.$$
(28.6)

Let

$$p_{b,y,t}(x) = \lim_{s \to +0} p_{b,y,t}(x+is).$$
(28.7)

Then

$$|p_{b,y,t}(x) + P(x-y,t)| \le P(x-y,t)e^{b(x)-(b*P(\cdot,t))(y)} \quad \text{a.e. } x, \qquad (28.8)$$

$$|p_{b,y,t}(x)| \le P(x-y,t)\left\{e^{b(x)-(b*P(\cdot,t))(y)} + 1\right\} \ (\in L^1(\mathbf{R}^1) \cap L^\infty(\mathbf{R}^1))$$

a.e. x.   (28.9)

Theorem 28.1. *Let μ be a finite complex measure on $\mathbf{R}_+^2$. Let*

$$\||\mu|\|_C \le 1.$$
(28.10)

Let

$$b_t(x) = -\iint_{\mathbf{R} \times (0,t]} P(x-u,s)d|\mu|(u,s).$$
(28.11)

Let $B_{b_t}(x+is)$, $p_{b_t,y,t}(x+is)$ and $p_{b_t,y,t}(x)$ be as in (28.4)–(28.5) and (28.7) with b_t in places of b. Let

$$h(x) = -\iint_{\mathbf{R}_+^2} p_{b_t,y,t}(x)d\mu(y,t),$$
(28.12)

$$k(x) = \iint_{\mathbf{R}_+^2} P(x-y,t)d\mu(y,t).$$
(28.13)

Then

$$h \in S_{H_{(r)}^*},$$
(28.14)

$$\|k-h\|_{L^\infty(\mathbf{R}^1,\mathbf{C})} \le C.$$
(28.15)

Proof. Note that

$$-(b_t * P(\cdot,t))(y) = \iint_{\mathbf{R} \times (0,t]} P(y-u,s+t)d|\mu|(u,s)$$

$$\le C \iint_{\mathbf{R} \times (0,t]} P(y-u,t)d|\mu|(u,s) \le C\||\mu|\|_C \le C. \qquad (28.16)$$

So,

$$\int |p_{b_t,y,t}(x)|\, dx \leq \int P(x-y,t)\{C+1\}dx \text{ by (28.9), } b_t < 0 \text{ and (28.16)}$$
$$\leq C.$$

Thus,

$$\int_{\mathbf{R}^1} dx \iint_{\mathbf{R}^2_+} |p_{b_t,y,t}(x)|\, d|\mu|(y,t) \leq \iint_{\mathbf{R}^2_+} d|\mu|(y,t) < +\infty.$$

So, the integral (28.12) converges at almost every $x \in \mathbf{R}^1$ and h belongs to $L^1(\mathbf{R}^1,\mathbf{C})$. It is easy to see that the integral of (28.13) converges at almost every $x \in \mathbf{R}^1$. Thus, the estimate (28.15) follows from

$$|k(x) - h(x)| = \left| \iint_{\mathbf{R}^2_+} (P(x-y,t) + p_{b_t,y,t}(x))\, d\mu(y,t) \right|$$

$$\text{by (28.12) and (28.13)}$$

$$\leq \iint P(x-y,t)e^{b_t(x)-(b_t*P(\cdot,t))(y)}d|\mu|(y,t) \text{ by (28.8)}$$

$$\leq C \iint P(x-y,t)e^{b_t(x)}d|\mu|(y,t) \text{ by (28.16)}$$

$$= C \iint P(x-y,t)e^{-\iint_{\mathbf{R}\times(0,t]} P(x-u,s)d|\mu|(u,s)}d|\mu|(y,t)$$

$$= C \int_{t\in(0,+\infty)} \left\{ e^{-\int_{s\in(0,t]} \int_{u\in\mathbf{R}} P(x-u,s)d|\mu|(u,s)} \right.$$

$$\left. \times \int_{y\in\mathbf{R}} P(x-y,t)d|\mu|(y,t) \right\}$$

$$\leq C.$$

(28.6) implies

$$p_{b_t,y,t}(\cdot) \in S_{H^*_{(r)}}(, \text{ where } \cdot \in \mathbf{R}^1.) \tag{28.17}$$

(28.10) and Lemma 1.13 imply $k \in \text{BMO}$, which combined with (28.15) implies $h \in \text{BMO}$. Thus, (28.14) follows from the formula (28.12) and (28.17).$\square$

Corollary 28.1. *Let $g \in \text{BMO}(\mathbf{R}^1,\mathbf{C})$ and let $\operatorname{supp} g$ be compact. Then there exists $h \in S_{H^*_{(r)}}$ such that*

$$\|g - h\|_{L^\infty} \leq C\|g\|_{\text{BMO}}.$$

Proof. Lemma 13.4 implies the existence of a finite complex measure μ on $\mathbf{R}^2_+$ that satisfies

$$\|\mu\|_c \leq C\|g\|_{\text{BMO}},$$

$$\left\| g(\cdot) - \iint_{\mathbf{R}^2_+} P(\cdot - y,t)d\mu(y,t) \right\|_{L^\infty} \leq C\|g\|_{\text{BMO}}.$$

Then the desired result follows from applying Theorem 28.1 to this measure μ. $\qquad\qquad\square$

Finally, using the idea of Theorem 28.1, we give another proof of the following, which is the one-dimensional case of Lemma 17.2.

Theorem 28.2. *Let*

$$f \in L^2(\mathbf{R}^1, \mathbf{C}), \quad \operatorname{supp} \mathcal{F}f \subset [0, +\infty), \qquad (28.18)$$
$$q > 0 \qquad (28.19)$$

and $(y, t) \in \mathbf{R}_+^2$. *Then*

$$|f * P(\cdot, t)(y)|^q \leq |f|^q * P(\cdot, t)(y). \qquad (28.20)$$

Proof. Let $\varepsilon > 0$ and

$$b(x) = -\log\left(\varepsilon + |f(x)|\right). \qquad (28.21)$$

Define $B_b(x)$ and $p_{b,y,t}(x)$ by the formulae (28.4), (28.5) and (28.7). Then

$$\left| \int_{-\infty}^{+\infty} f(x) P(x - y, t) dx \right| = \left| \int_{-\infty}^{+\infty} f(x) \left(P(x - y, t) + p_{b,y,t}(x) \right) dx \right|$$

$$\text{by (28.18) and (28.6)}$$

$$\leq \int_{-\infty}^{+\infty} |f(x)| \, P(x - y, t) e^{-\log(\varepsilon + |f(x)|)} e^{(P(\cdot, t) * \log(\varepsilon + |f|))(y)} dx$$

$$\text{by (28.8) with (28.21)}$$

$$\leq e^{(P(\cdot, t) * \log(\varepsilon + |f|))(y)} \quad \text{by } |f(x)| \, e^{-\log(\varepsilon + |f(x)|)} \leq 1.$$

Letting $\varepsilon \to +0$ gives

$$\left| \int_{-\infty}^{+\infty} f(x) P(x - y, t) dx \right| \leq e^{(P(\cdot, t) * \log|f|)(y)}$$

$$\left(\leq \{|f|^q * P(\cdot, t)(y)\}^{1/q} . \right)$$

$\qquad\qquad\square$

Notes. Constructive proofs of the Fefferman-Stein decomposition of $\mathrm{BMO}(\mathbf{R}^1, \mathbf{C})$ were first given by P. W. Jones [80a] and [83]. (Jones [83] was earlier than Uchiyama [82a, 82b].)

In the one-dimensional case, there are other proofs of the $H^1(\mathbf{R}^1)$-$\mathrm{BMO}(\mathbf{R}^1)$ duality theorem. R. Coifman–R. Rochberg–G. Weiss [76] gives a proof that uses the factorization theorem of $H^1(\mathbf{R}^1)$. As we mentioned in the note of Section 21, H. Helson–G. Szegö [60] combined with R. Hunt–B. Muckenhoupt–R. Wheeden [73] gives a proof that uses weight functions.

Appendix

Copying C. Fefferman–E. M. Stein [72] p.172 we give a proof to Theorem 6.A. Let

$$m \in \{2, 3, 4, \cdots\}.$$

Let $u(z)$ be harmonic on

$$B = \{z \in \mathbf{R}^m : |z| < 1\}.$$

For $r \in (0, 1)$ let

$$
m_p(r) = \begin{cases}
\left(\displaystyle\int_{\{z \in \mathbf{R}^m : |z| = r\}} |u(z)|^p \, d\sigma(z) \right)^{1/p} & \text{if } p \in (0, \infty), \\[2em]
\displaystyle\sup_{z \in \mathbf{R}^m : |z| = r} |u(z)| & \text{if } p = \infty,
\end{cases}
$$

where $d\sigma$ denotes the area of the shere $\{z \in \mathbf{R}^m : |z| = r\}$.

Proof of Theorem 6.A. We may assume

$$q \in (0, 1), \tag{29.1}$$

$$\int_B |u(z)|^q \, dz = 1. \tag{29.2}$$

We may assume

$$m_\infty(r) \geq 1 \text{ for all } r \in (0, 1) \tag{29.3}$$

because otherwise there is nothing to prove.

First note that if $0 < \rho < r < 1$, then

$$
\begin{aligned}
m_\infty(\rho) &= \sup_{|\zeta|=\rho} \left| \int_{|z|=r} C \frac{r^2 - \rho^2}{r|\zeta - z|^m} u(z) d\sigma(z) \right| \quad \text{by the harmonicity of } u \\
&\leq C(r - \rho)^{-m+1} m_1(r).
\end{aligned}
$$

Substituting $\rho = r^a$, where $a > 1$, gives

$$m_\infty(r^a) \leq C(r-r^a)^{-m+1}m_1(r) \leq C(r-r^a)^{-m+1}m_q(r)^q m_\infty(r)^{1-q}.$$

So,

$$m_\infty(r^a)m_\infty(r)^{q-1} \leq C(r-r^a)^{-m+1}m_q(r)^q.$$

Taking the logarithms of both sides gives

$$\log m_\infty(r^a) + (q-1)\log m_\infty(r) \leq C + (-m+1)\log(r-r^a) + q\log m_q(r).$$

Integrating both sides on the interval $(1/2, 1)$ gives

$$\int_{1/2}^1 \log m_\infty(r^a)dr/r + (q-1)\int_{1/2}^1 \log m_\infty(r)dr/r$$
$$\leq C(a,m) + q\int_{1/2}^1 \log m_q(r)dr/r$$
$$\leq C(a,m) \quad \text{by (29.2)}. \tag{29.4}$$

Therefore

$$(1/a + q - 1)\int_{1/2}^1 \log m_\infty(r)dr/r$$
$$= \int_{(1/2)^{1/a}}^1 \log m_\infty(r^a)dr/r + (q-1)\int_{1/2}^1 \log m_\infty(r)dr/r$$
$$\leq \int_{1/2}^1 \log m_\infty(r^a)dr/r + (q-1)\int_{1/2}^1 \log m_\infty(r)dr/r \quad \text{by (29.3)}$$
$$\leq C(a,m) \quad \text{by (29.4)}.$$

Taking $a > 1$ so that $1/a + q - 1 > 0$ gives

$$\int_{1/2}^1 \log m_\infty(r)dr/r \leq C(q,m),$$

which implies

$$|u(0)| \leq \inf_{r \in (1/2,1)} m_\infty(r) \leq C(q,m).$$

$\square$

Corollary 29.1. Let $p \in (0,1]$. Let $u(x,t)$ be harmonic on $\mathbf{R}_+^{n+1}$. Let

$$\sup_{t \in (0,\infty)} \|u(\,\cdot\,,t)\|_{L^p(\mathbf{R}^n)} \leq 1. \tag{29.5}$$

Then there exists $f \in \mathcal{S}(\mathbf{R}^n)'$ such that

$$\mathcal{F}f \in C(\mathbf{R}^n, \mathbf{C}), \tag{29.6}$$
$$|\mathcal{F}f(\xi)| \leq C(p,n)|\xi|^{n(1/p-1)}, \tag{29.7}$$
$$u(x,t) = \mathcal{F}^{-1}\left\{e^{-2\pi t|\xi|}\mathcal{F}f(\xi)\right\}(x). \tag{29.8}$$

Proof. Since

$$|u(x,t)| \ \leq \ C\left\{t^{-n-1} \iint_{(y,s)\in\mathbf{R}_+^{n+1}:|y-x|+|s-t|\leq t/2} |u(y,s)|^p\, dyds\right\}^{1/p}$$

by Theorem 6.A

$$\leq \ Ct^{-n/p} \text{ by (29.5)},$$

we have

$$\|u(\,\cdot\,,t)\|_{L^1(\mathbf{R}^n)} \leq \|u(\,\cdot\,,t)\|_{L^\infty(\mathbf{R}^n)}^{1-p}\, \|u(\,\cdot\,,t)\|_{L^p(\mathbf{R}^n)}^{p} \leq Ct^{(-n/p)(1-p)}.$$

Thus

$$\mathcal{F}u(\,\cdot\,,t) \in C(\mathbf{R}^n,\mathbf{C}), \tag{29.9}$$
$$|\mathcal{F}u(\,\cdot\,,t)(\xi)| \leq Ct^{-n(1/p-1)}, \tag{29.10}$$
$$u(x,s+t) = P(\,\cdot\,,t) * u(\,\cdot\,,s)(x) \text{ for all } x \in \mathbf{R}^n \text{ and all } s,t \in (0,\infty). \tag{29.11}$$

Let

$$q(\xi) = e^{2\pi t|\xi|}\mathcal{F}u(\,\cdot\,,t)(\xi). \tag{29.12}$$

(Note that (29.11) implies that g is independent of $t \in (0,\infty)$.) Then

$$\mathcal{F}u(\,\cdot\,,t)(\xi) = e^{-2\pi t|\xi|}g(\xi), \tag{29.13}$$
$$q \in C(\mathbf{R}^n,\mathbf{C}) \text{ by (29.9)}, \tag{29.14}$$
$$|g(\xi)| = \left|e^{2\pi}\mathcal{F}u(\,\cdot\,,1/|\xi|)(\xi)\right| \text{ by (29.12)}$$
$$\leq C|\xi|^{n(1/p-1)} \text{ by (29.10)}. \tag{29.15}$$

Let

$$f = \mathcal{F}^{-1}g \in \mathcal{S}'.$$

Then (29.6)–(29.8) follow from (29.13)–(29.15). $\square$

References

We include some papers which are not refered to in this book but related to our topics.

D. Adams and M.Frazier
[88] BMO and smooth truncation in Sobolev spaces, Studia Math. 89 (1988), 241–260.

E. Amar and A. Bonami
[79] Measures de Carleson d'order α et solutions au bord de l'equation $\bar{\partial}$, Bull. Soc. Math. France, 107 (1979), 23–48.

A. Baernstein II and E. T. Sawyer
[85] Embedding and multiplier theorems for $H^p(\mathbf{R}^n)$, Mem. Amer. Math. Soc. No. 318 (1985).

R. Bañuelos
[86] Brownian motion and area functions, Indiana Univ. Math. J., 35 (1986), 643–668.

H. Bateman
[53] Higher Transcendental Functions, Mcgraw-Hill, 1953.

A. Bernard
[79] Espaces H^1 de martingales à deux indices, Dualitè avec les martingales de type BMO, Bull. Sci. Math., 103 (1979), 297–303.

J. Bourgain
[86a] On high dimensional maximal functions associated to convex bodies, Amer. J. Math. 108 (1986), 1467–1476.

[86b] On the L^p-bounds for maximal functions associated to convex bodies in $\mathbf{R}^n$, Israel J. Math., 54 (1986), 257–265.

[86c] Averages in the plane over convex curves and maximal operators, J. Anal. Math., 47 (1986), 69–65.

H–Q. Bui
[83] On Besov, Hardy and Triebel spaces for $0 < p \leq 1$, Ark. Mat., 21 (1983), 169–184.

D. L. Burkholder
[75] One-sided maximal functions and H^p, J. Funct. Anal., 18 (1975), 429 –454.

D. L. Burkholder and R. F. Gundy
[72] Distribution function inequalities for the area integral, Studia Math., 44 (1972), 527–544.

D. L. Burkholder, R. F. Gundy and M. L. Silverstein
[71] A maximal function characterization of the class H^p, Trans. Amer. Math. Soc., 157 (1971), 137–157.

A. P. Calderón
[77] An atomic decomposition of distributions in parabolic H^p spaces, Adv. in Math., 25 (1977), 216–225.

A. P. Calderón and A. Torchinsky
[75] Parabolic maximal functions associated with a distribution, Adv. in. Math., 16 (1975), 1–63.

[77] ——————————————————————————, II, Adv. in Math., 24 (1977), 101–171.

A. P. Calderón and A. Zygmund
[64] On higher gradients of harmonic functions, Studia Math., 24 (1964), 211–226.

S. Campanato
[63] Prorietà di hölderianità di alcune classi di funzioni, Ann. Scoula Norm. Sup. Pisa, 17 (1963), 175–188.

A. Carbery
[86] An almost-orthogonal principle with applications to maximal functions associated to convex bodies, Bull. Amer. Math. Soc., 14 (1986), 269–273.

L. Carleson
[62] Interpolation by bounded analytic functions and the corona problem, Ann. of Math., 76 (1962), 547–559.

[70] The corona theorem, Lecture Notes in Math., Vol. 118 (1970), 121–132.

[76] Two remarks on H^1 and BMO, Adv. in Math., 22 (1976), 269–277.

[81] BMO-10 years' development, 18th Scandinavian Congress of Mathematicians, Progress in Math., 11, 3–21, Birkhäuser, Boston, Mass., 1981.

L. Carleson and J. Garnett
[75] Interpolating sequences and separation properties, J. Anal. Math., 28 (1975), 273–299.

S.-Y. Chang
[76] A characterization of Douglas subalgebras, Acta Math., 137 (1976), 81–89.

S.-Y. Chang and R. Fefferman
[80] A continuous version of duality of H^1 and BMO on the bidisc, Ann. of Math., 112 (1980), 179–201.

[85] Some recent developments in Fourier analysis and H^p-theory on product domains, Bull. Amer. Math. Soc., 12 (1985), 1–43.

S.-Y. Chang, J. M. Wilson and T. H. Wolff
[85] Some weighted norm inequalities concerning the Schrödinger operators, Comment. Math. Helvetici 60 (1985), 217–246.

J.-A. Chao
[74] H^p-spaces of conjugate systems on local fields, Studia Math., 49 (1974), 267–287.

[82] Hardy spaces on regular martingales, Lecture Notes in Math., 939 (1982), 18–28.

J.-A. Chao and M. H. Taibleson
[73] A subregularity inequality for conjugate systems on local fields, Studia Math., 46 (1973), 249–257.

[pre] A Hilbert transform for non-homogeneous martingales, preprint.

M. Christ
[84] Characterization of H^1 by singular integrals: necessary conditions, Duke Math. J., 51 (1984), 599–609.

M. Christ and D. Geller
[84] Singular integral characterizations of Hardy spaces on homogeneous groups, Duke Math. J., 51 (1984), 547–598.

R. Coifman
[72] Distribution function inequalities for singular integrals, Proc. Nat. Acad. Sci. USA, 69 (1972), 2838–2839.

[74] A real variable characterization of H^p, Studia Math., 51 (1974), 269–274.

R. Coifman and B. Dahlberg
[79] Singular integral characterizations of nonisotropic H^p spaces and the F. and M. Riesz theorem, Proc. Symp. Pure Math., 35 (1979), 231– 234.

R. Coifman and C. Fefferman
[74] Weighted norm inequalities for maximal functions and singular integrals, Studia Math., 51 (1974), 241–250.

R. Coifman, P. W. Jones and J. L. Rubio de Francia
[83] Constructive decomposition of BMO functions and factorization of A_p weights, Proc. Amer. Math. Soc., 87 (1983), 675–676.

R. Coifman, Y. Meyer and E. M. Stein
[82] Un nouvel espace fontionnel adapté à l'étude des opérateurs définis par des intégrales singulières, Lecture Notes in Math., 992 (1982), 1–15.

[85] Some new function spaces and their applications to harmonic analysis, J. Funct. Anal., 62 (1985), 304–335.

R. Coifman and R. Rochberg
[80] Another characterization of BMO, Proc. Amer. Math. Soc., 79 (1980), 249–254.

R. Coifman, R. Rochberg and G. Weiss
[76] Factorization theorems for Hardy spaces in several variables, Ann. of Math., 103 (1976), 611–635.

R. Coifman and G. Weiss
[70] On subharmonicity inequalities involving solutions of generalized Cauchy-Riemann equations, Studia Math., 36 (1970), 77–83.

[77] Extensions of Hardy spaces and their use in analysis, Bull. Amer. Math. Soc., 83 (1977), 569–645.

A. Córdoba and C. Fefferman
[76] A weighted norm inequality for singular integrals, Studia Math., 57 (1976), 97–101.

M. Cowling and G. Mauceri
[85] Inequalities for some maximal function, I, Trans. Amer. Math. Soc., 287 (1985), 431–455.

B. Dahlberg
[80a] Approximation of harmonic functions, Ann. Inst. Fourier, 30 (1980), 97–107.

[80b] Weighted norm inequalities for the Lusin area integral and the nontangential maximal functions for functions harmonic in a Lipschitz domain, Studia Math., 67 (1980), 297–314.

G. David and J. L. Journé
[84] A boundedness criterion for generalized Calderón-Zygmund operators, Ann. of Math., 120 (1984), 371–397.

J. Duoandikoetxea and J. L. Rubio de Francia
[85] Estimations indépendantes de la dimension pour les transformées de Riesz, C. R. Acad. Sc. Paris, Ser. I, Math., 300 (1985), 193–196.

P. Duren
[69] Extension of a theorem of Carleson, Bull. Amer. Math. Soc., 75 (1969), 143–146.

[70] Theorey of H^p spaces, Academic Press, 1970, New York.

P. Duren, B. Romberg and A. Shields
[69] Linear functionals on H^p spaces with $0 < p < 1$, J. Reine Angew. Math., 238 (1969), 32–60.

E. B. Fabes, R. L. Johnson and U. Neri
[76] Spaces of harmonic functions representable by Poisson integrals of functions in BMO and $\mathcal{L}_{p,\lambda}$, Indiana Univ. Math. J., 25 (1976), 159–170.

C. Fefferman
[71] Characterization of bounded mean oscillation, Bull. Amer. Math. Soc., 77 (1971), 587–588.

[74] Recent progress in classical Fourier analysis, Proc. of the International Congress of Math., Vancouver, 1974, 95–118.

[76] Harmonic analysis and H^p spaces, MAA Stud. Math., 13 (1976), 38–75.

C. Fefferman, N. Rivière and Y. Sagher
[74] Interpolation between H^p spaces, the real method, Trans. Amer. Math. Soc.,191 (1974), 75–81.

C. Fefferman and E. M. Stein
[71] Some maximal inequalities, Amer. J. Math., 93 (1971), 107–115.

[72] H^p spaces of several variables, Acta Math., 129 (1972), 137–193.

R. Fefferman, R. Gundy, M. Silverstein and E. M. Stein
[82] Inequalities for ratios of functionals of harmonic functions, Proc. Natl. Acad. Sci. USA, 79 (1982), 7958–7960.

R. Fefferman and E. M. Stein
[82] Singular integrals on product spaces, Adv. in Math., 45 (1982), 117–143.

G. B. Folland and E. M. Stein
[82] Hardy spaces on homogeneous groups, Math. Notes, 28 (1982), Princeton Univ. Press.

A. P. Frazier
[72] The dual space of H^p of the polydisc for $0 < p < 1$, Duke Math. J., 39 (1972), 369–379.

M. Frazier
[85] Subspaces of BMO($\mathbf{R}^n$), Trans. Amer. Math. Soc., 290 (1985), 101–125.

T. Gamelin
[80] Wolff's proof of the corona theorem, Israel J. Math., 37 (1980), 113–119.

A. Gandulfo, J. García-Cuerva and M. Taibleson
[76] Conjugate system characterization of H^1: counter examples for the Euclidean plane and local fields, Bull. Amer. Math. Soc., 82 (1976), 83–85.

J. García-Cuerva and J. L. Rubio de Francia
[85] Weighted norm inequalities and related topics, North-Holland, 1985.

J. B. Garnett
[79] Two constructions in BMO, Proc. Symp. Pure Math., 35 (1979), 295–302.

[81] Bounded analytic functions, Academic Press, 1981.

J. B. Garnett and P. W. Jones
[78] The distance in BMO to L^∞, Ann. of Math., 108 (1978), 373–393.

[82] BMO from dyadic BMO, Pacific J. Math., 99 (1982), 351–371.

J. B. Garnett and R. Latter
[78] The atomic decomposition for Hardy spaces in several complex variables, Duke Math. J., 45 (1978), 815–845.

A. Garsia
[73] Martingale inequalities, Benjamin, 1973.

D. Goldberg
[79] A local version of real Hardy spaces, Duke Math. J., 46 (1979), 27–42.

R. F. Gundy
[73] Inégalités pour martingales à un et deux indices: L'espace H^p, Lecture Notes in Math., 774 (1980), 251–334.

R. F. Gundy and E. M. Stein
[79] H^p theory for the poly-disc, Proc. Nat. Acad. Sci. USA, 76 (1979), 1026–1029.

R. F. Gundy and R. Wheeden
[74] Weighted integral inequalities for the nontangential maximal function, Lusin area integral and Walsh-Paley series, Studia Math., 49 (1974), 107–124.

M. de Guzmán
[75] Differentiation of Integrals in $\mathbf{R}^n$, Lecture Notes in Math., 481 (1975), Springer-Verlag.

G. H. Hardy and J. E. Littlewood
[32] Some properties of conjugate functions, J. für Mathematik, 167 (1932), 405–423.

V. P. Havin, S. V. Hruščëv and N. K. Nikol'skii
[84] Linear and complex analysis problem book, Lecture Notes in Math., 1043 (1984), Springer-Verlag.

H. Helson and G. Szegö
[60] A problem in prediction theory, Ann. Math. Pura Appl., 51 (1960), 107–138.

C. Herz
[74a] H_p-Spaces of martingales, $0 < p \leq 1$, Z. W., 28 (1974), 189–205.

[74b] Bounded mean oscillation and regulated martingales, Trans. Amer. Math. Soc., 193 (1974), 199–215.

L. Hörmander
[67] L^p estimates for (pluri-) subharmonic functions, Math. Scand., 20 (1967), 65–78.

R. Hunt
[72] An estimate of the conjugate function, Studia Math., 44 (1972), 371–377.

R. Hunt, B. Muckenhoupt and R. Wheeden
[73] Weighted norm inequalities for thr conjugate function and Hilbert transform, Trans. Amer. Math. Soc., 176 (1973), 227–251.

S. Igari
[63] An extension of the interpolation theorem of Marcinkiewicz II, Tôhoku Math. J., 15(1963), 343–358.

S. Janson
[76] On functions with conditions on the mean oscillation, Ark. Mat., 14 (1976), 189–196.

[77] Characterization of H^1 by singular integral transforms on martingales and $\mathbf{R}^n$, Math. Scand., 41 (1977), 140–152.

[78] Mean oscillation and commutators of singular integral operators, Ark. Mat., 16 (1978), 263–270.

F. John and L. Nirenberg
[61] On functions of bounded mean oscillation, Comm. Pure Appl. Math., 14 (1961), 415–426.

P. W. Jones
[78] Constructions with functions of bounded mean oscillation, Ph. D. Thesis, U. C. L. A., 1978.

[80a] Carleson measures and the Fefferman-Stein decomposition of BMO($\mathbf{R}$), Ann. of Math., 111 (1980), 197–208.

[80b] Factorization of A_p weights, Ann. of Math., 111 (1980), 511–530.

[80c] Estimates for the corona problem, J. Funct. Anal., 39 (1980), 162–181.

[83] L^∞ estimates for the $\bar{\partial}$-problem in a half plane, Acta Math., 150 (1983), 137–152.

[pre] Square functions, Cauchy integrals, analytic capacity and harmonic measure, preprint. (Harmonic analysis and partial differential equations (El Escorial, 1987), 24–68, Lecture Notes in Math., 1384 (1989), Springer-Verlag.)

J. L. Journé
[83] Calderón-Zygmund operators, Pseudo-differential operators and the Cauchy integral of Calderón, Lecture Notes in Math., 994 (1983), Springer-Verlag.

M. Kaneko
[87] Estimates of the area integrals by the nontangential maximal functions, Tôhoku Math. J., 39 (1987), 589–596.

P. Koosis
[78] Sommabilité de la fonction maximale et appartenance à H_1, C. R. Acad. Sc. Paris 286 (1978), Série A 1041–1043.

[79] Sommabilité de la fonction maximale et appartenance à H_1, plusieurs variables, C. R. Acad. Sc. Paris 288 (1979), Série A 489–492.

[80] Lectures on H^p spaces, London Math. Soc. Lecture Notes Series 40 (1980).

R. Latter
[78] A characterization of $H^p(\mathbf{R}^n)$ in terms of atoms, Studia Math., 62 (1978), 93–101.

R. Latter and A. Uchiyama
[79] The atomic decomposition of parabolic H^p spaces, Trans. Amer. Math. Soc., 253 (1979), 391–398.

R. Macías and C. Segovia
[79] A decomposition into atoms of distributions on spaces of homogeneous type, Adv. in Math., 33 (1979), 271–309.

M. P. Malliavin and P. Malliavin
[77] Intégrales de Lusin-Calderón pour les fonctions biharmoniques, Bull. Sc. Math., 2^cSérie 101 (1977), 357–384.

D. Marshall
[76] Subalgebras of L^∞ containing H^∞, Acta Math., 137 (1976), 91–98.

T. McConnell
[84] Area integrals and subharmonic functions, Indiana Univ. Math. J., 33 (1984), 289–303.

K. G. Merryfield
[85] On the area integral, Carleson measures and H^p in the polydisc, Indiana Univ. Math. J., 34 (1985), 663–685.

N. G. Meyers
[64] Mean oscillation over cubes and Hölder continuity, Proc. Amer. Math. Soc., 15 (1964), 717–721.

A. Miyachi
[83] Products of distributions in H^p spaces, Tôhoku Math. J., 35 (1983), 483–498.

[84] Weak factorization of distributions in H^p spaces, Pacific J. Math., 115 (1984), 165–175.

[87] Maximal functions for distributions on open sets, Hitotsubashi J. of Arts and Sciences, 28 (1987), 45–58.

[pre1] H^p spaces over open subsets of $\mathbf{R}^n$, preprint. (Studia Math., 95 (1990), 205–228.)

[pre2] Some Littlewood-Paley type inequalities and their application to the Fefferman-Stein decomposition of BMO, preprint. (Indiana Univ. Math. J., 39 (1990), 563–583.)

B. Muckenhoupt
[72] Weighted norm inequalities for the Hardy maximal function, Trans. Amer. Math. Soc., 165 (1972), 207–226.

[79] Weighted norm inequalities for classical operators, Proc. Symp. Pure Math., 35 (1979), 69–84.

B. Muckenhoupt and R. Wheeden
[74] Norm inequalities for the Littlewood-Paley function g_λ^*, Trans. Amer. Math. Soc., 191 (1974), 95–111.

[76] Weighted bounded mean oscillation and the Hilbert transform, Studia Math., 54 (1976), 221–237.

[78] On the dual of weighted H^1 of the half-space, Studia Math., 63 (1978), 57–79.

T. Murai
[88] A real variable method for the Cauchy transform and analytic capacity, Lecture Notes in Math., 1307 (1988), Springer-Verlag.

T. Murai and A. Uchiyama
[84] Good λ inequalities for the area integral and the nontangential maximal function, Studia Math., 83 (1986), 251–262.

E. Nakai and K. Yabuta
[85] Pointwise multipliers for functions of bounded mean oscillation, J. Math. Soc. Japan, 37 (1985), 207–218.

U. Neri
[75] Fractional integration on the space H^1 and its dual, Studia Math., 53 (1975), 175–189.

H. Reimann and T. Rychener
[75] Funktionen Beschränkter Mittlerer Oszillation, Lecture Notes in Math., 487 (1975), Springer-Verlag.

J. L. Rubio de Francia
[84] Factorization theory and A_p weights, Amer. J. Math., 106 (1984), 533–547.

D. Sarason
[73] Algebras of functions on the unit circle, Bull. Amer. Math. Soc., 79 (1973), 86–99.

[75] Functions of vanishing mean oscillation, Trans. Amer. Math. Soc., 207 (1975), 391–405.

[78] Function theorey on the unit circle, Notes for lectures at a conference at Virginia Poly. Inst. and State Univ. Blacksberg, Virginia, 1978.

D. Stegenga
[76] Bounded Toeplitz operators on H^1 and applications of the duality between H^1 and the functions of bounded mean oscillation, Amer J. Math., 98 (1976), 573–589.

E. M. Stein
[70] Singular integrals and differentiability properties of functions, Princeton Univ. Press, 1970.

[76] Maximal functions: Spherical means, Proc. Nat. Acad. Sci. USA, 73 (1976), 2174–2175.

[82] The development of square functions in the work of A. Zygmund, Bull Amer. Math. Soc., 7 (1982), 359–376.

[83] Some results in harmonic analysis in $\mathbf{R}^n$ for $n \to \infty$, Bull. Amer. Math. Soc., 9 (1983), 71–73.

E. M. Stein and J.-O. Strömberg
[83] Behavior of maximal functions in $\mathbf{R}^n$ for large n, Ark. Mat., 21 (1983), 259–269.

E. M. Stein and G. Weiss
[59] An extension of a theorem of Marcinkiewicz and some of its applications, J. Math. Mech., 8 (1959), 263–284.

[60] On the theorey of harmonic functions of several variables, Acta Math., 103 (1960), 26–62.

[68] Generalization of the Cauchy-Riemann equations and representa- tions of the rotation group, Amer. J. Math., 90 (1968), 163–196.

[71] Introduction to Fourier analysis on Euclidian spaces, Princeton Univ. Press, 1971.

J.-O. Strömberg
[79] Bounded mean oscillation with Orlicz norm and duality of Hardy spaces, Indiana Univ. Math. J., 28 (1979), 511–544.

J.-O. Strömberg and A. Torchinsky
[89] Weighted Hardy spaces, Lecture Notes in Math., 1381(1989), Springer-Verlag.

M. Taibleson
[75] Fourier analysis on local fields, Math. Notes. 15 (1975), Princeton Univ. Press.

M. Taibleson and G. Weiss
[80] The molecular characterization of certain Hardy spaces, Astérsique, 77 (1980), 67–149.

A. Torchinsky
[86] Real-variable methods in harmonic analysis, Academic Press, 1986.

A. Uchiyama
[78] On the compactness of operators of Hankel type, Tôhoku Math. J., 30 (1978), 163–171.

[80a] A remark on Carleson's characterization of BMO, Proc. Amer. Math. Soc., 79(1980), 35–41.

[80b] A maximal function characterization of H^p on the space of homogeneous type, Trans. Amer. Math. Soc., 262(1980), 579–592.

[81] The factorization of H^p on the space of homogeneous type, Pacific J. Math., 92(1981), 453–468.

[82a] The construction of certain BMO functions and the corona ploblem, Pacific J. Math., 99(1982), 183–204.

[82b] A constructive proof of the Fefferman-Stein decomposition of BMO on simple martingales, Conference on Harmonic Analysis in Honor of Antoni Zygmund, Wadsworth, 1982, 495–505.

[82c] A constructive proof of the Fefferman-Stein decomposition of $BMO(\mathbf{R}^n)$, Acta Math., 148(1982), 215–241.

[83] The singular integral characterization of H^p on simple martingales, Proc. Amer. Math. Soc., 88 (1983), 617–621.

[84] The Fefferman-Stein decomposition of smooth functions and its application to $H^p(\mathbf{R}^n)$, Pacific J. Math., 115 (1984), 217–255.

[85a] Characterization of $H^p(\mathbf{R}^n)$ in terms of generalized Littlewood- Paley g-functions, Studia Math., 81 (1985), 135–158.

[85b] Extension of the Hardy-Littlewood-Fefferman-Stein inequality, Pacific J. Math., 120 (1985), 229–255.

[86] On the radial maximal function of distributions, Pacific J. Math., 121 (1986), 467–483.

[87] On McConnell's inequality for functionals of subharmonic functions, Pacific J. Math., 128 (1987), 367–377.

[pre] On the characterization of $H^p(\mathbf{R}^n)$ in terms of Fourier multipliers, preprint. (Proc. Amer. Math. Soc.,109 (1990). 117–123.)

A. Uchiyama and J. M. Wilson
[83] Apporoximate identities and $H^1(\mathbf{R})$, Proc. Amer. Math. Soc., 88 (1983), 53–58.

N. Th. Varopoulos
[77] BMO functions and the $\bar{\partial}$ equation, Pacific J. Math., 71 (1977), 221–273.

[78] A remark on functions of bounded mean oscillation and bounded harmonic functions, Pacific J. Math., 74 (1978), 257–259.

T. Walsh
[73] The dual of $H^p(\mathbf{R}_+^{n+1})$ for $p < 1$, Canad. J. Math., 25 (1973), 567–577.

G. Weiss
[79] Some problems in the theory of Hardy spaces, Proc. Symp. Pure Math., 35 (1979), 189–200.

[89] Book review, Bull. Amer. Math. Soc., 20 (1989), 198–207.

J. M. Wilson
[82] A simple proof of the atomic decomposition of $H^p(\mathbf{R}^n)$ $0 < p \leq 1$, Studia Math., 74 (1982), 25–33.

[85] On the atomic decomposition for Hardy spaces, Pacific J. Math., 116 (1985), 201–207.

[88] Green's theorem and balayage, Michigan Math. J., 35 (1988), 21–27.

T. Wolff
[pre] Counterexamples to two variants of the Helson-Szegö Theorem, preprint.

[hand] Examples of harmonic gradients in $\mathbf{R}^3$, hand written paper.

K. Yosida
[68] Functional analysis, Springer-Verlag, 1968.

A. Zygmund
[59] Trigonometric series, Cambridge, 1959.

R. Bañuelos
[88] A sharp good λ inequality with an application to Riesz transforms, Michigan Math. J., 35 (1988), 117–131.

R. Bañuelos and C. N. Moore
[89a] Sharp estimates for the nontangential maximal function and the Lusin area function in Lipschitz domains, Trans. Amer. Math. Soc., 312 (1989), 641–662.

[89b] Laws of the iterared logarithm, sharp good λ inequalities and L^p-estimates for caloric and harmonic functions, Indiana Univ. Math., J. 38 (1989), 315–344.

M. Frazier and B. Jawerth
[pre] A discrete transform and decompositions of distribution spaces, preprint. (J. Funct. Anal., 93 (1990), 34–170.)

Index